# 10th Southeast Asia Astronomy Network 2018



Journal of Physics: Conference Series
Volume 1231

Bandar Lampung, Indonesia
18 – 21 October 2018

**Editors:**

**Hakim L. Malasan**
**Wahyu S. Putro**
**Robiatul Muztaba**

ISBN: 978-1-5108-9550-8
ISSN: 1742-6588

**Printed from e-media with permission by:**

Curran Associates, Inc.
57 Morehouse Lane
Red Hook, NY 12571

**Some format issues inherent in the e-media version may also appear in this print version.**



Institute of Physics
Dirac House, Temple Back
Bristol BS1 6BE UK

Phone: 44 1 17 929 7481
Fax: 44 1 17 920 0979

techtracking@iop.org

**Additional copies of this publication are available from:**

Curran Associates, Inc.
57 Morehouse Lane
Red Hook, NY 12571 USA
Phone: 845-758-0400
Fax: 845-758-2633
Email: curran@proceedings.com
Web: www.proceedings.com

# TABLE OF CONTENTS

## INTERDISCIPLINARY TOPICS RELATED TO EARTH AND PLANETARY SCIENCES



## OPTICAL ASTRONOMY

## RADIO ASTRONOMY



## SOLAR PHYSICS AND NEUTRINO



## SPACE SCIENCE EDUCATION INTER DISCIPLINE (EARTH AND DISASTER)



## THEORETICAL ASTRONOMY AND ASTROPHYSICS

# Preface

The 10th Southeast Asia Astronomical Network (SEAAN) Meeting 2018 served as a forum of astronomers related to Theoretical Astronomy and Astrophysics; Optical Astronomy; Solar Physics and Neutrino; Space Science Education Interdiscipline (Earth and Disaster); Interdiciplinary Topics Related to Earth and Planetary Sciences in Southeast Asia and its surrounding. It is intended to be used as a scientific forum for accommodating discussion among researchers in astronomy and related fields.

The 10th SEAAN Meeting 2018 is organized by Institut Teknologi Sumatera (ITERA) Astronomical Observatory, Department of Atmospheric Science and Planetary in conjuction with Department of Physics, ITERA, and Government of Province of Lampung, Indonesia. The meeting was held on 19 to 21 October 2018 at Bukit Randu Resort & Restaurant, Bandar Lampung, Indonesia. The 10th SEAAN Meeting 2018 is composed as a meta-conference and business session to admit presentations of activities in astronomy over South-east Asia with international research communities and deployment of ideas that extend the field. On 20 October 2018, participants were invited for a ground-breaking ceremony of Lampung Astronomical Observatory in Mt. Betung, Lampung.

There are 77 participants along with distinguised international speakers from regions. More than 33 papers are selected to go through a strict peer reviewed process.

We would like to thank the scientific organizing committee and the members of reviewers for their kind assistance in critically reading the papers. We are indebted to ITERA and NARIT for substantial support. We would also like to extend our gratitude for all plenary and invited speakers for their invaluable contribution and ideas shared in the conference.

The Editors
Hakim L. Malasan
Wahyu S. Putro
Robiatul Muztaba

IOP Conf. Series: Journal of Physics: Conf. Series **1231** (2019) 011001 doi:10.1088/1742-6596/1231/1/011001

# Message From SEAAN President

One of a key success factor for the achievement a leap growth in Astronomy is a strong collaboration and synergy among the member countries in the region to promote and develop an effective research platform for researchers and students. The Southeast Asia Astronomy Network (SEAAN) has been established since 2007 to nurture astronomy and establish an effective mechanism for promoting the development and experiences in astronomy research and education among South-east Asian countries. SEAAN has also been recognized by the ASEAN Committee on Science and Technology (ASEAN COST) in 2015.

SEAAN organized an annual meeting as a research and academic platform for the member countries and collaborative institutes outside Southeast Asia to report on the development and research results in Astronomy during the past year. The 10th SEAAN meeting was hosted by ITERA, Lampung, Indonesia during 18-20 October 2018 which academic and business meeting were organized. In this meeting, there were more than 40 abstracts submitted and all had been reviewed by the SOC before approving for both oral and poster presentations. There are 32 full accepted papers which had been carefully reviewed and checked to prevent plagiarism by the SOC and LOC to ensure a high quality proceedings published for the first time in the SEAAN meeting.

I would like to express my sincerely thanks ITERA for hosting the 10th SEAAN meeting and all the SOC and LOC for their hardworking in making SEAAN meeting success and a high quality standard.

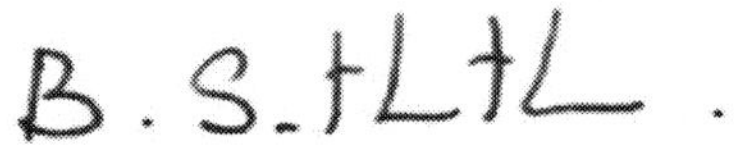

Prof. Boonrucksar Soonthornthum
President of SEAAN

IOP Conf. Series: Journal of Physics: Conf. Series **1231** (2019) 011001 doi:10.1088/1742-6596/1231/1/011001

## LIST OF COMMITTEE

**Conference Chair:**
Toto Winata – Institut Teknologi Sumatera (ITERA)

**International Advisory Boards:**
Boonrucksar Soonthornthum – Thailand
Shengbang Qian – China
Bambang Hidayat - Indonesia
Thomas Djamaluddin – Indonesia
Paul T. P. Ho – USA
Osamu Hashimoto-Japan

**Scientific Committee**
Hakim L. Malasan – Indonesia (Chair)
Thomas Djamaluddin - Indonesia
Boonrucksaar Soonthorntum – Thailand
Hasan Abu Kassim - Malaysia
Abel Yang - Singapore
Pham Ngoc Diep - Vietnam
Khin Swe Myint - Myanmar
Cynthia Celebre- Philippines
Chey Chan Oeurn - Cambodia

**Organizing Committee**
Robiatul Muztaba. – Institut Teknologi Sumatera (ITERA)
Nindhita Pratiwi – Institut Teknologi Sumatera (ITERA)
Okta Pilopa – Institut Teknologi Sumatera (ITERA)
Wahyu Sasongko Putro – Institut Teknologi Sumatera (ITERA)
Dimas Triyono – Institut Teknologi Sumatera (ITERA)
Destriyadi – Institut Teknologi Sumatera (ITERA)
Wirid Birastri – Institut Teknologi Sumatera (ITERA)
Annisa Novia Indra Putri – Institut Teknologi Sumatera (ITERA)
Ikah N.P. Permanasari – Institut Teknologi Sumatera (ITERA)
Vicka Tamaya – Institut Teknologi Sumatera (ITERA)
Danni Gathot Harbowo – Institut Teknologi Sumatera (ITERA)
Juwita Indrya Saputra – Institut Teknologi Sumatera (ITERA)
Muhammad Hajid An Nur – Institut Teknologi Sumatera (ITERA)
Heru Ruwandar – Institut Teknologi Sumatera (ITERA)
Rahmat Nawi Siregar – Institut Teknologi Sumatera (ITERA)
Bambang Prasetio – Institut Teknologi Sumatera (ITERA)
Susi Susyanti - Institut Teknologi Sumatera (ITERA)

IOP Conf. Series: Journal of Physics: Conf. Series **1231** (2019) 011001 doi:10.1088/1742-6596/1231/1/011001

## CONFERENCE PHOTOGRAPH

IOP Conf. Series: Journal of Physics: Conf. Series **1231** (2019) 011001 doi:10.1088/1742-6596/1231/1/011001

IOP Conf. Series: Journal of Physics: Conf. Series **1231** (2019) 011001 doi:10.1088/1742-6596/1231/1/011001

IOP Conf. Series: Journal of Physics: Conf. Series **1231** (2019) 011001 doi:10.1088/1742-6596/1231/1/011001

# Peer review statement

All papers published in this volume of *Journal of Physics: Conference Series* have been peer reviewed through processes administered by the proceedings Editors. Reviews were conducted by expert referees to the professional and scientific standards expected of a proceedings journal published by IOP Publishing.

# Pattern recognition as tools for understanding night sky brightness variabilities

**M Rezky**[1,*], **R Priyatikanto**[2], **A G Admiranto**[2] and **E Soegiartini**[1]

[1] Department of Astronomy, Institut Teknologi Bandung, Bandung, Jawa Barat 40132, Indonesia
[2] Space Science Centre, National Institute for Aeronautics and Space (LAPAN), Bandung, Jawa Barat 40173, Indonesia

Email: mrezky088@gmail.com

**Abstract.** Night sky brightness (NSB) research related to the artificial light pollution issues has been increasing all over the world, using various measurement techniques and tools. The research produces tens of thousands of data for each month such that proper handling, processing and analysing the data become challenging. In this article, we demonstrate an alternative method for processing the NSB data by utilising pattern recognition techniques: Canny edge detection and Hough transform. These techniques were applied to identify data and extract important parameter from the NSB density plot semi-automatically. Datasets collected from Bandung, Garut, Subang and Sumedang used as the test cases. Three time-segments (dusk, night and dawn) became the main focus of the analyses and our method successfully extracted following parameters: the rate of sky brightness change at dusk and dawn, the average NSB at night and the intersections which indicate transition time. This method, along with its many possible improvements, enables us to process data more effectively and encourage more observation campaigns to be conducted in the future.

***Keywords:*** *Night sky brightness (NSB) and Artificial Light Pollution*

## 1. Introduction

The artificial light pollution has become a growing issue on the 21st century, especially within urbanised communities who concerned about its negative impacts on the environment [1] and human health [2] ground-based astronomical observatories that affected by the brighter night sky. Night sky brightness (NSB) measurements are set up as a way to quantify light pollution and mitigate its impact. Since Cinzano et al. (2001) presented their first artificial NSB World Atlas [3] which became a prominent reference for many light pollution research, many in-situ measurements have been conducted all around the globe (e.g. [4], [5]). There is also continuing improvements in measurement methods and tools to gather more accurate and reliable data [6].

NSB measurements can generate thousands of datasets for every night. Depends on the objectives, continuous measurements may be conducted as a short-term or long-term activity. For achieving long-term data (e.g. [7]), researchers are facing challenges such as increasing datasets as well as varying meteorological conditions and ageing of the measurement tools. Posch et al. (2018) have developed several ways to visualise large datasets on many types of diagram [8]. In some cases, we need to investigate the NSB growth rate over time and detecting outliers that may occur during the measurement period. Added with the necessity to measure the sky brightness transition at dawn or dusk, a researcher may need additional tools for differentiating important data from outliers and also clustering datasets in related time segments (dawn, night and dusk). With many datasets involved, pattern recognition in data computation became a powerful instrument to implement this task.

In this article, we demonstrate an alternative method for NSB data processing by using pattern recognition techniques to extract some features needed to produce a result that corresponds with the growing trend of light pollution phenomenon. This method will allow us to treat the datasets as values that represent particular concentrations rather than as individual points, so we could focus on meaningful data points and extract information about NSB change rate trends.

## 2. Data source and measurement methods

The NSB datasets analysed in this paper have been carried out at LAPAN's four observation stations in West Java (see table 1). Measurements were conducted using Unihedron Sky Quality Meter (SQM) – LU type that was placed inside weather-proof housing and directed to the zenith. This type of photodiode-based instruments has an effective field of view of ~20° and produce sky brightness data in mag/arcsec$^2$. All SQM was set to record zenithal sky brightness magnitude from 5 PM to 7 AM local time with a sampling rate of one minute, regardless of weather conditions and moon phases. Datasets used in this study were obtained from April to July 2018. These raw datasets archived at LAPAN Space Science Database in Bandung.

**Table 1.** Geographical coordinates and conditions of 4 NSB observation stations.

| Stations | | Longitude | Latitude | Types |
|---|---|---|---|---|
| Bandung | (BDG) | 107° 40' 40.21" E | 6° 55' 32.03" S | Urban |
| Subang | (SBG) | 107° 46' 07.43" E | 6° 33' 44.99" S | Suburban |
| Sumedang | (SMD) | 107° 50' 13.97" E | 6° 54' 47.08" S | Suburban |
| Garut | (GRT) | 107° 41' 31.97" E | 7° 39' 00.22" S | Rural |

## 3. Data processing

The purpose of utilising pattern recognition techniques for NSB data processing is to identify features, especially linear trends, from the scotogram (see figure 1(a)). The scotogram is a density plot of sky brightness (or inversely sky darkness) magnitude values as a function of time. The scotograms in this article have a horizontal resolution of 5 minutes and a vertical resolution of 0.1 mag/arcsec$^2$. It becomes the primary input of the pattern recognition algorithm which consists of two main parts: (1) Canny edge detection and (2) Hough transform. We will discuss the main idea of each algorithm and its role in the scotogram analysis in the next subsections. We are using OpenCV modules [9] on Python 3.5.2 version to execute both of the methods.

### *3.1. Canny edge detection*

Canny edge detection [10] is a computational procedure developed by John Canny to determine arbitrary edge profiles on the desired image. It was designed to detect a single boundary between contrasting signal-to-noise region in the image. The scotogram image is applied with a 5 × 5 Gaussian blur kernel to smoothen the image and reducing noise and 3 × 3 Sobel operation to determine segments or boundary lines which are represented by pixel intensity gradient in the horizontal and vertical direction of the image ($G_x$ and $G_y$). Both values determine the overall intensity gradient $G$ and direction $\theta$,

$$G = \sqrt{G_x^2 + G_y^2} \quad (1)$$

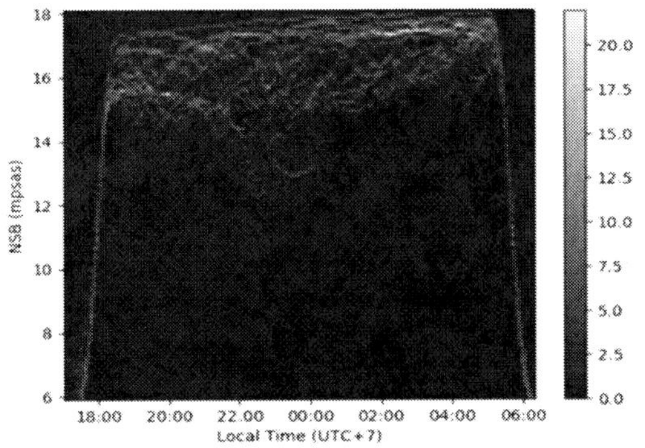


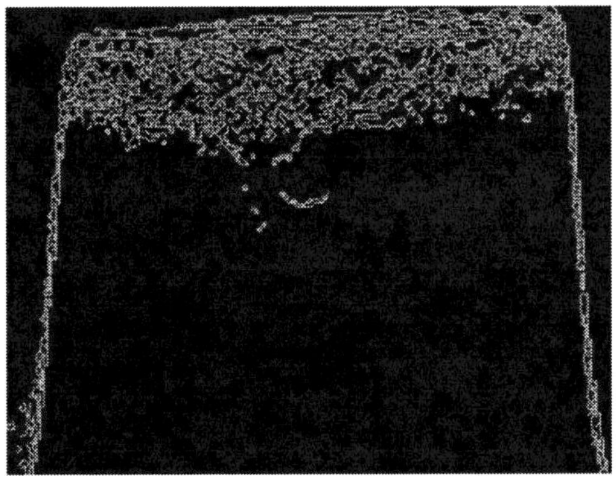

$$\theta = tan^{-1}\left(\frac{G_y}{G_x}\right). \tag{2}$$

Those boundary lines then undergo an edge detection process, which involves two user-defined thresholds (upper and lower), called hysteresis thresholding. Those thresholds play a role in filtering which local maximum pixels can construct an edge line. Pixels with higher intensity than upper limit are automatically accepted as a real local maximum. Pixels with intensity level below the lower limit are automatically denied. Pixels with intensity level between those limits could be accepted as a real local maximum only if such pixel can connect to a nearby real local maximum within few steps, or else it is denied. Those real local maximum pixels will construct a real edge line for the image as shown in figure 1(b). In this article, we used 14-15 variations of threshold range to examine how threshold ranges affect the final results.

### *3.2. Hough transform*

To detect straight or curved lines in the image, Hough transform [11] is employed. This procedure involves pixel coordinate transformation between Cartesian space $(x, y)$ and parameter space $(\theta, \rho)$ which described as equation

$$x \cos\theta + y \sin\theta = \rho. \tag{3}$$

The transformed parameters space with $\theta$ and $\rho$ as its axis is called the Hough space. Each active pixel on the original image is represented as a sinusoidal curve in Hough space. Then the algorithm will produce a list of $(\theta, \rho)$ coordinates that correspond to nodes of sinusoidal curves, which defined as local maxima on Hough space. Each node contains information about the total of curves accumulated on the node, determines its weight (counts) against other nodes.

Nodes with the highest weights in the Hough space are selected and converted back into Cartesian space as a representative trend line of the datasets. Additionally, users can also assign how many nodes or how low the counts threshold of the nodes that can be issued as the output of this process. In the context of this article, this means that we could treat all NSB datasets in a Hough space, shown in figure 1(c), and then extract a representative trend line for each of the time segments without split them manually, as seen in figure 1(d).

## 4. Results and Discussion

Our data processing algorithm had successfully extracted linear feature for each time segment, e.g. dusk, night and dawn. The segments were articulated as three clusters of nodes in Hough space at which the most significant nodes were extracted. For every line, the gradient and y-intercept were stored as the output for further analysis.

Linear features obtained by Hough transform can be translated into several parameters. First, line gradients at dusk (or at dawn) are the gradual change of sky brightness over time as the Sun goes away from (or to) horizon. The average value of sky brightness change at four observing sites is 0.23 mag/arcsec$^2$/min which is in agreement with the result reported by Tyson & Gal [12], but lower than the value obtained at Paranal, Chile, which is 0.26 mag/arcsec$^2$/min for V band [13]. There is no systematic difference between the changes at dusk and dawn, but brightness changes observed at SBG and GRT stations which are close to sea level are steeper compared to the one from BDG station (~750 masl). All values are summarised in table 2. Since the changes of surface brightness at these time segments are related to the scattering of sunlight in the atmosphere, it is plausible to correlate the difference of sky brightness changes to the difference in atmospheric pressures or to the atmospheric column density.

**Table 2.** Sky brightness parameters obtained by the algorithm for different observing stations, including associated errors (in bracket). The gradients for the night segment (rightmost column) express the whole changes overnight. A negative value means suspicious brightening. The NSB values were not corrected by -0.25 mag/arcsec$^2$ offset that caused by instrument cover.

| Stations | Dusk segment | Dawn segment | Night segment | |
|---|---|---|---|---|
| | Gradient (mag/arcsec$^2$/min) | Gradient (mag/arcsec$^2$/min) | Average NSB (mag/arcsec$^2$) | Gradient (mag/arcsec$^2$/min) |
| BDG | 0.22 (0.00) | -0.21 (0.00) | 17.25 (0.08) | 0.77 |
| SBG | 0.23 (0.04) | -0.23 (0.00) | 18.55 (0.08) | 0.21 |
| SMD | 0.21 (0.04) | -0.25 (0.00) | 20.39 (0.34) | -0.69 |
| GRT | 0.27 (0.06) | -0.22 (0.00) | 20.86 (0.26) | -0.01 |

The second parameter produced by the algorithm is related to the NSB itself. In this context, the NSB value is the most statistically significant value, associated with a high density at scotogram. NSB values during cloudy and overcast condition and NSB values with moonlight contamination were excluded by the algorithm. The extracted line indicates the most representative model of NSB during the measurement period. However, there were indeed variations in the obtained NSB values and trends due to the tuning of input parameters required by the algorithm, which we discussed later in this section.

In general, urban area (BDG) has the brightest night sky that reaches 17.25 mag/arcsec2 while GRT becomes the station with the darkest sky (20.86 mag/arcsec2). It is important to note that the obtained values were not corrected by -0.25 mag/arcsec2 offset due to the housing window. Slight changes of NSB were observed, especially at the BDG station where the sky at early night was approximately 2 times brighter than the sky before dawn (0.77 mag/arcsec2 difference). Positive gradient means that the sky is getting darker over time, either by outdoor light reduction that is associated with human activities at night or by alteration of atmospheric condition. The average NSB for any station is defined as the midpoint of night segment trend line between dusk-to-night (d-n) and night-to-dawn (n-d) intersections.

The last parameters to be emphasised are the intersections between the segments. The intersections have notable importance in astronomical observation scheduling and also in the determination of prayers time for Muslims. In this article, the intersections are denoted in the local time since the analysis was done to measurements data obtained in a relatively short duration. During this period, there was a slight change in the setting and rising time of the Sun (~15 minutes). As an insight, the obtained time of d-n and n-d transitions are presented in table 3. For more proper analysis, utilisation of scotogram with Sun altitude as the horizontal axis is highly recommended, though the algorithm explained in this article is still applicable.

**Table 3.** Transition time range on dusk-to-night (d-n) and night-to-dawn (n-d) time segment (in 24-hour format, UTC+7).

| Stations | d – n transition | n – d transition |
|---|---|---|
| BDG | 18:38 – 18:40 | 05:14 – 05:19 |
| SBG | 18:27 – 18:35 | 05:13 – 05:14 |
| SMD | 18:25 – 18:35 | 05:07 – 05:12 |
| GRT | 18:24 – 18:32 | 05:00 – 05:03 |

Parameters extracted by the algorithm from the scotogram do not always represent the current conditions of light pollution in a particular area, but the models could be updated as time goes by. These models could be our basis to understand the improvement/worsening trend of the pollution periodically. It is worth noted that this kind of evaluation must be treated carefully to produce reliable conclusions as there are many contributing factors affects accuracy, such as the climate of the measurement station or the degrading measurement instrument sensitivity. For example, we could compare the datasets that produced at a particular season that year with the datasets from the similar season a year before. Combining datasets from different seasons will not generate a proper model because of the difference in night duration (especially in temperate regions) or cloud coverage.

It is also important to note that the outputs generated by the algorithm are subject to change due to the hysteresis threshold variations used. As mentioned in subsection 3.1, the variation of input parameters on the Hough transform algorithm (the thresholds) may produce various results. In this case, there are 14-15 different results obtained on each station dataset. Varied input parameters could produce different level of spreads or errors in the output parameters. For NSB, the range was less than 0.3 mag/arcsec$^2$ or 2% relative error. For the brightness gradient at dusk, the maximum spread was 0.06 mag/arcsec2 or approximately 20% relative error. This considerably large error could be associated with the nature of data submitted into the algorithm. The utilisation of scotogram as a function of Sun altitude is expected to suppress the error at dusk and dawn segments.

By comparing each varying results, we believe that there is no clear relation between the hysteresis threshold variations and the resulting data from it. The reason is that the lower and upper threshold affected differently on the pixels evaluated in images and the fact that thresholds determination is still subjective, which means it depends on the user's judgement to assign specific values on the thresholds that satisfy their needs. This subjectivity could hamper the results produced by the Hough transform, because its input already suffers from Canny edge algorithm's poor anti-noise ability [14] and makes possible for the noises to be slipped off the results, increasing its deviation value. It was also found that the lower threshold determination influences the results more than the upper threshold because it sets the tolerance for a pixel in the detection process to be considered as building pixels of the real edge.

However, this data processing algorithm has a potential to be developed further into a semi-supervised machine learning algorithm. By modifying the current methods and made the program learns into and evaluate the results, we could gather more information about threshold variabilities and its effect on the trend lines, which will be given to the algorithm to produce the most acceptable set of rules on threshold determination. Medina-Carnicer et al. even suggest that it is possible to modify it into unsupervised learning [15]. Another problem that could be solved is the boundaries between each time segment. For this article, we set the boundary values based on visual information on the Hough space, which will generate some trouble when we are dealing with more data points. By letting the algorithm clustering the nodes by itself, it will ease the voting procedures when we need to produce trend lines for each time segment.

## 5. Conclusion

By utilising pattern recognition methods like Canny edge detection and Hough transform to the sky brightness measurement datasets, information about NSB variabilities (trend lines and its characteristic values) for dusk, night and dawn time segment can be extracted. Bandung has the brightest sky (NSB = 17.26 mag/arcsec2) among four stations evaluated, while Garut has the darkest one (NSB = 20.86 mag/arcsec2). Besides, typical decrement/increment of sky brightness at dusk/dawn was detected with an average change of 0.23 mag/arcsec2/min. This information could be used to examine the changing environment effects on light pollution problems in general and long-term scale.

Problems that occurred during implementation of this methods, like the hysteresis threshold on Canny edge detection and boundary value determination on Hough space were identified. Further development and modification on the algorithm into semi-supervised algorithm are expected to resolve these problems.

**Acknowledgement**

We acknowledge the observers at LAPAN's observation stations and Nana Suryana at Subang for acquiring the NSB datasets in their respective station. We also acknowledge the help of Siti Maryam and Elyyani for maintaining the archive of NSB data at LAPAN Space Science Center. This article is a product of short-term internship program (Kuliah Kerja Astronomi) supervised by Dr. Hesti Retno Tri Wulandari. We also acknowledge financial support by LAPAN and FMIPA ITB for this publication.

## References

[1] Longcore T and Rich C 2004 *Front. Ecol. Environ.* **2** 191–8

[2] Reiter R J, Gultekin F, Manchester L C and Tan D-X 2006 *J. Pineal Res.* **40** 357–8

[3] Cinzano P, Falchi F and Elvidge C D 2001 *Mon. Notices Royal Astron. Soc.* **328** 689–707

[4] Puschnig J, Schwope A, Posch T and Schwarz R *J. Quant. Spectrosc. Radiat. Transf.* **139** 76–81

[5] Bará S 2016 *R. Soc. Open Sci.* **3** 160541

[6] Hänel A, Posch T, Ribas S J, Aubé M, Duriscoe D, Jechow A, Kollath Z, Lolkema D E, Moore C, Schmidt N, Spoelstra H, Wuchterl G and Kyba C C 2018 *J. Quant. Spectrosc. Radiat. Transf.* **205** 278–90

[7] Ochi N and Wuchterl G 2015 *Journal of Toyo University* **58** 1–12

[8] Posch T, Binder F and Puschnig J 2018 *J. Quant. Spectrosc. Radiat. Transf.* **211** 144–65

[9] Bradsky G 2000 *Dr. Dobb's Journal of Software Tools*

[10] Canny J 1986 *IEEE Trans. on Pattern Anal. Mach. Intell.* **PAMI-8** 679–98

[11] Duda R O and Hart P E 1972 *Commun. ACM* **15** 11–5

[12] Tyson N D and Gal R R 1993 *Astron. J.* **105** 1206–12

[13] Patat F, Ugolnikov O S and Postylyakov O V 2006 *Astron. Astrophys.* **455** 385–93

[14] Feng Y, Zhang J and Wang S *AIP Conf. Proc.* **1890** 040011

[15] Medina-Carnicer R, Munoz-Salinas R, Yeguas-Bolivar E and Diaz-Mas L *Pattern Recognit.* **44** 1201–11

# Spatial Analysis of Light Pollution Dynamics Around Bosscha Observatory and Timau National Observatory Based on VIIRS-DNB Satellite Images

**H A Prastyo**[1,*], **D Herdiwijaya**[2]

[1]Master Program in Astronomy, Faculty of Mathematics and Natural Sciences, ITB, Bandung, Indonesia
[2]Astronomy Research Group and Bosscha Observatory, Faculty of Mathematics and Natural Sciences, ITB, Bandung, Indonesia

Email: prastyohendra@gmail.com

**Abstract.** Bosscha Observatory is the largest and oldest observatory in Indonesia that has been actively doing astronomical research since 1928. Bosscha Observatory was originally very suitable for observing astronomical objects because of its ideal night sky conditions and covered most of the northern and southern sky areas. However, starting at the end of 1980, the quality of the night sky at Bosscha Observatory has decreased along with the development of cities around the observatory. Therefore, the construction of a new observatory in Indonesia is currently underway, i.e. the Timau National Observatory. One of the main causes of the decreasing quality of the night sky at Bosscha Observatory is light pollution. Light pollution is one form of pollution caused by excessive light outside the room released into the sky, causing an increase in brightness of the night sky. The brightness increasing of the night sky causes relatively faint astronomical object, e.g. Milky Way Galaxy that is difficult to observe. Based on the analysis of the dynamics of light pollution in a radius of 20 km from Bosscha Observatory and at a radius of 90 km from the Timau National Observatory in 2013-2017 using VIIRS-DNB satellite imagery, it is known that there is a wide change in light pollution for the very low, low, medium, high, and very high. The average area of light pollution at a radius of 20 km from Bosscha Observatory for the category of very low, low, medium, high, and very high categories experienced a rate of $-41.3 \pm 135.2$ km$^2$ / year, $5.9 \pm 97.2$ km$^2$ / year, $15.1 \pm 14.6$ km$^2$ / year, $18.0 \pm 42.6$ km$^2$ / year, and $2.1 \pm 10.2$ km$^2$ / year. The rate of increase in the extent of light pollution for the medium, high, and very high categories occurs south of the Bosscha Observatory, which is towards the city of Bandung. The average area of light pollution in a 90 km radius from the Timau National Observatory for the very low, low, medium, and high categories experienced a rate of $20.8 \pm 43.4$ km$^2$ / year, $-22.1 \pm 40.9$ km$^2$ / year, $3.0 \pm 7.3$ km$^2$ / year, and $-1.7 \pm 7.3$ km$^2$ / year. Therefore, the night sky in the 128-268 degrees azimuth direction at Bosscha Observatory is relatively not ideal as a location for observing astronomical objects, and the Timau National Observatory is still relatively ideal as a location for observing astronomical objects, but it is necessary to monitor the dynamics of light pollution in the direction of 207-215 degrees azimuth which leads to Kupang City.

***Keywords:*** *Light Pollution Dynamics, Bosscha Observatory, Timau National Observatory, and VIIRS-DNB Satellite Images*

## 1. Introduction

Since the invention of the electric lamp in 1879 by Thomas Alva Edison, the use of artificial light for outdoor lighting at night has increased. Gradually, along with increasing population, an increase in the amount of artificial light as outdoor lighting causes an increase in the brightness of the night sky and produces light pollution. Light pollution is one form of environmental pollution due to artificial light at night (ALAN), where the artificial light at night is increasingly out of control, both in terms of number and method of installation. Light pollution is defined as artificial light that increases the brightness of the sky by 10% of the brightness of the natural sky above a height of 45° [1]. According to the International Dark Sky Association (IDSA) [2], light pollution has several types, i.e. sky glow - visible light in the urban sky due to artificial light, glare - dazzling artificial light, light trespass - the entry of light that is not desirable and not needed from the outside into a person's home resulting in difficulty sleeping, and light clutter - a group of artificial light sources that causes visual disturbances. Astronomers are the first realizing that the dark sky has been lost due to light pollution [3].

Light pollution is caused by excessive use of artificial outdoor lightings such as street lighting, fluorescent markers, and luminous advertising board lights [4]. Light pollution is increasing along with the development of an area, especially urban areas. The California Energy Commission or the California Energy Commission in 2005 stated that in the United States, about 6% of 4,054 million megawatt hours (MWh) of electricity was used for outdoor lighting and an estimated 30% of outdoor lighting was a source of light pollution [5]. Light pollution prevents astronomers from getting a clear night sky and reducing the visibility of the stars [6]. Based on the four main components classified as light pollution, sky glow is the main cause of the decline in the quality of the night sky around the observatory. Sky glow is formed from the reflection and scattering of light by molecules and aerosols in the atmosphere [7] and many cities that produce sky glows that can be observed as far as 100 miles [5]. This shows that light pollution is the main problem currently faced by observatories around the world, one of which is the Bosscha Observatory in Indonesia.

Bosscha Observatory is one of the observatories in Southeast Asia and is the largest observatory in Indonesia. The Bosscha Observatory was established in 1923 and in 1928, the Zeiss double refractor telescope was installed at the Bosscha Observatory. The existence of the Zeiss double refractor telescope in 1928 became the beginning of the Bosscha Observatory in conducting astronomical observations and research. At the beginning of the establishment of the Bosscha Observatory, Lembang was chosen as an observatory location because, at that time, Lembang was a hilly area surrounded by tea plantations and had a dark night sky, making it ideal for astronomical observations. In addition, the location of the Bosscha Observatory is a strategic location because it covers most of the northern and southern sky areas, considering only a few observatories in the world in the equatorial region.

However, starting in late 1980, the night sky conditions at Bosscha Observatory experienced a decline. This was due to the development of cities around the Bosscha Observatory, i.e. the City of Lembang and the City of Bandung. The development of these cities led to a change in land use around the Bosscha Observatory, i.e. the North Bandung Region from the forest, plantation, and livestock areas to the area of built land (settlements and hospitality). The developed land area is expanding and increasing the use of artificial light for lighting at night. An increase in the amount of artificial light for lighting at night causes light pollution and produces a light dome above the cities around the Bosscha Observatory. Therefore, the construction of a new observatory in Indonesia is currently underway, i.e. the Timau National Observatory. The problem of light pollution around the Bosscha Observatory is a spatial problem, so to overcome it, it is necessary to analyze the dynamics of light pollution spatially, one of them is by using VIIRS-DNB satellite imagery and as a comparison, a similar study was carried out around the national observatory of Timau.

VIIRS-DNB (Visible Infrared Imaging Radiometer Suite-Day / Night Band) satellite imagery is a night-time satellite image managed by the Earth Observation Group, National Oceanic and Atmospheric Administration (NOAA), where this satellite image measures the radiance of night-time lighting on the Earth surface in the new Moon phase [7]. Nightly satellite imagery of VIIRS-DNB has visual and infrared wavelength channels for radiometric measurement of land, atmosphere, cryosphere, and oceans. Nightly satellite imagery data of VIIRS-DNB is used to measure cloud and aerosol objects, ocean color, sea and land surface temperature, ice movement and temperature, hotspots, and earth albedo. Climatology experts use the VIIRS-DNB data to increase understanding of climate change globally. VIIRS-DNB satellite imagery provides cloud imagery during the day and some image data in the infrared channel for cloud imaging applications day and night. In addition, VIIRS-DNB image data is also used to measure weak light radiations for studies of lighting energy estimates [8]. VIIRS-DNB satellite imagery was integrated with sky brightness measurement data using CCD and SQM to make light pollution atlases around the world [9]. Thus, VIIRS-DNB satellite imagery can be used to analyze the dynamics of light pollution at a radius of 20 km from Bosscha Observatory and at a radius of 90 km from the Timau National Observatory.

## 2. Methodology

The object of this research is light pollution around Bosscha Observatory (107° 37' E; 6° 49' S; altitude 1300 m above sea level) and Timau National Observatory (123° 56,794' E; 9° 35,77' S; altitude 1281 m above sea level) with 20 km radius and 90 km from the observatory. The data used in this study is the VIIRS-DNB nighttime satellite data in 2013-2017. The classification of light pollution levels is divided into 5 classes (very low, low, medium, high, and very high). The classification process from VIIRS-DNB at night satellite imagery uses two classification methods, the unsupervised classification method for mapping the level of light pollution around Bosscha Observatory and the supervised classification method for mapping the level of light pollution around the Timau National Observatory.

**Table 1.** Classification of Light Pollution

| Class | Digital Number Value ($10^9$ nW/cm$^2$/sr) |
|---|---|
| Very Low | -0.13 – 0.90 |
| Low | 0.91 – 3.40 |
| Medium | 3.41 – 8.37 |
| High | 8.39 – 18.16 |
| Very High | 18.20 – 86.01 |

### *2.1. Unsupervised Classification*

The process of making a map of light pollution around the Bosscha Observatory with a radius of 20 km is to classify the level of light pollution based on satellite imagery at night VIIRS-DNB. The method used in the classification process is the unsupervised method with the K-Means algorithm. The unsupervised classification process is carried out without user intervention in selecting a group of data sets that will be used as a reference for classification [10]. The user role is only limited to determine the number of classes or clusters and the number of iterations. K-Means algorithm is implemented by moving a set of clusters (center) recursively using the closest distance to mean approach until the center location of the cluster does not change, or until changes from one iteration to the next iteration are less than some predetermined thresholds [10]. Thus, the classification process with this algorithm produces a relatively accurate classification. The classification process using the K-Means algorithm is iterated 10 times to get a stable classification result.

### *2.2. Supervised Classification*

The process of making a map of light pollution around the National Timau Observatory with a radius of 90 km is to classify the level of light pollution based on satellite imagery at night VIIRS-DNB. The method used in the classification process is a supervised method with a Minimum Distance algorithm. The supervised method was chosen because most of the area around the Timau National Observatory to a radius of 90 km is relatively dark. If the classification process uses the unsupervised method, the classification process is not in accordance with the actual conditions. Therefore, the supervised method is selected and the range of image pixel values for each class of light pollution is adjusted to the range of image pixel DN values around the Bosscha Observatory. Then, the Minimum Distance algorithm is selected because this algorithm performs a classification process similar to the K-Means algorithm. The Minimum Distance Algorithm classifies based on the closest distance of the sample pixels to cluster center points.

IOP Conf. Series: Journal of Physics: Conf. Series **1231** (2019) 012002 doi:10.1088/1742-6596/1231/1/012002

## 3. Result

### *3.1. Light Pollution Map 20 km From Bosscha Observatory*

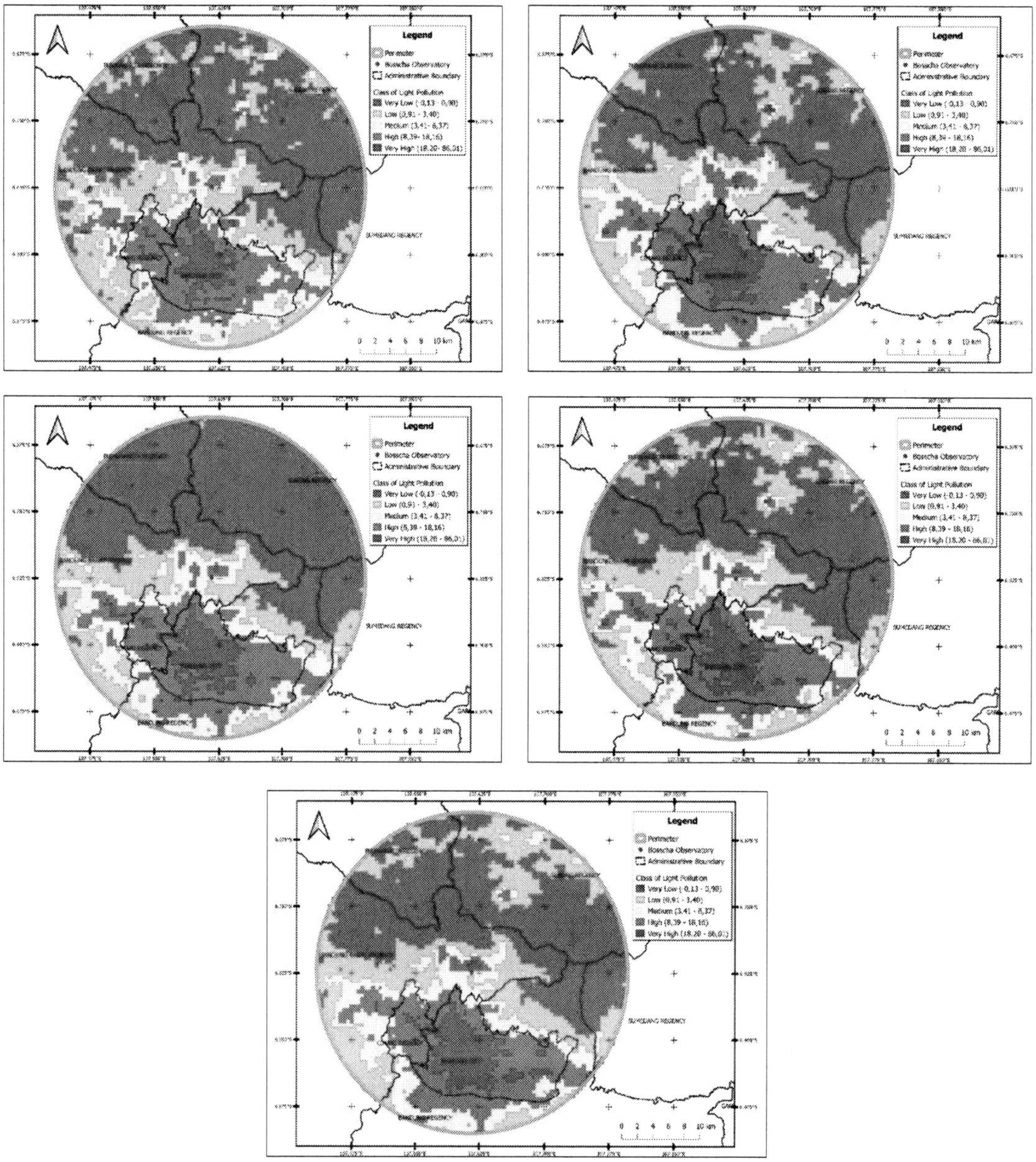


**Figure 1.** Light Pollution Map Around Bosscha Observatory in 2013-2017 (left to right from up to down)

### *3.2. Light Pollution Map 90 km From Timau National Observatory*

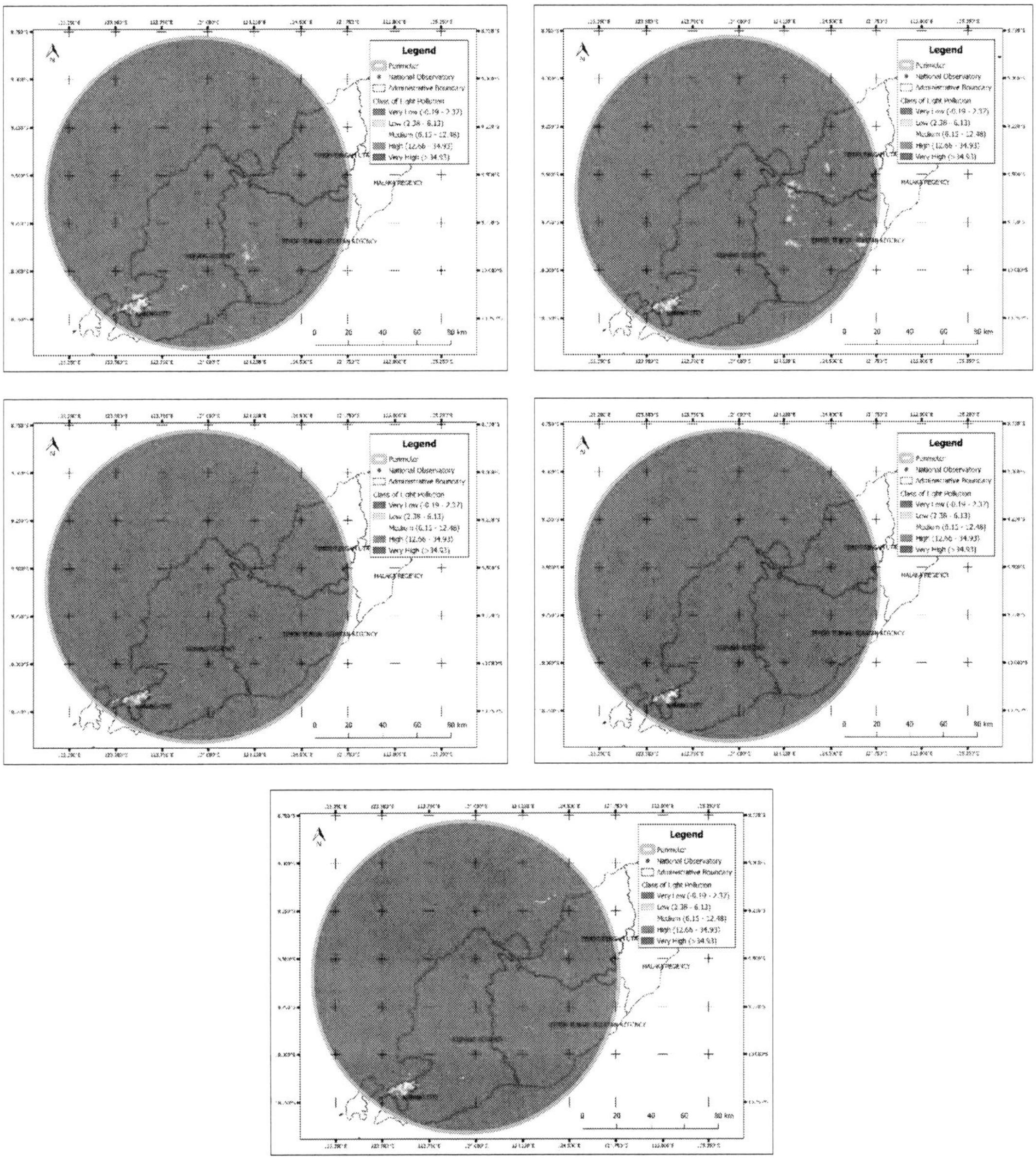


**Figure 2.** Light Pollution Map Around Bosscha Timau National Observatory in 2013-2017 (left to right from up to down)

## 4. Discussion

### *4.1 Light Pollution Around Bosscha Observatory*

Based on the light pollution level map around the Bosscha Observatory (figure 1), it is known that the distribution of the level of light pollution around the Bosscha Observatory is relatively uneven. The level of light pollution for the very low and low class is mostly located in the northern region of Bosscha Observatory, while the level of light pollution for the medium, high, and very high class is mostly in the southern region of Bosscha Observatory or in the 129-268 degrees azimuth.

Agglomeration the level of light pollution around the Bosscha Observatory is strongly influenced by its geographical conditions. The northern region of Bosscha Observatory is a mountainous and hilly area, while the southern region of Bosscha Observatory is a basin area and is an urban area. Based on figure 1, the city of Bandung and Cimahi are areas that have high levels of light pollution. The dynamics of light pollution around the Bosscha Observatory and Timau National Observatory can be analyzed based on the area changes of each class of light pollution levels in 2013-2017. Based on the VIIRS-DNB image data in 2013-2017, it is known that there is area change of each class. The following is a graph of area changes in the extent of each category of light pollution levels around the Bosscha Observatory.

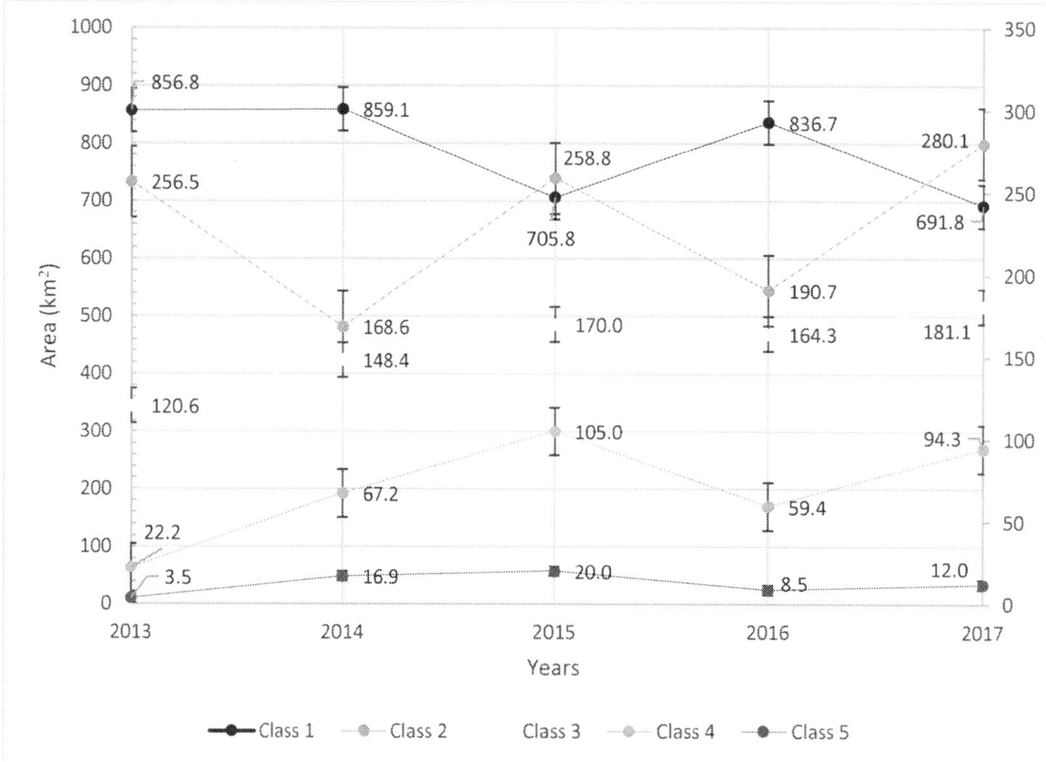


**Figure 3.** Extensive Area Changes in Each Class of Light Pollution Level Within Radius 20 km Centered from Bosscha Observatory.

Based on Figure 3, it is known that the area for very low classes did not experience very significant changes. In 2015, the area for very low classes decreased from 859.1 km$^2$ to 705.8 km$^2$. But in 2015 it increased to 836.7 km$^2$, and then in 2017, it decreased to 691.8 km. The area in 2017 is the lowest area for very low classes. Then, the area for the low class fluctuated. In 2014 and 2016 the area was decreased, i.e. 168.6 km$^2$ and 190.7 km$^2$. The area for the class is showing an increase from 2013-2017. Average changes in area per year for each class start from very low to very high classes of -41.3 ± 135.2 km$^2$ / year, 5.9 ± 97.2 km$^2$ / year, 15.1 ± 14.6 km$^2$ / year, 18.0 ± 42.6 km$^2$ / year, and 2.1 ± 10.2 km$^2$ / year. The average area change of light pollution for the high class has the highest value compared to the other classes. This shows that there is an increase in light pollution in the area around the Bosscha Observatory within a radius of 20 km. The average change in light pollution for the high category will certainly continue to increase along with the development of urban areas in the south and north of the Bosscha Observatory.

Based on Figure 1, the direction of the development of light pollution levels for the medium, high, and very high class in the south of the Bosscha Observatory tends to move eastward. This shows that the development of industrial areas in the eastern city of Bandung contributes to an increase in light pollution in the city of Bandung. Then, the development of the level of light pollution for the high category in the north of the Bosscha Observatory shows a relatively insignificant change in the area. However, because of its proximity to the Bosscha Observatory (<1 km), light pollution in the City of Lembang also affects the observation of astronomical objects in the northern sky even though it is not as severe as in the southern sky. This is because the City of Lembang has a much smaller administrative area compared to the cities of Bandung and Cimahi. Nevertheless, it is necessary to periodically monitor the dynamics of light pollution in the northern region of Bosscha Observatory.

### *4.2 Light Pollution Around Timau National Observatory*

Based on the light pollution level map around the observatory (figure 2), it is known that within a 90 km radius, the Timau National Observatory area is mostly in very low class. Low, medium and high classes can only be found around Kupang City. This shows that the Timau National Observatory area is still relatively ideal as an observatory location. The following picture is a chart of changes in the extent of each category of light pollution levels around the Timau National Observatory.

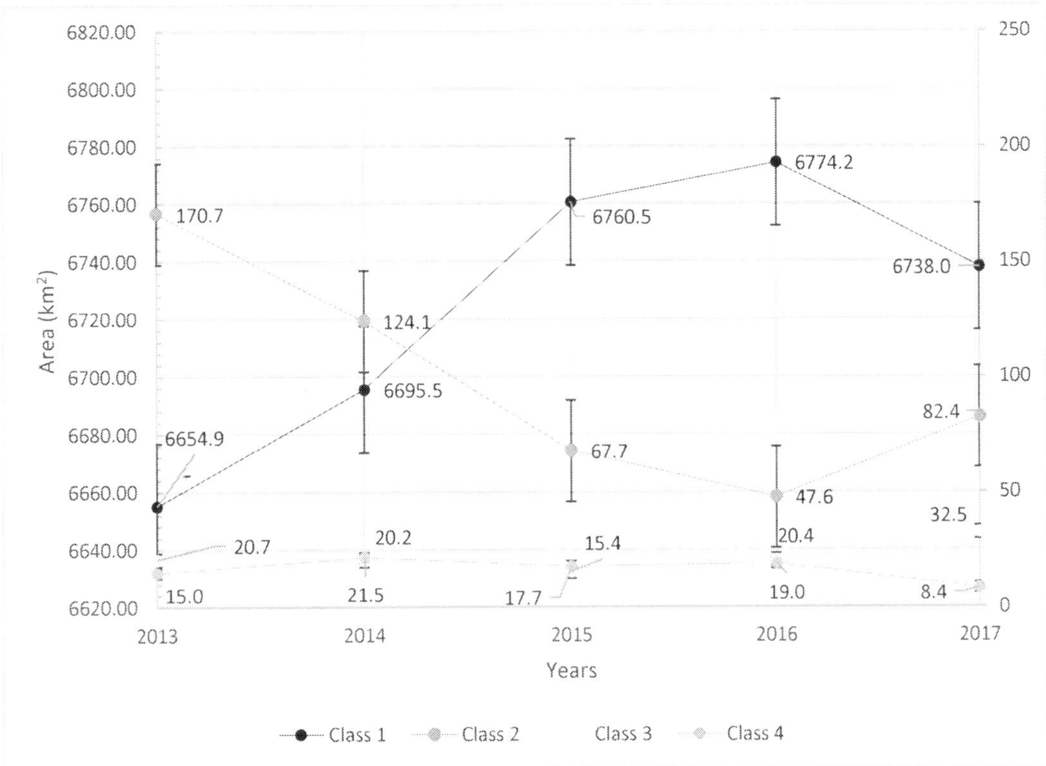


**Figure 4.** Extensive Area Changes in Each Class of Light Pollution Level Within Radius 90 km Centered from Timau National Observatory.

Based on Figure 4, it is known that very low classes area dominates the area around the Timau National Observatory. The very low classes area is also relatively unchanged which is indicated by a flat graphical form. Then, the low classes area experienced a decline from 2013-2016, from 170.7 $km^2$ in 2013 to 47.6 $km^2$ in 2016, but experienced an increase in 2017, which was 82.4 $km^2$. The middle and high classes area has a relatively similar value in 2014-2016, but in 2017 the class area is experiencing an increase to 32.5 $km^2$ from the previous in 2016 amounted to 20.4 $km^2$, while the high classes area has decreased to 8.4 $km^2$ from before in 2016 amounted to 19.0 $km^2$. The average change in area for each class is from very low to high at 20.8 ± 43.4 $km^2$ / year, -22.1 ± 40.9 $km^2$ / year, 3.0 ± 7.3 $km^2$ / year, and -1.7 ± 7.3 $km^2$ / year. The average change in the area of each class shows that the very low class does not experience an extensive decline, while the low class has a decrease of 22.1 ± 40.9 $km^2$ / year. This shows that the Timau National Observatory area is still relatively ideal as an observatory location. However, the development of the area south of the Timau National Observatory needs to be monitored periodically, especially in the 207-215 degrees azimuth range. This is because the development of the area in the south of the Timau National Observatory is predicted to grow further with the development of road access. If the development of the region in the south is not controlled, then the possibility of the condition of the southern sky at the Timau National Observatory will experience a very significant increase in light pollution.

## 5. Conclusion

The conclusion that can be obtained based on the analysis of the dynamics of light pollution around the Bosscha Observatory and Timau National Observatory is as follows:

a) The level of light pollution for the very low and low class is mostly located in the northern region of Bosscha Observatory, while the level of light pollution for the medium, high, and very high class is mostly in the southern region of Bosscha Observatory or in the 129-268 degrees azimuth. Grouping the level of light pollution around the Bosscha Observatory is strongly influenced by its geographical conditions. The northern region of Bosscha Observatory is a mountainous and hilly area, while the southern region of Bosscha Observatory is a basin area and is an urban area. Based on figure 1-5, the city of Bandung and Cimahi are areas that have high levels of light pollution.

b) The average area change of each class shows that the very low class does not experience an extensive decline. This shows that the Timau National Observatory area is still relatively ideal as an observatory location. However, the development of the area south of the Timau National Observatory needs to be monitored periodically, especially in the 207-215 degrees azimuth range. This is because the development of the area in the south of the Timau National Observatory is predicted to grow further with the development of road access. If the development of the region in the south is not controlled, then the possibility of the condition of the southern sky at the Timau National Observatory will experience a very significant increase in light pollution.

**Acknowledgments**

We thank to Magister Study Program in Astronomy, ITB for supporting this work. DH thanks to ITB's P3MI research fund.

## References

[1] Hu Z, Hu H and Huang Y 2018 *Environmental Pollution* **239** 30
[2] Wu B and Wong H 2012 *ISPRS Annals of the Photogrammetry, Remote Sensing and Spatial Information Sciences* **I-2** 171
[3] Aubé M and Roby J 2014 *Journal of Quantitative Spectroscopy & Radiative Transfer* **139** 52
[4] Pun C S J, So C W, Leung W Y and Wong C F 2014 *Journal of Quantitative Spectroscopy & Radiative Transfer* **139** 90
[5] Gallaway T, Olsen R N and Mitchell D M 2010 *Ecological Economics* **69** 658
[6] Elsahragty M and Kim J L 2015 *Procedia Engineering* **118** 480
[7] Duriscoe, D M, Anderson S J, Luginbuhl C B and Baugh K E 2018 *Journal of Quantitative Spectroscopy & Radiative Transfer* **214** 135
[8] Cao C and Bai Y 2014 *Remote Sensing* **6** 11915
[9] Falchi F, Cinzano P, Duriscoe D, Kyba C C M, Elvidge C D, Baugh K, Portnov B A, Rybnikova N A and Furgoni R 2016 *Science Advances* **2:e1600377** 1
[10] Mather P M and Tso B 2009 *Classification Methods for Remotely Sensed Data* CRC Press Boca Raton **2** 54

IOP Conf. Series: Journal of Physics: Conf. Series **1231** (2019) 012003 doi:10.1088/1742-6596/1231/1/012003

# Night sky brightness measurement during the March 2018 Earth Hour in Yogyakarta

**M Sukma[1,*],Y Pramudya[2],Muchlas[2] and Okimustava[2]**

[1] Graduate Physics education studies, Universitas Ahmad Dahlan IV, Jl. Ring Road Selatan, Tamanan, Baguntapan, Bantul, Yogyakarta, Indonesia
[2] Postgraduate studies of Physics education, Universitas Ahmad Dahlan II, Jl. Pramuka No. 42 sidikan Yogyakarta, Indonesia

Email: mega1500007072@webmail.uad.ac.id

**Abstract.** Switching off the light for one hour is annual global event to celebrate the Earth Hour. At that time, there is also campaign to observe the night sky with minimal light pollution. Hence, during this event, it is important to measure the level of light pollution using the Sky Quality Meter (SQM) device that was directed to the zenith. The research is intended to obtain comparison of the night sky brightness (NSB) between reguler days (23[th] and 25[th] of March) and Earth Hour day (24[th] of March). The measurement was performed in Yogyakarta that is located in -7.798893° south and 110.364992° East using the SQM-LU device. The location is the city center where the artificial light comes from the building, advertising boards, city lights and vehicles. The data distribution analysis was performed to obtain the mode value of NSB for each day. The mode value of NSB on 23[th], 24[th], 25[th] of March 2018 are 15.5 mag/arcsec$^2$ , 14.61 mag/arcsec$^2$ 15.02 mag/arcsec$^2$ and the mode value on the Earth Hour event is 17.4 mag/arcsec$^2$. Hence, it can be concluded that the night sky in Yogyakarta has a high level of light pollution. By turning off the light, the light pollution can be reduced.

***Keywords:*** *Night sky brightness and East Yogyakarta*

## 1. Introduction

The sources of light pollution are the natural lighting and excessive artificial lighting. The light pollution will make the sky glow thus the number of visible stars will be decreasing. The street lighting, garden lighting, billboard lighting and the stadium lighting are some example of the artificial light that potentially contributes to the light pollution. Research in the city of Hong Kong about the protection of light in several places including cities and rural areas including differences in the sky in these places. The measurement results in the city have a value of 15.0 mag/arcsec$^2$ while the city has a value of 20.1 mag/arcsec$^2$ [1]. the impact of this light pollution is not only to the astronomy but also in the other aspect. Light pollution also affects animal behavior. The glatik (great tits, Parus Major) who sleeps in a bed exposed to LED light all night long wakes up from its nest than birds that sleep when it's dark. The birds exposed to these lights earlier start the day out of the clover. The impact of the environment, especially for animals that are active at night (nocturnal), such as owls, one time the owls will migrate to a distant place and will return to place of origin. These birds use light from the moon as well as constellations to be used as a guide in migration. The astronomers are the most disadvantaged people on light pollution. They have difficulties to obtain the celestial object data since the sky is too bright the number of visible light is decreasing. As the pollution is getting worse, the bird migration got disrupted. The light pollution is bad for human health, especially for the hormonal system because the brightness of the lights will change the cycle of human hormones.Hence,changes in health and psychology in humans [2].

On March 24$^{th}$, 2018, there was an event of Earth Hour. On that event, the light was switching off for one hour. Therefore, all this time, it is very important to measure the level of light using the Sky Quality Meter (SQM) device which is directed to the zenith. This research was conducted to obtain accurate information of night sky brightness (NSB) between reguler days (23$^{th}$ and March, 25$^{th}$) and Earth Hour (March, 24$^{th}$). Hence, the effect of earth hour to the light pollution can be investigated. The figure 1 shows. The Light pollution is very high because the city center is full of artificial lighting

**Figure 1.** Location of data collection is in the cityater. The light pollution is high. (*https://www.lightpollutionmap.info*).

## 2. Methodology

Observation has been conducted at coordinates of -7.798893° South and 110.364992° East in Yogyakarta. The location is the city center. Hence, the light pollution is high due to lighting from building. The moon phase at the time of data collection was the first quarter. The SQM was pointed to the zenith as shown in figure 2. The data collection for 3 days on 23$^{rd}$, 24$^{th}$ and 25$^{th}$ of March 2018 with data collection time from 6:00 pm to 10:00 pm local time. The time interval for data collection is every 10 seconds

**Figure 2.** The design of data collection. The Sky Quality Meter (SQM) was directed to the zenith http://unihedron.com.

IOP Conf. Series: Journal of Physics: Conf. Series **1231** (2019) 012003 doi:10.1088/1742-6596/1231/1/012003

## 3. Data and Analysis

NSB on Earth Hour day and the day before and after Earth Hour day are shown in figure 3. On March 23 $^{rd}$ , the data collection was stop at 8:00 p.m due to rain. Figure 3 show that there is jump on NSB value on 24$^{th}$ of March. The NSB increase about 2 magnitude/ arcsec$^2$. This is due to the light being swithed off.

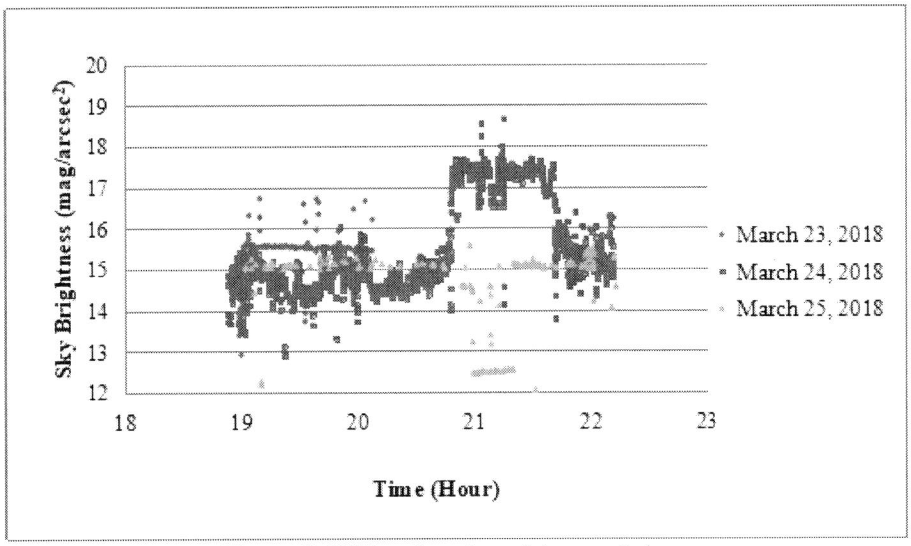


**Figure 3.** NSB as function of time on 3 days.

The figure 4 shows using statistical approach, the number of data at particular interval of NSB can be calculated. The mode of NSB data on March 23$^{th}$ is in the interval of 15.51 - 15.60 mag/arcsec$^2$. The mode of data collected on the 23$^{rd}$ of March is 2665 data at a value of sky brightness of 15.51 - 15.60 mag / arcsec$^2$. This shows the normal brightness value of the sky, the brightness value of the sky due to the amount of light pollution in the location.

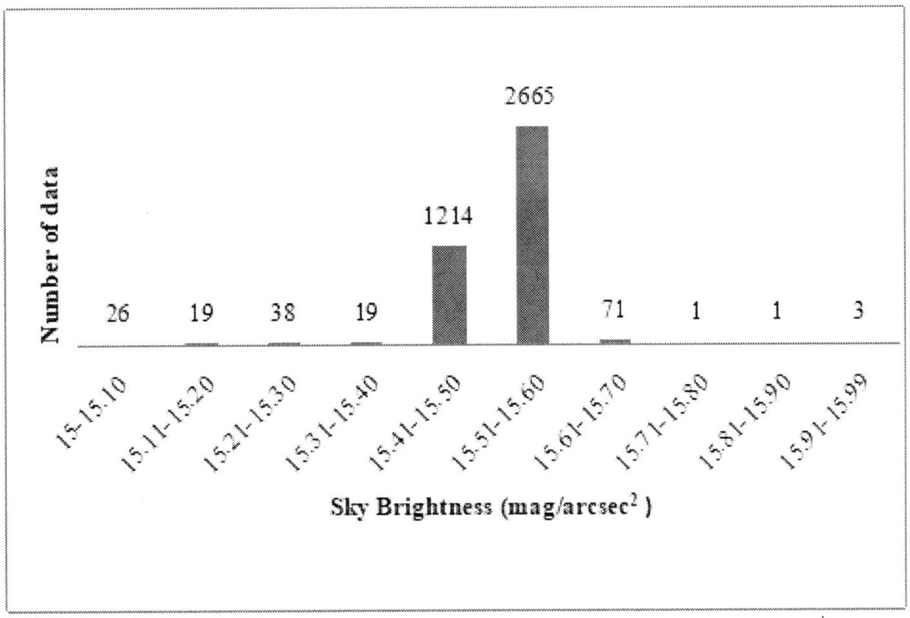


**Figure 4.** The number of data at particular interval NSB on March 23 $^{th}$ .

The figure 5 shows the mode of NSB at March 25$^{th}$ is in the interval of 12.41 - 12.50 mag/arcsec . The cloudy condition at March 23$^{th}$ caused the difference of mode between day before and after the Earth Hour. In addition, both days can be categorized as the regular day. The cloud make the sky darker

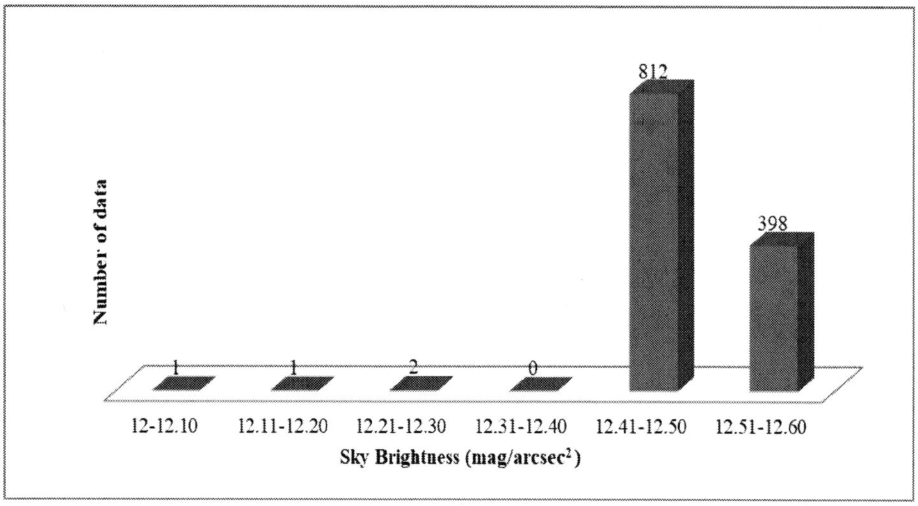


**Figure 5.** The number of data at particular NSB on March 25$^{th}$.

The Earth Hour day, the figure 6 indicated that the mode occurred in the interval of 14.81 - 14.90 mag/arcsec$^2$ as show in figure 6. To emphasize the effect of Earth Hour celebration by turning light off, the figure 6 show the shifting of the mode by turning light off. The NSB increasing to the value of 17 mag/arcsec$^2$. This value is also higher compared to the regulerl days NSB.

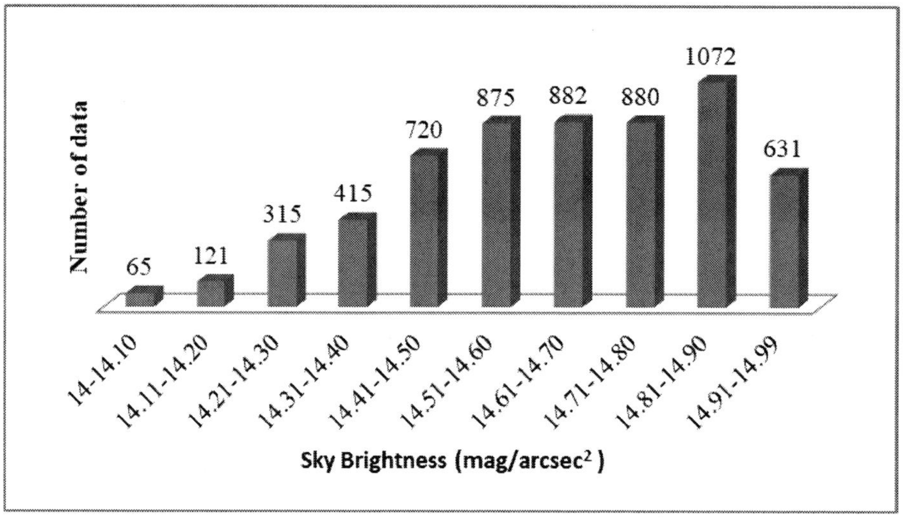


**Figure 6.** The number of data at particular NSB data on March 24$^{th}$ excluding the Earth Hour

The figure 7 shows. The data collection on sky brightness on the 24$^{th}$ Earth Hour is different from reguler and event sky brightness during earth hour. At normal sky brightness of 14 mag / arcsec$^2$ with 5973 data while when earth hour sky brightness is 17 mag / arcsec$^2$ with 2725 data amount this is because when the Earth Hour event at that location simultaneously turns off the lights resulting in dark lagging and the SQM measurement values are increasing.

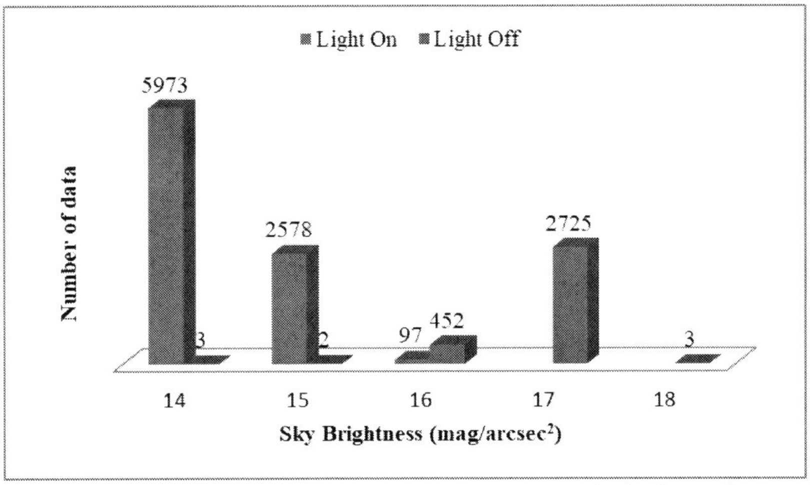


**Figure 7.** The comparison between the number of data at particular interval on Earth Hour day at the time of light was tuning on and period of turning light off only.

## 4. Discussion

The data distribution analysis was performed to obtain the mode value of NSB for each day. The mode value of NSB on 23$^{th}$, 24$^{th}$, and 25$^{th}$ of March 2018 are 15.5 mag/arcsec$^2$, 14.61 mag/arcsec$^2$ 15.02 mag/arcsec$^2$. The sky condition on the 23$^{rd}$ is cloudy resulting in scattering of light that makes the sky bright. The results of eart hour measurements are in line with the research conducted [3] that is, clouds will cause light reflection, the brightness of the sky on the 23$^{rd}$ is better than the 25$^{th}$ and at the Earth Hour on the 24$^{th}$ the sky brightness value of the mode is 17.4 mag/arcsec$^2$ at the event all the lights are turned off but there are still interference from artificial light such as traffic signs and lights from vehicle lights around the Earth hour event. Hence, it can be concluded that the night sky in Yogyakarta has a high level of light pollution. By turning off the light, the light pollution can be reduced.

## Acknowledgment

This research was supported by the research funding from Lembaga Penelitian dan Pengabdian Masyarakat Universitas Ahmad Dahlan.

## References

[1] Pun C S J, Chu W S and Chung F T W 2012 The Night Sky Monitoring Network in Hongkong. *Highlights of Astronomy* **16** 740
[2] IDA. 2010, International Dark-Sky Association, http://www.darksky.org. Accessed 29 August 2010
[3] Lolkema, D.E., Haima, M., O uter, P.N. den., & Spoelstra, H. Effects of Atmospheric Conditions on Night Sky Brightness. *National Institute for Public Health and the Environment*
[4] Pramudya, Y., & Arkanuddin, M. (2016). The Sky Brightness Measurement During the 2016. *Journal of Physics: Conference Series 771*, 1-4.
[5] Raisal, A. Y., Pramudya, Y., Okimustava, & Muchlas. (2017). The Moon Phases Influence on the Beginning of Astronomical Dawn Determination in Yogyakarta. *International Journal of Science and Applied Science: Conference Series*, 1-7.
[6] (2011). *SQM-LU Operator's Manual*. http://unihedron.com.

# Preliminary study on relation between temperature, humidity and Night Sky Brightness in Yogyakarta

**Y Pramudya[1,*], K S Budi[1], Okimustava[2], Muchlas[1]**

[1]Magister Pendidikan Fisika, Universitas Ahmad Dahlan, Yogyakarta, Indonesia
[2]Pendidikan Fisika, Universitas Ahmad Dahlan, Yogyakarta, Indonesia

Email: yudhiakto.pramudya@pfis.uad.ac.id

**Abstract.** The meteorological data is important to be collected especially at the observatory. Temperature and humidity are two parameters that can be monitored as part of meteorological data collections. The observatory site also need to be protected from light pollution. The light pollution level is usually measured by the value of Night Sky Brightness (NSB). In this paper, the relation between temperature, humidity and NSB will be studied. The temperature and humidity were measured by DHT11 Sensor that is assembled and monitored using Internet of Thing method. The NSB was measured using Sky Quality Meter. The measurements have been done in Observatorium UAD, Yogyakarta. The correlation calculation was employed to perform relationship analysis. The correlation value of temperature on NSB are in the range of -0.767 to 0.576. Hence, there is no particular pattern for the relationship between temperature and NSB. The correlation value of humidity on NSB are in the range of -0.812 to 0.977. Here, the humidity tend to have more number positive relationship to NSB. As the humidity increased, the NSB is increasing. However, the investigation on temperature and humidity data changes in particular duration of time measurement will give complete information about the relationship between temperature, humidity, and NSB. The conclusion can be used to create prediction model of NSB monitoring system.

***Keywords:*** *Night Sky Brightness, temperature, and humidity parameter*

## 1. Introduction

The monitoring of night sky brightness in observatory is important for observatory protection. Since, the light pollution is main threat for observation activity. The Mount Wilson Observatory, National Observatory at Kilt Peak and many other new telescopes in China, Arizona, Hawaii, and the Canary Islands are constantly got threat by increasing level of light pollution [1]. Although the night sky brightness is decreasing as the moon is above the horizon [2]. Night sky brightness can be measured by Sky Quality Meter (SQM). Night sky observation is also facing the problem of cloud coverage and seeing. Hence, the night sky brightness as the measurement of sky quality for observation should be supported by the meteorological data. Hence, seeing, night sky brightness and meteorological information are three important factors for observation, especially at observatory [3]. The stability of temperature and humidity are needed to obtain the high quality data on SQM [4]. Temperature and humidity can be used to forecast the weather condition. They are also important to be measured to

IOP Conf. Series: Journal of Physics: Conf. Series **1231** (2019) 012004 doi:10.1088/1742-6596/1231/1/012004

assess the condition of sky quality meter that is located outside of the building. Cloud coverage, cloud height and aerosol parameter were measured at CESAR to investigate the effect of atmospheric conditions on the night sky brightness [5]. This research is intended to measure the relationship between the temperature and humidity to the nigh sky brightness. Hence, modelling for night sky brightness can be completed with the meteorological data as input parameter.

## 2. Methodology

The research has been done using the observational method. The location of measurement is located at 7°48' South and 110°22'59" East. It is located at the compound site of Controversial Ahmad Lahnda Observatory. The duration of data collection is January 2018 until April 2018. Optometry method for measuring the NSB was employed. The NSB was measured by SQM LE that was connected to the computer. The SQM LE is the type of SQM that is connected to the computer via Ethernet. The SQM is powerful and versatile instrument that can be used not only for measuring the NSB but also the sky brightness changing during the day. The measurement of sky brightness changing during the total solar eclipse has been done using SQM [6]. Temperature and humidity data were collected by DHT11 Sensor. The sensor is controlled by Wemos D1 Mini to act as microcontroller and access point to internet. The Real Time Clock (RTC) is also connected to the Wemos D1 Mini as shown in Figure 1.

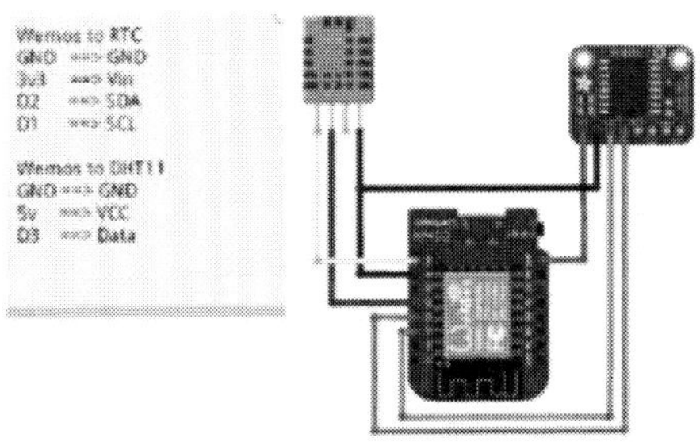


Figure 1. The electrical circuit of the temperature and humidity measurement and data transmision. The Wemos D1 Mini is connected to the DHT11 Sensor and the Real Time Clock (RTC).

The NSB as a function of time will be plotted together with temperature and humidity as function of time. The correlation coefficient ($r$) is calculated using the Pearson Product Moment. The correlation coefficient value is between -1 and +1. The positive coefficient means both variable are increasing together or proportionally relation. While one variable increasing and the other decreasing, it means that the coefficient is negative. With the value of $X$ as an independent variable and $Y$ as an independent variable, the coefficient can be calculated as

$$r = \frac{\sum(X,Y)}{\sqrt{S_X^2 S_Y^2}} \tag{1}$$

The value of $\Sigma(X,Y)$ can be calculated as

$$\sum(X,Y) = \frac{\sum(X-\overline{X})(Y-\overline{Y})}{n-1} \tag{2}$$

The variant of $X$ data can be calculated as

$$S_X^2 = \frac{\sum(X-\overline{X})}{n-1} \tag{3}$$

The variant of $Y$ data can obtained by calculating

$$S_Y^2 = \frac{\sum(Y-\overline{Y})}{n-1} \tag{4}$$

Those equation are necessary to have the average value of X and Y data. The average of X data can be calculated as

$$\overline{X} = \frac{X}{n} \tag{5}$$

The average of Y data is

$$\overline{Y} = \frac{Y}{n} \tag{6}$$

which $n$ is the number of data.

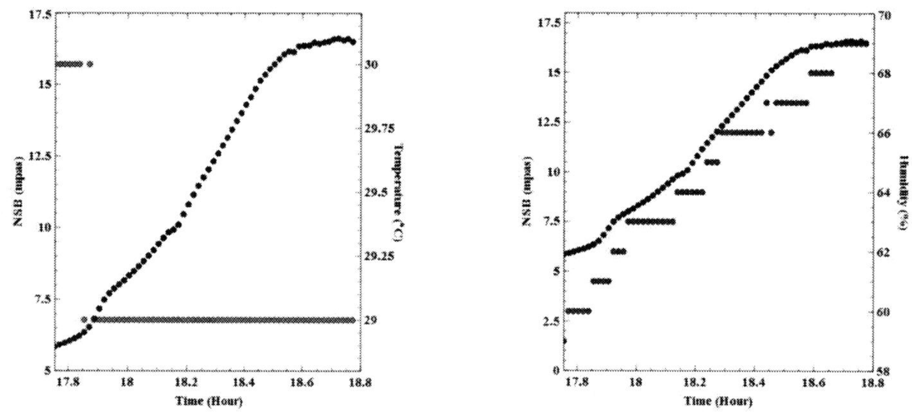


(a) (b)

Figure 2. The black dots, red dots and blue dots are the data of NSB, temperature and humidity, respectively. The time is local time (UTC + 07 Hour). (a) The graph of NSB and temperature versus time on March, 9th 2018 on increasing NSB. (b) The graph of NSB and humidity versus time on March, 9th 2018 on increasing NSB.

## 3. Result and Discussion

The result of the observation is explained in this section. The data need to be divided into 3 parts since the characteristic of the NSB changing is different. They are region of increasing NSB, constant NSB and decreasing NSB. The division is basically similar with the division of dawn, night and dusk.

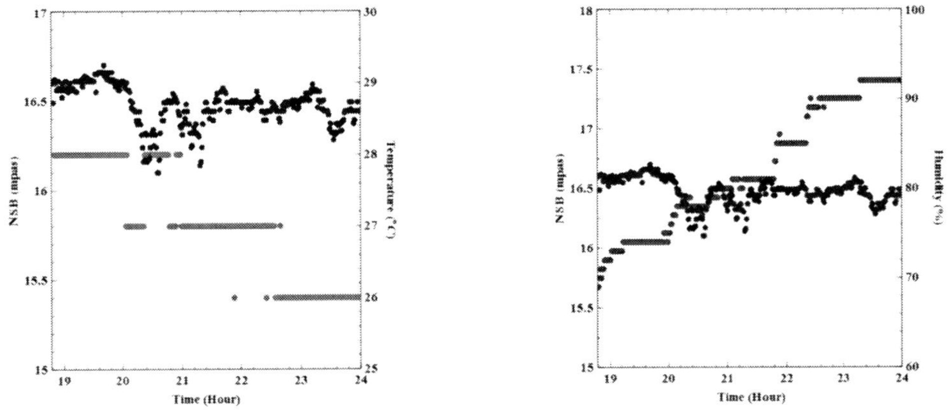


(a) (b)

Figure 3. (a) The graph of NSB and temperature versus time on March, 9th 2018 on constant NSB. (b) The graph of NSB and humidity versus time on March, 9th 2018 on constant NSB.

The trend on temperature and humidity as function of time are plotted in figure 2 for increasing NSB value. Figure 2a shows the graph of NSB and temperature as function of time. The graph indicate that there is no significant correlation between the NSB changing and the temperature. However, figure 2b shows the humidity changing in similar trend as NSB changing. Analysis can be done also for the constant NSB. Figure 3a shows that the temperature and humidity insignificantly correlate with the NSB. NSB relatively stable regardless the changing of temperature and humidity. Figure 2 and figure 3 show different sky condition. The dusk is represented by figure 2. The figure 3 is the representation of evening or night. The atmospheric condition relatively changing especially the humidity due to the changing of water vapor content in the atmosphere. This is related to the sun light intensity that fall on the earth surface.

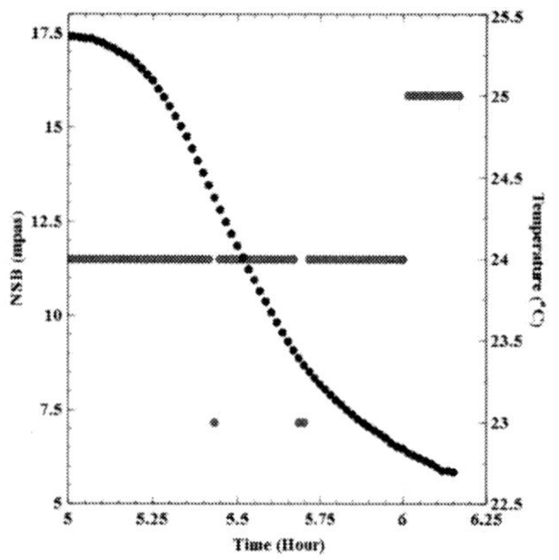


(a)

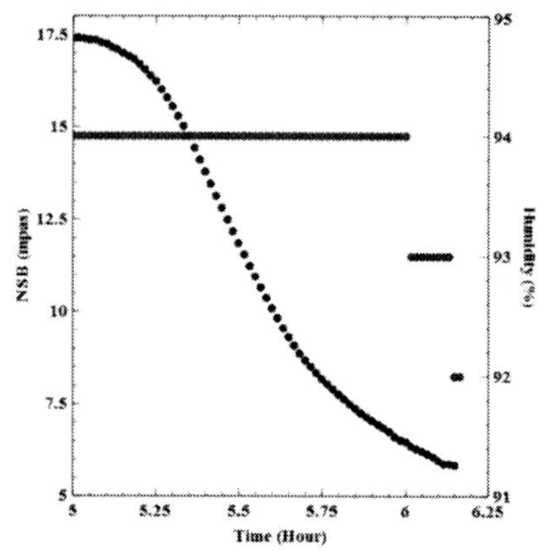


(b)

Figure 4. (a) The graph of NSB and temperature versus time on January, 28th 2018 on decreasing NSB. (b) The graph of NSB and humidity versus time on January, 28th 2018 on constant NSB.

At dawn, the water vapor supposed to be have similar behaviour. However, the changing of humidity did not correlate significantly with the NSB changing. Positive correlation is found on this particular date and condition. As the NSB decrease, the humidity also decrease but it was insignificant. The temperature and NSB behave indifferent way. As the sun rising from the eastward direction, NSB decrease and the temperature start to rise. More data is required to obtain the complete description of correlation between NSB and temperature and humidity.

The correlation coefficient between NSB and temperature and humidity are shown in table 1. The table shows the coefficient for each observation date as well as the division of NSB characteristics. The correlation coefficient between NSB and temperature for increasing NSB show 4 negative values from 4 data sets. The minimum is -0.229 and the maximum is -0.767. It means that as the sky getting darker, the temperature is getting colder. The correlation coefficient between NSB and humidity for increasing NSB show 4 positive values from 4 data sets. The minimum is 0.713 and the maximum is 0.904. As the sky getting darker, the air become more humid. The dependencies NSB to the humidity is stronger compared to the temperature.

The correlation coefficient between NSB and temperature for decreasing NSB show 4 negative values from 4 data sets. The minimum is -0.229 and the maximum is -0.767. It means that as the sky getting darker, the temperature is getting colder. The correlation coefficient between NSB and humidity for increasing NSB show 4 positive values from 4 data sets. The minimum is 0.713 and the maximum is 0.904. As the sky getting darker, the air become more humid. The dependencies NSB to the humidity is stronger compared to the temperature.

Table 1: Correlation coefficient between NSB and temperature ($r_T$) and NSB and humidity ($r_H$)

| Date | Time | Correlation Coefficient | |
|---|---|---|---|
| January, 12th | 17:46 - 19:44 | $r_T$ | -0.229 |
| | (increasing NSB) | $r_H$ | 0.904 |
| | 21.32 - 22.35 | $r_T$ | -0.197 |
| | (constant NSB) | $r_H$ | 0.156 |
| February, 27th | 17:53 - 19:00 | $r_T$ | -0.767 |
| | (increasing NSB) | $r_H$ | 0.858 |
| | 20:15 - 24:00 | $r_T$ | 0.367 |
| | (constant NSB) | $r_H$ | -0.316 |
| February, 28th | 00:00 - 04:59 | $r_T$ | -0.842 |
| | (constant NSB) | $r_H$ | 0.878 |
| | 05:00 - 06:09 | $r_T$ | -0.419 |
| | (decreasing NSB) | $r_H$ | 0.533 |
| March, 9th | 17:45 - 18:46 | $r_T$ | -0.537 |
| | (increasing NSB) | $r_H$ | 0.978 |
| | 18:47 - 23:59 | $r_T$ | 0.288 |
| | (constant NSB) | $r_H$ | -0.356 |
| March, 10th | 00:00 - 01:58 | $r_T$ | 0.576 |
| | (constant NSB) | $r_H$ | -0.322 |
| April, 4th | 18:17 - 18:59 | $r_T$ | -0.418 |
| | (increasing NSB) | $r_H$ | 0.713 |
| April, 5th | 00:00 - 03:24 | $r_T$ | 0.485 |
| | (constant NSB) | $r_H$ | -0.812 |
| | 03:24 - 05:00 | $r_T$ | -0.631 |
| | (increasing NSB) | $r_H$ | 0.488 |
| | 05:00 - 06:01 | $r_T$ | -0.816 |
| | (decreasing NSB) | $r_H$ | 0.406 |

There is 1 data of increasing NSB that was occurred on April, 5th in the interval of 03:24 - 05:00, as shown in table 1. This is a peculiar data since it was occurred during the night. The temperature correlation coefficient and humidity coefficient correlation are negative and positive, respectively. However, both values are in the medium range, since they are close to -0.5 and 0.5. At this interval, the analysis can be composed in the specific way since probably there are some complexity conditions on the weather and astronomical phenomena occurred. The following discussion is about the coefficient temperature at the interval of constant NSB. There are 6 data sets. There are 2 negative values and 4 positive values of temperature coefficient. The negative values are -0.197 and -0.842. The positive values are 0.288, 0.367, 0.485, 0.576. From these results, temperature can be stated that it did insignificantly correlate the NSB. The humidity correlation coefficients are -0.812, -0.316, -0.322 , -0.356, 0.156, 0.878. Hence, there is no conclusive the dominant correlation coefficient. Since, the NSB is constant or very small changing. Any changing of the humidity and temperature did not effect the NSB. There are 2 data set of decreasing NSB. The temperature correlation coefficients are-0.419 and -0.816. The humidity correlation coefficients are 0.533 and 0.406. Hence, the humidity correlation coefficients are more consistent compared to the temperature correlation coefficient. As the sky getting brighter, the humidity decrease since the sun rising. The variation of correlation coefficient can be explained by different sky condition, for example cloud coverage and moon phase. The lunar phase on January 12th, 2018 was the third quarter. Hence, the illumination by the moon will be significant from the beginning of midnight to dawn. The similar lunar phase also occur on March 9th and April, 4th and 5th, 2018. However, the lunar phase on February 27th and 28th, 2018 was almost full moon. Hence, the moon will be above the horizon from dusk till dawn. The lunar phase should be included into the analysis since it can effect the NSB especially to the beginning of the astronomical twilight [2][7].

## 4. Conclusion

The NSB correlation coefficient to the temperature and humidity are vary with the time. However, the humidity tend to have positive relation to the NSB. It means, the changing of water vapor concentration correspond to the changing of the sky quality. Hence, the model of NSB can be composed by including the humidity factor.

## Acknowledgments

The research was supported by the internal research grant of Lembaga Penelitian dan Pengabdian Masyarakat Universitas Ahmad Dahlan

## References

[1] Luginbuhl C B, Walker C E, Wainscoat R J, 2009 *Physics Today* **62 (12)** p 32-37
[2] Raisal A Y, Pramudya Y, Okimustava, Muchlas, 2017 *International Journal of Science and Applied Science : Conference Series* **2 (1)** p 1-7
[3] Zhang J, Ge L, Lu X, Cao Z, Chen, X, Mao Y, Jiang X, 2015 *Publications of The Astronomical Society of the Pacific* **127** p 1292-1306
[4] Zainuddin M Z, Haron S, Niri M A, Ahmad N, Nawawi M S A M, Man S, Rodzali M Z, Ramli R, Wahab R A, Ismail K, Zaki N H A, 2013 *Middle-East Journal of Scientific Research* **13 (2)** p 220-223
[5] Lolkema D E, Haaima M, den Outer P N, Spoelstra H, 2010 *RIVM Report* 680151002 p 23-30
[6] Pramudya Y, Arkanuddin M, 2016 *Journal of Physics: Conference Series* **771 (2016)** 012013
[7] Kyba CCM, Ruhtz T, Fischer J, Hölker F (2011) Cloud Coverage Acts as an Amplifier for Ecological Light Pollution in Urban Ecosystems. *PLoS ONE* **6(3)**: e17307

# Optical spectroscopy of nova ASASSN-17hx at Bosscha Observatory

**A Adhyaqsa[1,*], A N Istiqomah[1,2], D G Ramadhan[1,2], I Imaduddin[1,2], H L Malasan[1,2,3], A Arai[4], and H Kawakita[4]**

[1] Astronomy Study Program, Faculty of Mathematics and Natural Sciences, Institut Teknologi Bandung, Bandung 40391, Indonesia
[2] Bosscha Observatory, Institut Teknologi Bandung, Lembang, Indonesia
[3] ITERA Astronomical Observatory, Lampung, Indonesia
[4] Kyoto Sangyo University, Motoyama, Kyoto 603-8555, Japan

Email: azlizanadhyaqsa224@gmail.com

**Abstract.** We report optical spectroscopic observations in 2017 at Bosscha Observatory on an interesting nova, ASASSN-17hx (Nova Sct 2017). The nova ASASSN-17hx was observed on July 26th, July 29th, July 31st, and August 14th. Based on the light curve, we propose that our spectroscopic observation of ASASSN-17hx was conducted during "iron curtain" phase. The observed spectra showed a strong H$\alpha$ emission line and multiple emission lines of Fe II multiplets, which is typical for a Fe II type nova. However, since ASASSN-17hx was classified as a He/N type only few days after its outburst, it is suggested that the nova is actually a hybrid type nova. We will discuss the characteristic of ASASSN-17hx by analysing its expansion velocity, electron density, and evolution of the emission lines.

***Keywords:*** *Optical Spectroscopy and Nova ASASSN-17hx*

## 1.Introduction

Novae are stellar explosions caused by a thermonuclear runaways on the surface of a white dwarf in a close binary system, which exhibit a sudden rise in optical brightness within a period of time. Since only several novae outbursts reported every year, they are very interesting to be studied. Spectroscopic phases of novae can be explained by examining its light curve and emission lines. The elements of emission lines in the spectra can also be used to classify the nova, which consist of three types : He/N type, Fe II type, and hybrid type. The outburst of nova ASASSN-17hx was discovered with All Sky Automated Survey for SuperNovae (ASASSN) and announced in Astronomer's Telegram on June 23rd 2017 [8]. We aim to determine the characteristic of ASASSN-17hx by analysing the changes of expansion velocity, electron density, and evolution of the emission lines.

## 2. Observations

The optical spectroscopic observations were conducted on four nights in July and August 2017 at Bosscha Observatory. We employed two spectrographs. They are a low resolution spectrograph NEO-R1000 ($R$~1000, grating=600 grooves/mm) attached on Celestron C-11 28-cm Schmidt-Cassegrain telescope ($f/10$) with SBIG ST-8 CCD camera and a high resolution spectrograph LHIRES III

IOP Conf. Series: Journal of Physics: Conf. Series **1231** (2019) 012006 doi:10.1088/1742-6596/1231/1/012006

($R$~5000, grating=1200 grooves/mm) attached on Meade 25-cm Schmidt-Cassegrain telescope ($f/10$) with SBIG ST-402 CCD camera. The nova ASASSN-17hx was observed using NEO-R1000 on July 26$^{th}$, July 29$^{th}$, and July 31$^{st}$ and using LHIRES III on August 14$^{th}$. We used Image Reduction and Analysis Facility (IRAF) to perform image reductions and calibrations. FeNeAr spectral atlas was used to perform wavelength calibration for the low resolution images and Neon-Argon spectral atlas for the high resolution image. We employed HR7001 as a spectrophotometric standard star to conduct flux calibrations. We also used line list provided by Williams [10] to identify the spectral lines of the nova. Although city light lines were detected on the low resolution spectra, we could still identify the lines from the nova.

## 3. Spectra of ASASSN-17hx

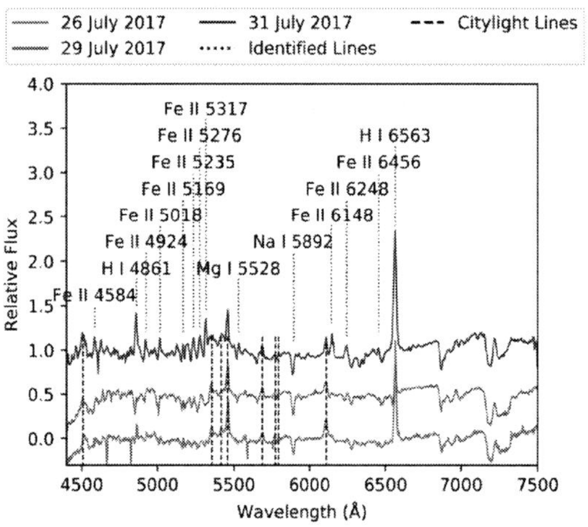


**Figure 1.** Spectra of ASASSN-17hx obtained with low resolution spectrograph NEO-R1000.

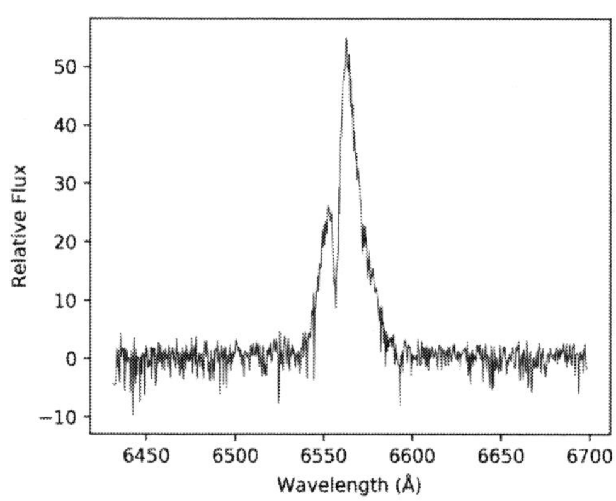


**Figure 2.** The Hα line profile of ASASSN-17hx obtained with high resolution spectrograph LHIRES III on August 14$^{th}$.

The wavelength calibrated and normalised spectra of ASASSN-17hx can be seen in Figure 1 and Figure 2. We detected emission lines of Fe II, Mg I, Hα, and Hβ. An absorption line of Na I was also detected. During three nights of observations, we could see the evolution of intensity of the emission lines. On July 26$^{th}$, the Hβ line showed a P-Cygni profile, the metal lines were not clearly visible yet, and we could see a strong Hα emission. In the next observation on July 29$^{th}$, the intensity of Hα and Hβ were weakened, with P-Cygni profile could be seen on the Hα line. The intensity of metal lines were increasing, but still relatively weak. The evolution of the spectra could be distinctly seen on July 31$^{st}$. The emission lines of metal became opaque, with a significant increase of Hα and Hβ intensity. In this spectrum, the P-Cygni profiles were observable in the metal lines. The high resolution spectra of Hα obtained on August 14$^{th}$ showed an asymmetric double-peak emission line. After barycentric correction, we measured that the radial velocities corresponding to blue-shifted peaks and central absorption were 485.672 km/s, 23.403 km/s, and 287.131 km/s respectively.

## 4. Discussion

The chemical elements and characteristics of the spectral lines were used to classify the nova into Fe II type, He/N type, or even hybrid type [10]. Based on our observations of ASASSN-17hx, the Fe II lines were dominating the spectra with visible P-Cygni profiles in every spectrum. These signatures are usually belongs to the Fe II type nova. However, previous observation a few days after the outburst showed that ASASSN-17hx had the characteristics of He/N type [5]. Similar case was already been seen on the recurrent nova T Pyx [3]. It is suggested that the changing type of nova is common mechanism but rarely observed because lack of data at different phases [10].

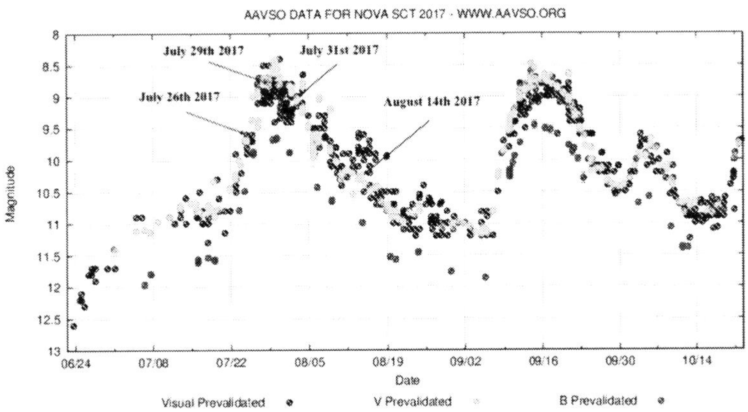


**Figure 3.** Light curve of ASASSN-17hx from American Association of Variable Star Observers (AAVSO) database.

Based on Figure 4, ASASSN-17hx was observed during its final rise and initial decline phase. In these phases, our observed spectra represented the shell of nova. Physical interpretation of shell can be better understood if we examine the spectral evolution by measuring expansion velocities and electron densities at different times. To determine the expansion velocity, we employed the equation (1), which was previously used by Azaliah *et al.* [1] to analyse nova spectra.

$$v_{expansion} = \frac{\lambda_{absorption} - \lambda_{emission}}{\lambda_0} \tag{1}$$

After obtaining the expansion velocities, we measured the electron densities. Electron densities were derived using equation (2) by Doroshenko [2]. Since it required the intensity of Hβ, electron densities were only measured on the low resolution spectra.

$$n_e^2 = 0.54 \times 10^{69} \left[ \frac{Intensity_{H\beta}}{(\overline{v_{exp}} \times t_{sinceoutburst})^3} \right] \tag{2}$$

To determine the uncertainty of our measurements, we used the method of error propagation. The uncertainty of expansion velocity were derived from the uncertainty of observation parameters, which were the wavelength value of absorption and emission of the P-Cygni profiles. Then, for the uncertainty of electron density, we propagated the uncertainty of $H\beta$ intensity and the mean expansion velocity. The results of our measurements of expansion velocities and electron densities could be seen in Table 1.

**Table 1.** Expansion velocity and electron density of ASASSN-17hx

| Date | July 26th 2017 | July 29th 2017 | July 31st 2017 | August 14th 2017 |
|---|---|---|---|---|
| **Expansion velocity (*km/s*)** | 867.639 ± 17.041 | 966.361 ± 23.883 | 921.169 ± 28.802 | 226.197 ± 6.687 |
| **Electron density ($\times 10^{10}$ $cm^{-3}$)** | 12.002 ± 0.285 | 5.350 ± 0.139 | 11.897 ± 0.281 | - |

The values from Table 1 indicated that the shell of nova was expanding with varying velocity and electron density. During the final rise, the shell's velocity increased until it reached its maximum on July 29th. During this time, the electron density had the lowest value. The less dense medium suggested that the shell was undergoing a cooling stage and therefore allowed the recombination process of metal lines. The process caused metal lines to be more observable on July 31st, during the early initial decline. The emerging metal lines caused the shell to be denser, with lower expansion velocity. This fact suggested that the nova was observed at the spectral "iron curtain" phase [7]. Two weeks after the start of initial decline, the shell was still expanding but with significantly lower velocity. The asymmetric double-peak of Hα implied that the shell had inhomogeneous density at the time, which may be caused by intrinsic nature of the nova or the effect of inclination angle between the ejecta and our line of sight.

## 5. Summary

Our observations of ASASSN-17hx in July and August 2017 showed a distict spectral evolution. We could see that the intensities of the emission lines varied between the nova phases. Based on the analysis of expansion velocities and electron densities, the metal lines in the shell of ASASSN-17hx were formed after the peak brightness. It was implied that the low density at the peak brightness trigged the recombination of the metal lines. The appearance of metal lines in the spectra of ASASSN-17hx is also interesting since the nova was classified first as He/N type. Therefore, this nova can be classified as a hybrid type. Since hybrid nova is rarely observed, the case of ASASSN-17hx is important to be studied, particularly about the transition of novae classification and its mechanism.

## Acknowledgements

We are grateful to the staff of Bosscha Observatory for the time and assistance in the use of NEO-R1000 and LHIRES III spectrographs. This research is carried out under the framework of nova spectroscopic network between Kyoto Sangyo University and Institut Teknologi Bandung based on a Memorandum of Understanding in 2015.

## References

[1] Azaliah *et al.* 2015 *Publications of the Korean Astronomical Society* **30** 251-254
[2] Doroshenko V T 1968 *Astronomischeskii Zhurnal* **45** 121-132
[3] Ederoclite A 2013 *Highlights of Spanish Astrophysics VII, Proceedings of the X Scientific Meeting of the Spanish Astronomical Society (SEA)* 539-542
[4] Kogure T and Kam-Ching L 2007 *The Astrophysics of Emission-Line Stars* (New York: Springer)
[5] Kurtenkov *et al.* 2017 *The Astronomer's Telegram* **10527**
[6] Muztaba *et al.* 2016 *Journal of Physics : Conference Series* **771** 012054
[7] Shore S N 2012 *Bulletin of the Astronomical Society of India* **40** 185
[8] Stanek *et al.* 2017 *The Astronomer's Telegram* **10523**
[9] Williams R 1992 *The Astronomical Journal* **104** 725-733
[10] Williams R 2012 *The Astronomical Journal* **144** 98

# The robotic telescopes system for GRB optical follow-up in Timau National Observatory

**M D Danarianto**[1,*] **and M B Saputra**[1]

[1]Space Science Center, National Institute of Aeronautics and Space (LAPAN), Bandung, West Java, Indonesia 40173

Email: dio.danarianto@lapan.go.id

**Abstract.** The occurrence of Gamma Ray Bursts (GRBs) is yet unpredictable. Thus, follow-up observations to find its optical counterpart should be made as quickly as possible after the first detection, as well as kept being monitored on the observable area. In Southeast Asia region, two new robotic 50-cm telescopes system will be placed in Timau National Observatory, East Nusa Tenggara, Indonesia, which has about 70% of clear sky fraction with low-latitude sky coverage. In this manuscript, the possibility of implementing the system for GRB follow-up is discussed, followed by its future prospects as a multi-messenger optical follow-up robotic telescope. The current properties and status of the system are also described.

***Keywords:*** *GRB optical, robotic telescopes system, and Timau National Observatory*

## 1. Introduction

The Gamma Ray Bursts (GRBs) are the most energetic electromagnetic flashes from celestial origin. The "prompt" gamma-ray radiation is usually detected by space-based detectors (e.g. *Swift* and *Fermi* satellite), which lasts for seconds to few minutes and is usually followed by an "afterglow" in X-ray, optical, and radio wavelengths, that could last for a few minutes to months. Roughly, there are two classes of burst based on its prompt duration: long-GRBs (>2s) and short-GRBs (<2s). Long-GRBs are usually thought as a result of a massive stellar core collapse. Meanwhile, short-GRBs are linked with the merger of compact objects, which is confirmed by multi-messenger observations, e.g. the gravitational wave GW170817 detection [1].

The coordination among independent space-based detectors and ground-based optical follow-up observatories plays a crucial role to perform early afterglow observation. Due to the nature of the random and short-lived emission, rapid follow-up of GRB satellite detections by ground-based telescopes provides important data to understand the nature of its emission. In terms of location, the observations should be performed uniformly on Earth to chase the early afterglow in the observable area [2]. A new observatory is planned to be built in Timau, Timor island, East Nusa Tenggara, Indonesia (9°35′50.2″S, 123°56′48.5″E). Based on 5-year satellite data, Timau National Observatory site has about 70% of yearly clear sky fraction [3]. Beside the main 3.8-m telescope, two 50-cm robotic telescopes are planned to be placed at the site for survey purpose (Figure 1). Because of the flexibility of robotic telescope, here we discuss the potential of the 50-cm robotic telescopes system as GRB follow-up telescopes.

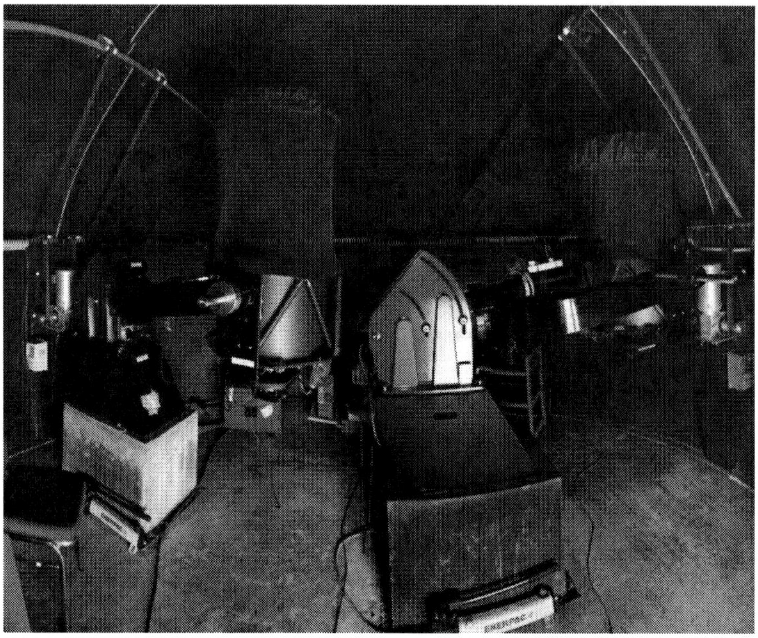

**Figure 1.** Two robotic telescopes systems. Officina Stellare (OS) ProRC 500 f/8 set (left) and RiFast 500 f/3.8 set (right) inside a 5-m AstroHaven dome.

## 2. Hardware Properties

Imaging properties of the system are described in Table 1. The system consists of two individual 50 cm telescopes inside a 5-m AstroHaven clamshell enclosure. The optical assembly of the f/8.0 telescope is Officina Stellare (OS) Ritchey-Chrétien Cassegrain reflector (ProRC 500) and the f/3.8 is OS proprietary aspheric primary mirror reflector (RiFast 500). Each telescope is mounted on the Paramount Taurus equatorial fork mounting. The cameras are FLI PL16803 with a 4k×4k *ON Semi*-KAF 16803 sensor for the f/8.0 and FLI PL4240 with a 2k×2k *e2v* CCD42-40-1-368 sensor for the f/3.8. Both cameras are attached to FLI CL1-10 filter wheels with standard Bessel BVRI and LRGB filters.

**Table 1.** Timau National Observatory 50-cm Telescopes Specification.

| | OS RiFast 500 | OS ProRC 500 |
|---|---|---|
| Primary Mirror Diameter | 500 mm | 500 mm |
| Focal Length | 1900 mm | 4000 mm |
| F-Number | 3.8 | 8 |
| Band | BVRI and LRGB | |
| Maneuver Speed | <3.5°/s | |
| FOV | 50'×50' | 32'×32' |
| Pixel scale | 1.46"/pixel | 0.46"/pixel |

The hardware are designed to be controlled autonomously. The components of the telescopes are controlled by various programs on Windows-based PC–mostly via ASCOM interface. The imaging system (cameras, filter wheels, and focusers) are controlled by MaximDL. The mountings are controlled by The SkyX software. The whole components (including the enclosure and the weather station) are planned to be integrated and controlled by ACP Observatory Control Software. Table 2 shows the comparison between Timau National Observatory 50-cm robotic telescopes with other established GRB follow-up telescopes.

**Table 2.** Comparison with other GRB follow-up telescopes.

| | Timau National Observatory's 50 cm F/3.8 | Timau National Observatory's 50 cm F/8 | OAO 50 cm [4] | Akeno 50 cm [4] | WRT [5] | Zadko Telescope [6] |
|---|---|---|---|---|---|---|
| Mirror Diameter | 500 mm | 500 mm | 500 mm | 500 mm | 400 mm | 1007 mm |
| F-Number | 3.8 | 8 | 6.5 | 6.0 | 14.25 | 4.01 |
| Band | BVRI and LRGB | | $g' R_c I_c$ | $g' R_c I_c$ | V R *g'r'i'* *Hα OIII* | |
| Maneuver Speed | <3.5°/s | | 4°/s | 9°/s | | 3 °/s |
| FOV | 50'×50' | 32'×32' | 26'×26' | 28'×28' | 14.5'×14.5' | 23'×23' |

In order to observe the GRB quickly, the system should be able to receive the GRB alert from satellite and perform observation automatically. One of the possible mechanisms is by subscribing into Gamma-ray Coordination Network (GCN) socket via a VOEvent protocol. The notification from VOEvent will be parsed by VOEvent Receiver in ACP Scheduler. The local position of the event then will be calculated. Then, if the event fulfills the constraints (i.e. weather, observability, lunar illumination, etc.), ACP Scheduler will execute the observation. Details of the command pipeline are described in Figure 2.

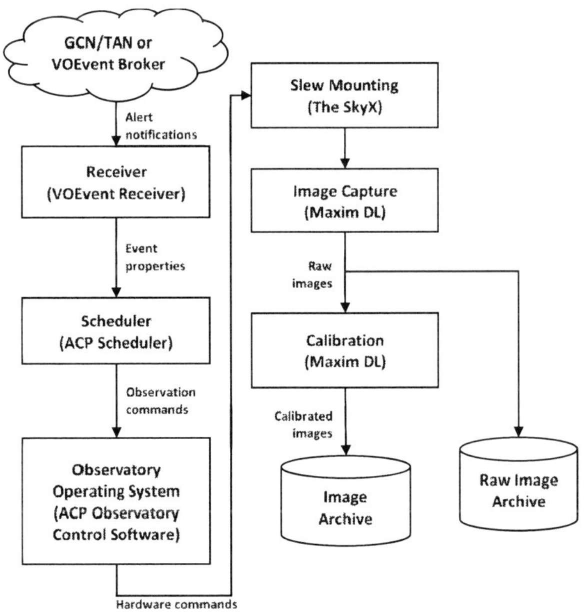


**Figure 2.** Proposed automatic GRB follow-up observation pipeline on Timau National Observatory 50 cm Telescope.

The telescopes are planned to be used for many purposes, such as near-Earth objects monitoring and extrasolar planet searching as the main priorities. Thus, the implementation of the rapid GRB follow-up robotic system should be made without interfering other observations. Satellite alert could be filtered out by selecting only confirmed and sufficiently bright event to be observed.

## 3. Scientific Potential

The slew duration of the telescope is about 30 to 60 seconds, depending on the initial position. By considering their response time, it is obvious that the system could not potentially observe prompt phase of GRB emission which lasts for only about 20 seconds in average [7]. However, the afterglow (or the transition phase between prompt and afterglow) could potentially be observed. Considering the system capabilities for GRB follow-up, there are some science opportunities and prospects that could be studied by using this system.

### *3.1. Testing fireball model*

In the fireball model, the standard picture of GRB emission is an expanding fireball of relativistic particle [8]. The expansion is driven by the central engine in the middle of the fireball. The super-Eddington energy release is thought to be responsible for the prompt emission and create a forward shock. The shock dissipation between high-energy particles and the surrounding medium are responsible for the afterglow emission. It is thought that reverse shock also exists, propagates back to the central engine and emits bright but short optical flashes. The forward and reverse shock properties could provide information about the explosion, ejecta, geometry, and properties of the surrounding medium [9,10].

### *3.2. High-z GRBs*

As one of the most luminous electromagnetic sources, radiation from high redshift GRBs could be bright enough to be detected. Analysis of GRB's optical emission could potentially probe stellar or interstellar matter properties at high-z. The high-z GRBs are also thought to originate from Population III stars [11]. Thus, optical follow-up for high-z GRBs could also probe the properties of Population III stars explosion scenario.

### *3.3. Multi-messenger follow-up*

There are two standard scenarios to explain GRB progenitor: core-collapse of massive stars and binary merger of compact objects. The second scenario was confirmed by coordinated observation of kilonova following GW 170817 detection [1]. The GRBs are also thought as one of the few candidates of ultra-high energy cosmic-ray. Multi-messenger observations could be useful to probe high-energy particles acceleration from GRBs [12].

If the system is capable to respond to GRB alerts automatically, the telescope may also able to receive alerts from other types of messenger, such as gravitational wave or high-energy cosmic neutrino. This could be implemented by receiving the alert via GCN/TAN notices or VOEvent protocol.

## 4. Current Status

The telescopes are currently located in National Institute of Aeronautics and Space (LAPAN), Bandung, West Java, Indonesia. The control system is currently being developed. Hardware components are also being tested to know the characteristics and performance of each part.

**Figure 3.** The telescopes are temporarily installed in LAPAN Space Science Center, Bandung.

**References**

[1] Abbott B P *et al* 2017 Multi-messenger Observations of a Binary Neutron Star Merger *Astrophys. J.* **848** L12
[2] Sakamoto T, Wallace C A, Donato D, Gehrels N, Okajima T and Ukwatta T N 2011 Goddard Robotic Telescope - Optical follow-up of GRBs and coordinated observations of AGNs *Adv. Sp. Res.* **47** 1444–50
[3] Hidayat T, Mahasena P, Dermawan B, Hadi T W, Premadi P W and Herdiwijaya D 2012 Clear sky fraction above Indonesia: An analysis for astronomical site selection *Mon. Not. R. Astron. Soc.* **427** 1903–17
[4] Kotani T *et al* 2007 MITSuME--Multicolor Imaging Telescopes for Survey and Monstrous Explosions *Nuovo Cim.* **C28** 755–8
[5] Melady G *et al* 2009 Watcher robotic telescope follow-ups of GRBs *AIP Conf. Proc.* **1133** 67–9
[6] Coward D M *et al* 2010 The Zadko telescope: A southern hemisphere telescope for optical transient searches, multi-messenger astronomy and education *Publ. Astron. Soc. Aust.* **27** 331–9
[7] Kouveliotou C, Meegan C A, Fishman G J, Bhat N P, Briggs M S, Koshut T M, Paciesas W S and Pendleton G N 1993 Identification of two classes of gamma-ray bursts *Astrophys. J.* **413** L101
[8] Piran T 1999 Gamma-ray bursts and the fireball model *Phys. Rep.* **314** 575–667
[9] Sari R and Piran T 1999 Predictions for the Very Early Afterglow and the Optical Flash *Astrophys. J.* **520** 641–9
[10] Middleton M J *et al* 2017 Paving the way to simultaneous multi-wavelength astronomy *New Astron. Rev.* **79** 1339–51
[11] Nakauchi D, Suwa Y, Sakamoto T, Kashiyama K and Nakamura T 2012 Long-duration X-ray flash and X-ray-rich gamma-ray bursts from low-mass population III stars *Astrophys. J.* **759** 128
[12] Murase K, Ioka K, Nagataki S and Nakamura T 2008 High-energy cosmic-ray nuclei from high- and low-luminosity gamma-ray bursts and implications for multimessenger astronomy *Phys. Rev. D - Part. Fields, Gravit. Cosmol.* **78** 1–22.

# Two 0.5-m robotic telescopes for Timau National Observatory in eastern Indonesia

**M T Bumbungan[1,*], R P Priyatikanto[1], M D Danarianto[1], M B Saputra[1], F Rohmah[1], M D Murti[1], F Mumtahana[1], M Z Nurzaman[1] and N D Legita[2]**

[1] Space Science Center, National Institute of Aeronautics and Space (LAPAN), Bandung, West Java, Indonesia 40173
[2] National Institute of Science and Technology, Jakarta, Indonesia 12630

Email: manoressy.tobias@lapan.go.id

**Abstract.** Timau National Observatory will be the new Indonesia's astronomical observatory in East Nusa Tenggara, planned for 2020. Besides the main 3.8-m optical telescope, it will host two 0.5-m "off-the-shelf" f/3.8 and f/8.0 telescopes, equipped with two CCD cameras and two filter wheels. The system will be controlled with a robotic system for autonomous operation. In this paper, we present the system description and current development progress, as well as the future plan for the facility, including its two potential scientific studies: near-Earth objects and exoplanetary transits.

***Keywords:*** *Robotic telescopes and Timau National Observatory*

## 1. Introduction

Since 1923, Bosscha Observatory in Lembang (107°36'E, 6°49'S), West Java, is the only major astronomical observatory operating in Indonesia. It hosts a 0.6-m double refractor telescope (installed in 1928) and a 0.7-m Schmidt reflector telescope (installed in 1960), which are still operating, but no longer sufficient to keep up with the latest development and progress in modern astronomy that are made possible by bigger and more advanced telescopes and instruments. Moreover, the sky quality over Lembang, near Bandung, has deteriorated, with worsening air and light pollution decreasing the number of photometric nights suitable for observation. New observing facilities are needed, and the idea has come up within a framework of national observatory. Meteorological satellite data spanning over 15 years (1996-2010) suggested that Timau, in the region of East Nusa Tenggara, south-eastern part of Indonesia, could be the most suitable location for an astronomical observatory [1]. Compared to other sites, Timau has more photometric nights and drier climate, with yearly usable night sky fraction of ~70%. The main planned facility for Timau National Observatory will be a 3.8-m optical telescope, currently under development in cooperation with Kyoto University. Other smaller telescopes are also included in the planning, including two units of 0.5-m robotic telescopes for astronomical surveys. Robotic telescopes are useful to minimize operation costs and number of required personnel. Their system can be designed to manage complex observational schedules based on the most appropriate targets and switch immediately to observe sudden events that require fast response. One of the largest examples is the 2-m Liverpool telescope, designed for study of unpredictable phenomena, survey and follow up, and monitoring of variable objects. The next generation of robotic telescopes includes the Liverpool Telescope 2, a 4-m telescope with the capability for faint-objects spectroscopy in the optical/infrared [2].

Here, we describe the 0.5-m robotic telescopes system to be established in Timau National Observatory (figure 1). We provide the description of the hardware (Section 2), system control (Section 3), preliminary testing (Section 4), plan for the weather system (section 5), and the scientific potential of the telescopes (Section 6).

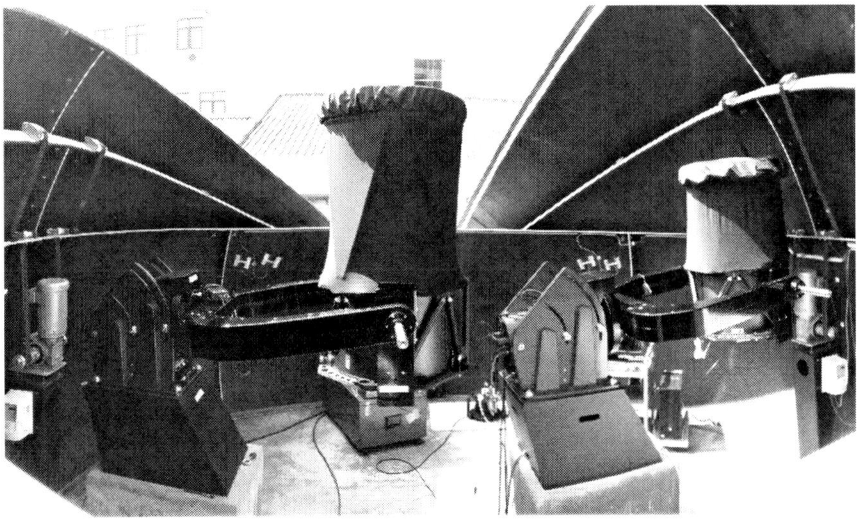

**Figure 1**. The two 0.5-m telescopes with the dome opened, temporarily installed in Space Science Center LAPAN, Bandung.

## 2. Hardware

Here, we provide the general description of the hardware for the planned robotic telescopes system.

### *2.1 Dome*

The two telescopes are currently enclosed in a 5-m Astrohaven clamshell dome. Such design is advantageous for rapid follow-up observations, since it is not necessary to rotate the dome to adjust with the telescope. The dome and the telescopes are currently installed in Space Science Center LAPAN, Bandung. The current major issue is the positioning of the two telescopes, which are aligned in north-south direction inside one dome, which is not adequate for observation since there is less space for pointing the telescopes. The reason for this were mainly administration and safety of the purchased equipment. We are currently planning to purchase a similar dome to be installed in Timau site, so each telescope will have its own enclosure. The primary power for the dome may be either 230 VAC three-phase or 230 VAC single-phase. The two shutters on each side are lifted by a pair of electric motors with a series of belts and pulleys. Each shutter requires 22 seconds to fully open or close.

### *2.2 Mounting*

The mounts for both telescopes are Software Bisque's Paramount Taurus Model 500 of equatorial fork design. It provides through-the-meridian tracking, with adjustable polar axis from $0^{\circ}$ to $58^{\circ}$. The maximum slew speed is $3.5^{\circ}$/sec. for both axes. The factory default is 80% of the maximum; this may depend on the payload and the ambient temperature. The mount is equipped with an on-axis encoder, providing $\leq$ 10" RMS all-sky pointing accuracy. Theoretically, the pointing's accuracy may reach the encoder's limit under 1". In practice, without the encoder, the achieved accuracy is $\leq$ 20" RMS and the periodic error for the RA gear is $\leq$ 7". With correction, the periodic error is less than the local seeing.

*2.3 Telescopes*

The two 0.5-m OTAs (optical tube assembly) are off-the-shelf Officina Stellare's RiFast 500 (f/3.8) and ProRC 500 (f/8.0), made of carbon and aluminium with low thermal expansion. Both OTAs are equipped with primary mirror heaters to prevent condensation, three ventilation fans for cooling the mirror, and four primary mirror shutters to protect the mirror. Each telescope's mass is 85 kg. RiFast employs spherical primary mirror and aspherical secondary mirror, with three corrector lenses (two are spherical) near the focus to minimize chromatic aberration. ProRC is a Ritchey-Chrétien telescope. Both OTA's secondary mirrors are motorized to adjust the focus. The linear obstruction of RiFast and ProRC are 55% and 47%, with back focus correction of 0.24 m and 0.3 m respectively. The telescopes use rotofocusers for the cameras.

*2.4 CCD Cameras and Filter Wheels*

The two CCD cameras are Finger Lakes Instrumentation's (FLI) PL4240 Midband for RiFast and PL16803 for ProRC (figure 2). Both cameras utilize thermoelectric cooling system capable for working temperature of 55°C below ambient temperature. The detectors require ~15 minutes to reach the working temperature between -30°C and -20°C from the ambient temperature of 22°C. Both cameras are equipped with FLI CenterLine CL-1-10 filter wheels, employing two research grade 50S UBVRI and LRGB filter sets. The detailed specifications are shown in Table 1.

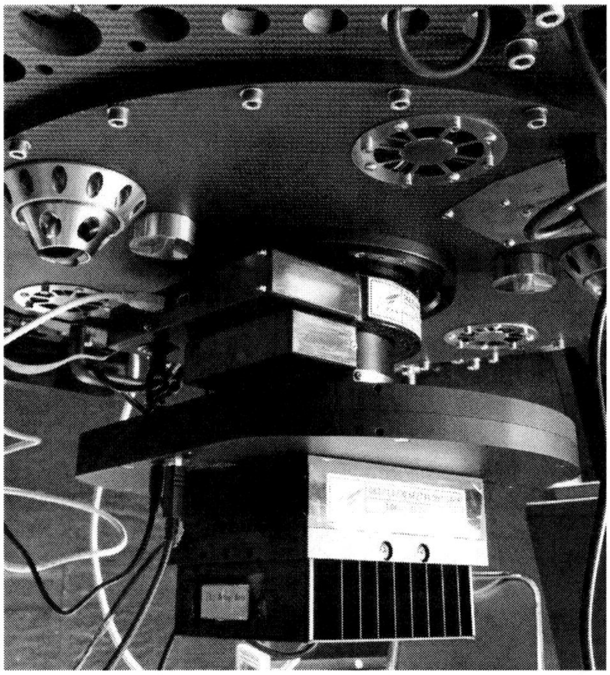

**Figure 2**. CCD camera with the filter wheel at the back plate of the telescope.

**Table 1.** CCD cameras specification.

| Spec. | PL 4240 | PL 16803 |
|---|---|---|
| Sensor type | e2v CCD42-40-1-368 | ON Semi KAF-16803 |
| Sensor size | 2048 x 2048 pixels; 27.65 x 27.65 mm$^2$ | 4096 x 4096 pixels; 36.86 x 36.86 mm$^2$ |
| CCD type | back illuminated | front illuminated |
| Pixel size | 13.5 micron | 9 micron |
| Full well capacity | $10^5$ $e^-$ | $10^5$ $e^-$ |
| Anti blooming | - | > 100x saturation exposure |
| Color | monochrome | monochrome |
| Pixel scale | 1.47"/pixel | 0.46"/pixel |
| CCD field of view | 49.8' x 49.8' | 31.8' x 31.8' |
| Readout mode | 500 kHz & 2 MHz | 1 MHz & 8 MHz |
| Readout noise ("slow" & "fast" readout) | | |
| • Binning 1x1 | 6.33 $e^-$ & 8.42 $e^-$ | 6.37 $e^-$ & 8.86 $e^-$ |
| • Binning 2x2 | 6.77 $e^-$ & 8.71 $e^-$ | 9.78 $e^-$ & 11.40 $e^-$ |
| • Binning 3x3 | 7.27 $e^-$ & 8.94 $e^-$ | 13.71 $e^-$ & 14.63 $e^-$ |
| Dead time ("slow" readout) | | |
| • Binning 1x1 | 11 seconds | 19 seconds |
| • Binning 2x2 | 6 seconds | 10 seconds |
| • Binning 3x3 | 4 seconds | 7 seconds |
| 100-s dark current ("slow" readout at -30°C) | < 10.0 $e^-$/pixel | < 2 $e^-$/pixel |
| Gain (from manufacturer, "slow" readout) | 1.38 $e^-$/ADU | 1.46 $e^-$/ADU |
| Filter | CenterLine CL-1-10 | |
| Number of filters | 2 (wheels) x 5 (slots) | |
| Size of filter slot | rectangular 50 mm | |
| Filter set | UBVRI and LRGB | |

## 3. System Control

As for now, each subsystem is controlled using separate software. Images are grabbed using MaxIm DL Pro 5 in its Camera Control; the CCD and its filter are configured in this menu. Movement of telescope on its mounting is controlled from TheSkyX. Officina Stellare ATC Remote is used for controlling the primary mirror shutter, monitoring the status of primary and secondary mirror temperature, as well as backfocus control. Environmental data such as temperature, pressure, relative humidity, and dew point from the sensor attached on the OTA are also showed in the OS ATC Remote software. Images taken are focused using RotoFoc 3" V2 that can be controlled by RotoFoc Control Panel Software. RotoFoc focuser can be moved from zero position to 25,000 μm.

All of these systems are planned to be controlled from one integrated software such as ACP Observatory Control. This is a part of the integrated robotic system that will be implemented to the telescopes and is currently under development.

## 4. Preliminary Testing

This section describes the initial tests that have been conducted so far, mainly on the detectors used for the telescopes and the optical aberration of the telescope.

### *4.1 Cameras*

The techniques used to characterize the CCD cameras here follow those of Abbott [3]. Bias and dark testing have been conducted for both cameras, while flat fielding analysis is still in progress.

A median of 9 bias images was taken for each camera in order to characterize the statistical and spatial distribution of the readout noise [4]. In figure 3 and 4, we plot the mean rows and columns to check the spatial distribution of the bias from PL4240 and PL16803. The count is represented in ADU (analog-to-digital unit).

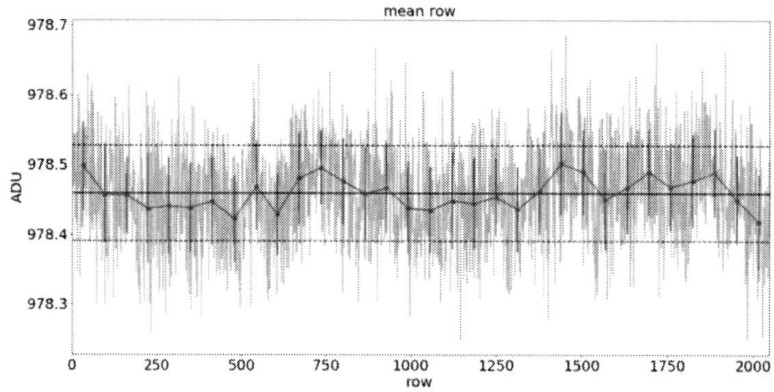


(a)

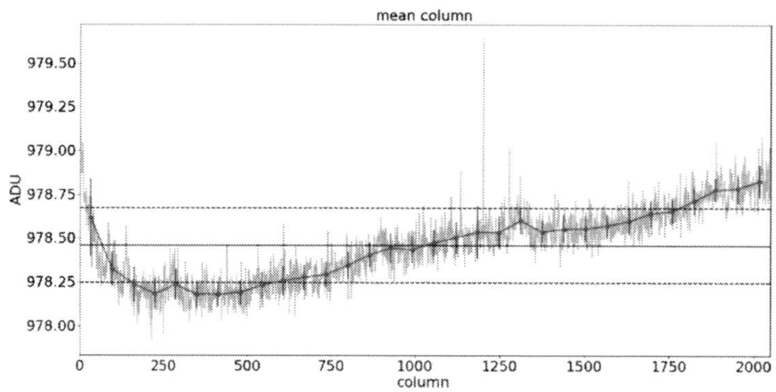


(b)

**Figure 3**. The mean (a) row and (b) column of the unpinned (1x1) median bias image of PL4240, "slow" readout mode. Grey indicates the mean values in each row and column. Red curves are the binned mean values with their error bars. The overall mean and the 1-sigma are shown with blue solid and dashed lines, respectively.

IOP Conf. Series: Journal of Physics: Conf. Series **1231** (2019) 012008 doi:10.1088/1742-6596/1231/1/012008

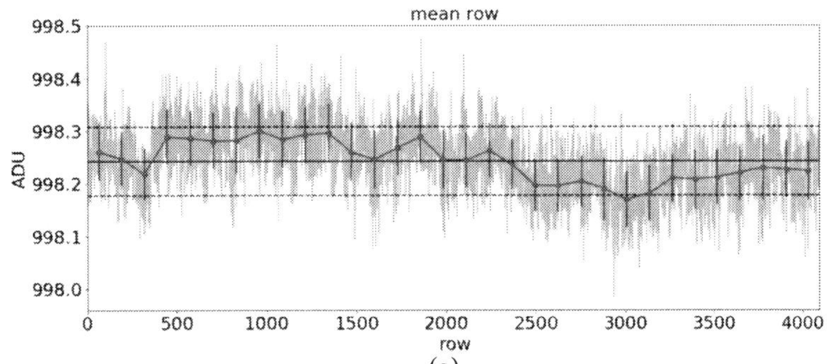

(a)

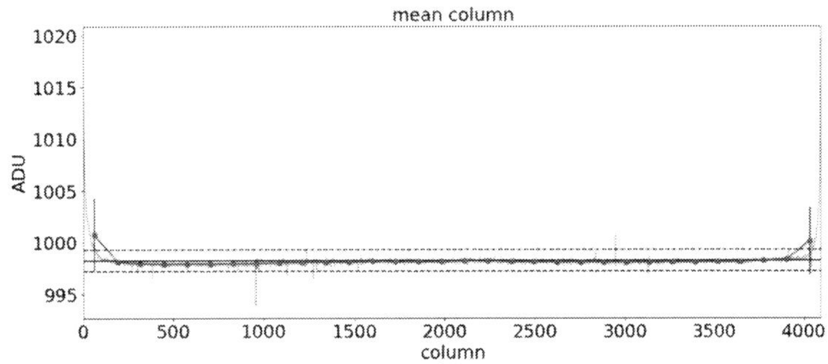

(b)

**Figure 4**. Same as Figure 3, but for PL16803.

It can be seen that the mean along the row direction for both cameras are relatively stable within ± 1 ADU. A gradient exists along the column direction for PL4240, with higher counts found at the "right" of the detector, but this is still within ± 1 ADU. Along the column direction of PL16803, the counts fluctuate within ± 1 ADU, with high jumps can be seen at both sides of the frame. We found similar patterns in each frame, irrespective of binning size and readout modes. Such patterns in both detectors were also found in previous testing and are relatively fixed, hence can be subtracted from the science image through debiasing.

Inspection of the statistical distribution of the bias for both cameras are shown in figure 5. The probability density functions were counted, as well as the mean ($\mu$) and the standard deviation ($\sigma$), which is a good measure of the readout noise. Slight deviation can be seen in PL4240's histogram, indicating the presence of systematics that may require further investigation. A skewed right distribution in PL16803 has been expected, considering the presence of significantly-higher counts at both sides of the frame. The readout noise increases with the binning factor and the readout rate as shown in Table 1. The gain, as provided by the manufacturer, decreases by only ~1.45% for faster readout modes for both detectors. We used the theoretical gain values to estimate the readout noises; further flat testing should provide a more realistic estimate for the gains, hence readout noises.

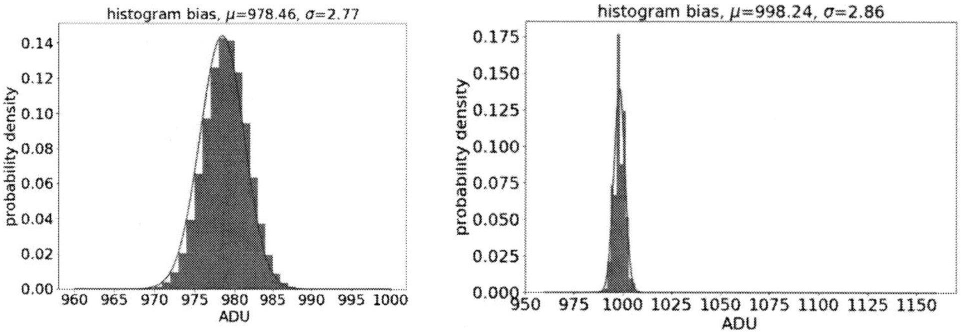


**Figure 5**. Statistical distribution of bias counts for PL4240 (left) and PL16803 (right), "slow" readout mode. The probability density function is shown in grey, while red curves show the calculated normal probability density function.

The dark current were measured for unbinned frames at "slow" readout mode as functions of temperature (-30° – 20°C, 5-second exposure time) and exposure time (2 – 1800 seconds at -30°C). The dark variability of both cameras are shown in figure 6 and 7, showing exponential and linear profiles for temperature and exposure time variability, respectively. The values of 100-s dark current of PL4240 and PL16803 are shown in Table 1. Our measured value for PL4240 is lower than that given by the manufacturer, but slightly higher for PL16803. More data are still required and further analysis will be conducted.

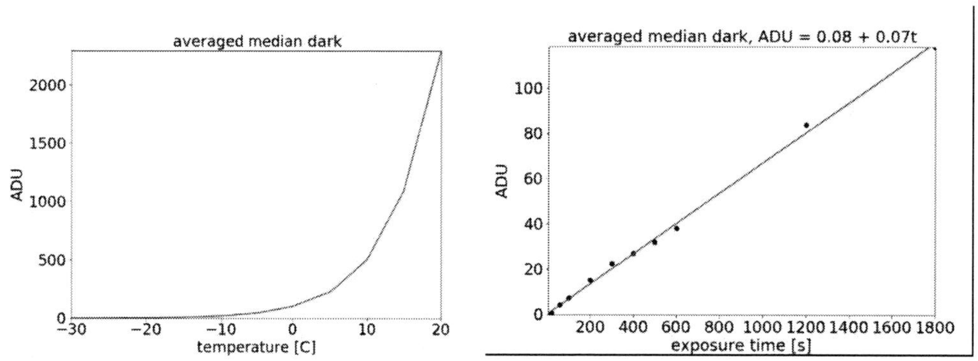


**Figure 6**. Dark variability of PL4240 as a function of temperature (left) and exposure time (right).

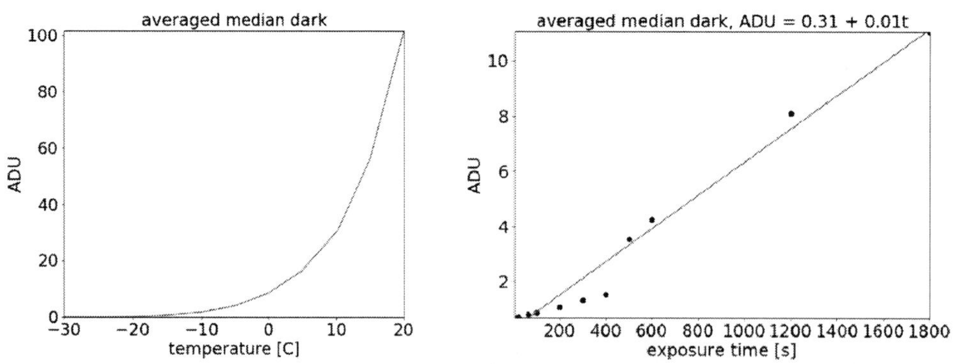


**Figure 7**. Same as Figure 6, but for PL16803.

IOP Conf. Series: Journal of Physics: Conf. Series **1231** (2019) 012008 doi:10.1088/1742-6596/1231/1/012008

An important factor to be taken into account is the humidity inside the dome, which needs to be maintained below a safe observing limit [5]. Some initial images from PL4240 show the presence of ice crystals that begin to form below -20°C (figure 8), suggesting high moisture around the detector. Putting the camera into a dry cabinet is one way to solve the issue. We found no similar problems with PL16803 so far.

For now, the temperature and humidity inside the dome is controlled with a portable air conditioner. We found that it is necessary to cool down the detector and begin calibration before the dome is open, maintaining lower internal humidity before the system is exposed to outer environment. The whole preparation and calibration procedure should take no more than 35 minutes. With one simple try-out, we managed to reduce frost forming on the detector. For future operation, dehumidifier will be installed in the dome and integrated into the weather system.

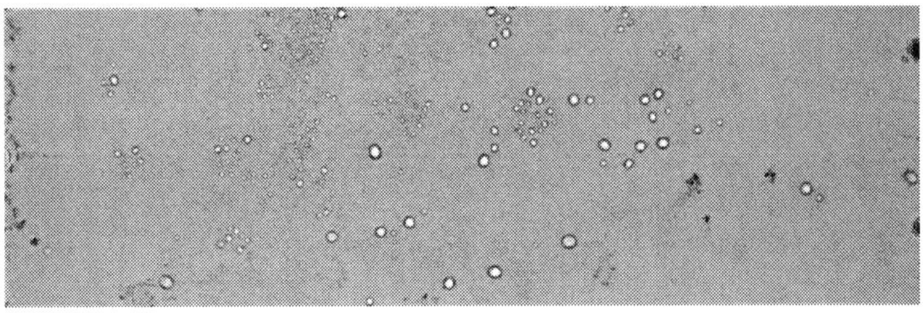

**Figure 8**. Subsection of a frame from PL4240. Signs of condensation and ice crystals are clearly seen. Image is shown with SAOImageDS9, *zscale*, *heat* color.

*4.2 Optics*

RiFast's optics is currently being investigated for off-axis aberration, particularly astigmatism. Such aberration is due to misalignments or residual aberrations and can be identified by defocusing punctual objects. The external contour of the defocused object is analysed to determine the equation of ellipse, in which the defocusing and astigmatism are represented by Zernike coefficients $Z_{20}$ and $Z_{22}$ [6][7].

One of the data taken from preliminary observation can be seen in figure 9, showing an out-of-focus image of Jupiter. Fitting can be done by measuring the gradual decrease of radial counts and fitting the outer border by an ellipse by least squares [8]. The position of the secondary mirror can be varied in steps to see how the geometry changes.

**Figure 9**. Defocused, off-axis image of Jupiter, taken with RiFast and PL4240, shown with SAOImageDS9, *zscale*, *cool* color.

IOP Conf. Series: Journal of Physics: Conf. Series **1231** (2019) 012008 doi:10.1088/1742-6596/1231/1/012008

## 5. Weather System

For the safety of the telescopes against bad weather, at least three weather stations are planned to be installed around the dome and the observatory site. The weather stations will monitor sky temperature, ambient temperature, humidity, rain clouds, and wind velocity around the area.

Currently, one available equipment is the AAG CloudWatcher. The sensor can be calibrated by using the cloud monitoring data from the same site. Similar device has been installed installed for pt5m telescope on La Palma, where it is calibrated with the images from the Gran Telescopio Canarias all-sky camera [5]. The other sensor being considered is the Boltwood Sensor II, which is able to send command to automatically close the dome when the weather is categorized as bad, even if the PC controlling the system is not functioning. This device is used for the Baker Robotic Autonomous Telescope in southern Missouri [9] and the Skynet Robotic Telescope Network at Dark Sky Observatory, USA [10].

## 6. Scientific Potential

### *6.1 Near-Earth Objects (NEO)*

Observation of Near-Earth Objects (NEO) is necessary due to the threats and potential destruction posed by them. The challenges in observing NEO include their possible appearance and fast movement throughout the whole celestial sphere, and near real time inspection of the data is required. Short focal length Cassegrain and Newtonian telescopes with CCD cameras have been useful for NEO observation [11]. The two robotic telescopes for Timau have this potential, though further test for its adjustment is certainly still needed.

### *6.2 Exoplanets*

Current examples of 0.5-m class telescopes for photometry of bright stars with transiting planets are DEMONEX [12], TRAPPIST [13], and pt5m [5]. DEMONEX is an f/8 Meade 0.5-m RCX400 telescope equipped with FLI Fairchild CCD3041 CCD camera (figure 8), while pt5m is an f/10 Officina Stellare ProRC 500 telescope with QSI-532ws camera. Both telescopes have been used for the study of hot Jupiters, which provides constraints on theories regarding planetary and disk formation, as well as planetary migration [14, 15]. TRAPPIST, as an f/8 0.6-m telescope, also conducts photometric search for transiting planets detected by radial velocity. The study of NIR emission from an extremely-irradiated planet that is blocked out during occultation is also one goal of this telescope.

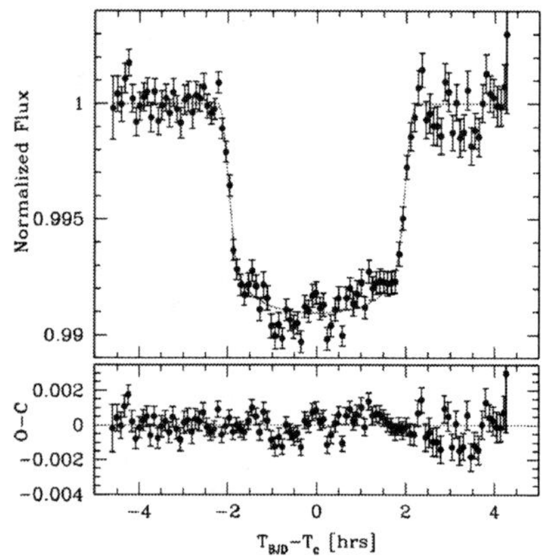


**Figure 10**. Left: DEMONEX telescope at Winer Observatory in Arizona, USA [12]. Right: science example with DEMONEX, combined light curve from 7 transits of hot Jupiter XO-4b [16].

## 7. Summary and Future Work

We have presented the brief description of the planned robotic telescopes system for Timau National Observatory. The system is currently undergoing preliminary testing, primarily for the detector and optics. More comprehensive testing will be conducted in 2019, before its installation in Timau site, planned in 2020. Hardware testing will be the most important process of preparation, especially for the weather system and the dome control. The software control will be tested after the hardware tests, to make sure that the hardware and the software work properly. The integration of the entire system along with the preparation of the automatic image reduction and scheduling pipelines will be conducted afterward.

## Acknowledgment

The robotic telescopes project for Timau National Observatory is a collaborative effort of Space Science Center LAPAN, Bosscha Observatory, and Department of Astronomy ITB.

## References

[1] Hidayat T, Mahasena P, Dermawan B, Hadi T W, Premadi P W and Herdiwijaya D 2012 Clear sky fraction above Indonesia: an analysis for astronomical site selection *Mon. Not. R. Astron. Soc.* **427** 1903–17
[2] Copperwheat C M, Steele I A, Bates S D, Smith R J, Bode M F, Baker I, Peacocke T and Thomson K 2014 Liverpool telescope 2: a new robotic facility for time domain astronomy in 2020+ *Proceeding of SPIE* vol **9145** p 914511
[3] Abbot T M C 1995 In situ CCD testing
[4] Bernardes D V, Martioli E and Rodrigues C V 2018 Characterization of the SPARC4 CCDs *Publ. Astron. Soc. Pacific* **130** 095002
[5] Hardy K, Butterley T, Dhillon V S, Littlefair S P and Wilson R W 2015 pt5m - a 0.5m robotic telescope on La Palma *Mon. Not. R. Astron. Soc.* **454** 4
[6] Luna E, Salas L, Gutiérrez L, Nuñez J M, Nacional U, México A De, Astronomía I De and Nacional O A 2007 Geometric method to measure astigmatism aberration at astronomical telescopes *Appl. Opt.* **46** 3439–43
[7] Noll R J 1976 Zernike polynomials and atmospheric turbulence *J. Opt. Soc. Am.* **66** 207–11
[8] Luna E, Cordero A, Valdez J, Gutiérrez L and Salas L 1999 Telescope alignment by out- of-focus stellar image analysis *Publ. Astron. Soc. Pacific* **111** 104–10
[9] Hicks L L, Reed M D, Thompson M A and Gilker J T 2012 The Baker Observatory Robotic Autonomous Telescope *Astron. Soc. India Conf. Ser.* **7** 123
[10] Smith A B, Caton D B and Hawkins R L 2016 Implementation and operation of a robotic telescope on Skynet *Publ. Astron. Soc. Pacific* **128** 1–21
[11] Boattini A 2001 The role of small telescopes in the discovery of follow-up of near-Earth objects *ASP Conf. Ser.* **246** 233–9
[12] Eastman J D, Gaudi B S and DePoy D L 2008 DEMONEX: the dedicated monitor of exotransits *Proceedings of the International Astronomical Union* vol **4** pp 408–11
[13] Gillon M, Jehin E, Magain P, Chantry V, Hutsemekers D, Manfroid J, Queloz D and Udry S 2011 TRAPPIST: a robotic telescope dedicated to the study of planetary systems *Eur. Phys. J. Web Conf.* **11**
[14] Fabrycky D and Tremaine S 2007 Shrinking binary and planetary orbits by Kozai cycles with tidal friction *Astrophys. J.* **669** 1298–315
[15] Nagasawa M, Ida S and Bessho T 2008 Formation of hot planets by a combination of planet scattering, tidal circularization, and the Kozai mechanism *Astrophys. J.* **678** 498–508
[16] Villanueva S, Eastman J D and Gaudi B S 2016 The DEdicated MONitor of EXotransits (DEMONEX): seven transits of XO-4b *Astrophys. J.* **820** 87

# Diffuse Interstellar Bands in the High-Resolution GAOES Data

**L Puspitarini [1,*], H L Malasan[1,2], O Hashimoto[3]**

[1] Astronomy Research Group, FMIPA Institut Teknologi Bandung and Bosscha Observatory, Indonesia
[2] ITERA Astronomical Observatory, Indonesia
[3] Gunma Astronomical Observatory, Japan

E-mail: lucky.puspitarini@as.itb.ac.id

**Abstract.** Diffuse interstellar bands (DIBs) are an enigmatic set of interstellar absorption features, observed at optical and near infrared wavelengths in the spectra of reddened stars in our Galaxy also in other galaxies. Identifying the carriers of DIBs has been a longstanding challenge and is still very much a work in progress. In recent years, surveys of DIBs have been conducted, but mostly focus on strong bands. In this work, we extracted and measured DIBs recorded in the high-resolution stellar spectra of ε Auriga (ε Aur) and IQ Per observed by using Gunma Astronomical Observatory Echelle Spectrograph (GAOES) (spectral resolution R~65000). Observations were aimed at obtaining high resolution spectra of double-lined detached eclipsing binaries. For each selected DIB, we performed spectral fitting of a combination of a smooth stellar continuum (polynomial function), an empirical DIB model, and a synthetic telluric transmission. The high-quality data allow us to measure not only strong and well-known DIBs, but also weaker DIBs.

***Keywords:*** *Diffuse Interstellar Bands and GAOES Data*

## 1. Introduction

Interstellar matter (ISM) is a "rich laboratory". Interstellar gas and dust in are found in a wide range of temperatures and densities providing numerous different interstellar species (atoms or molecules). Many of them are still unidentified. Diffuse interstellar bands (DIBs) are an enigmatic set of absorption features, they behave like interstellar lines with regard to occurrence, strength, and Doppler shift. Instead of being narrow like the well-known interstellar lines, their profiles are widened and irregular. The first observational record of DIBs was made in 1922, whereas systematic study has started since 1934 [17], however to date the carrier(s) of the DIBs are still unclear, except for the two infrared DIBs [3]. Identifying DIBs carriers is a subject of active research for many years (see [7],[12], [22] and references therein). Despite the unclear carrier(s), DIBs can be a promising ISM tracer since (1) DIBs are ubiquitous, around 500 DIBs are identified in optical and IR domain (see [15], [13], [16]), (2) DIBs are not easily saturated thus they can be used to trace distant Galactic IS structures [19], (3) DIBs are correlated with interstellar extinction but with some dispersion (see [18], [19] and references therein). The relationships can be used to estimate interstellar extinction and/or to provide information about interstellar physical conditions for the reason that some DIBs are sensitive to the stellar radiation field. Massive measurements of DIBs can be used to study the nature of DIBs (i.e. identifying DIB families [2]), but also to build spectroscopic database for ISM mapping

## 2. Data

We used high resolution stellar spectra observed by using Gunma Astronomical Observatory Echelle Spectrograph (GAOES) (see [9], [10], [11]). GAOES was designed and built for the GAO 150-cm telescope. Our data are spectra of two stars i.e., ε Aur (RA= 05 01 58.13, Dec= 43 49 23.87) and IQ Per (RA= 03 59 44.68, Dec=48 09 04.49) (see Fig. 1 and Table 1). Our data consists of one spectrum of ε Aur and two spectra of IQ Per with spectral resolution R or $\Delta\lambda/\lambda$ ~65000 and spectral range of 4779.54 Å to 6691.74 Å. Observations were primarily aimed at obtaining high resolution spectra of double-lined detached eclipsing binaries. But since the two stars are located behind interstellar clouds, this allows us to study DIBs. Stellar parallaxes were obtained from Gaia DR2 (see [5] and [6]). Interstellar extinction data were obtained by using a computerized model by [8] and from Gaia DR2 when available. We have selected 10 DIBs which consists of not only strong type of DIBs, but also weaker ones, i.e., 6283.84, 5780.48, 5797.06, 6613.62, 5849.81, 6204.49, 6203.05, 5705.08, 6195.98, 6445.28, and 6089.85 Å. The criteria of the selection are their equivalent width and FWHM, also the selected DIBs should be able to be distinguished from stellar features.

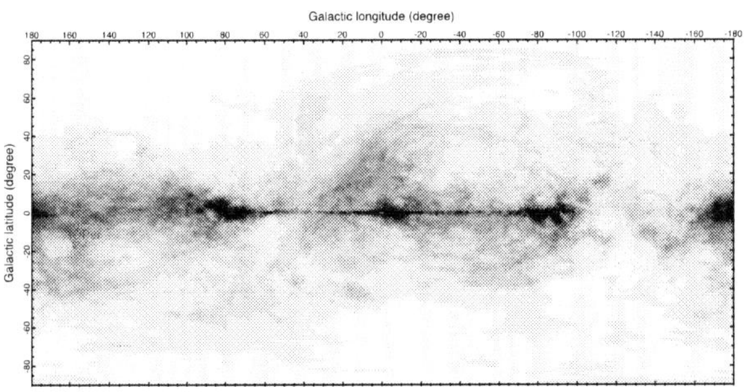


**Figure 1.** Distribution of the target stars (ε Auriga and IQ Per) as seen in the HI map.

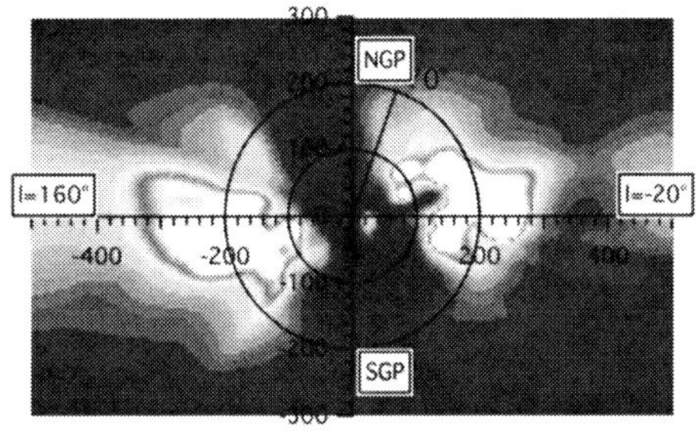


**Figure 2.** Vertical slices of the 3D IS dust map along the longitude of the target stars [14]. Interstellar clouds are seen at a distance of ~100-400 pc.

**Table 1.** Properties of the target stars.

| *Name* | Stype | Lon (°) | Lat (°) | p +/- Δp (mas) | d (pc) | $d_{min}$ (pc) | $d_{max}$ (pc) | Av DR2 | Av Model [8] |
|---|---|---|---|---|---|---|---|---|---|
| ε *Aur* | A9Ia | 162.79 | 1.18 | 2.41 +/- 0.51 | 414.18 | 341.73 | 525.62 | - | 0.77 +/- 0.32 |
| *IQPer* | B9 | 152.24 | -3.71 | 3.26 +/- 0.05 | 306.80 | 302.03 | 311.71 | 0.52 +/- 0.26 | 0.91 +/- 0.39 |

## 3. Method

We aim at extracting and measuring DIB equivalent width (W). The equivalent width is defined as the area between the line profile and the continuum, proportional to the measure of the strength of spectral features. We applied automated DIB fitting to a narrow region of spectrum which contains of DIB. We fitted the spectrum with a combination of smooth continuum (polynomial function), an individual DIB model, and a telluric transmission model simultaneously. The fitting used Levenberg-Marquardt algorithm in IGOR Pro 8 software (WaveMetrics, Inc.). The DIBs models are empirical models/profiles from series of observations of nearby stars previously determined (see [18]). To determine the model, spectra of few reddened stars were selected. Then, we normalized, scaled, and averaged the spectra. For a specific DIB, we took the new averaged-profile as the individual DIB model. The synthetic telluric transmission was computed by using the line-by-line radiative transfer model (LBLRTM) code and the HIghResolution-TRANsmission molecular absorption (HITRAN) spectroscopic database (see [4] and [21]) and is available online in the Transmissions of the Atmosphere for Astronomical data (TAPAS) web-based service [1].

## 4. Results

Some examples of DIB fitting can be found in Fig. 3 to 8. Table 2 shows result of DIB measurements. Since we have two independent spectra for the IQ Per, we compared DIB measurements from the first and the second spectrum (Fig. 9), it shows consistent results within the uncertainties. In addition, with the high-resolution spectra, we possibly detect sub-structure variability in the DIB feature (see Fig. 10). However, we caution that the target stars are binary type and variable star. Thus, further careful investigation needs to be done.

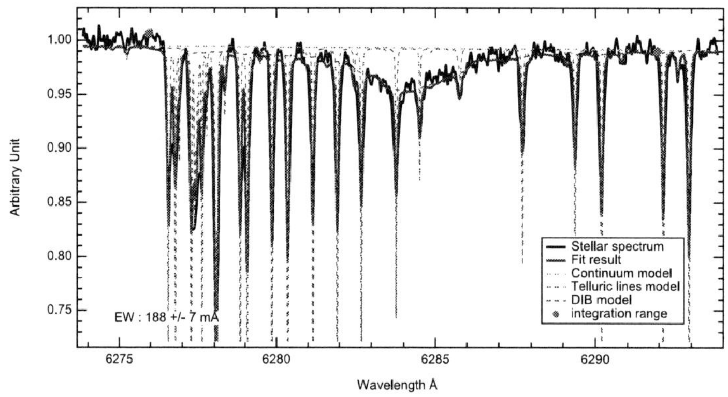


**Figure 3.** An example of automated fitting for 6283Å DIB in the IQ Per (1) spectrum. The 6283Å DIB is one of the strongest DIBs. The black line is the spectrum of target star and the blue line is the fitting result. Dashed lines show three models (DIB model + synthetic telluric transmission model + continuum).

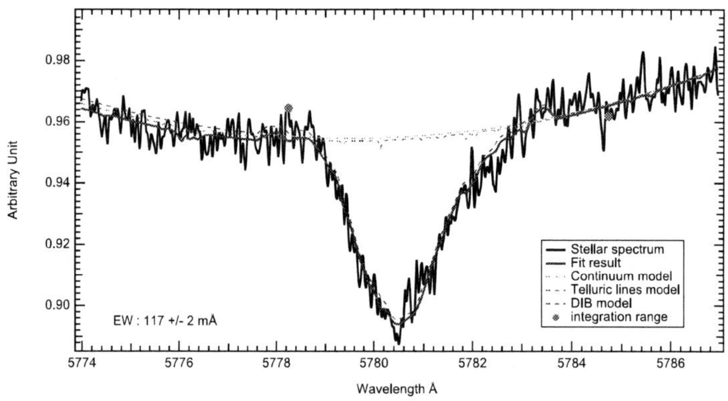


**Figure 4.** An example of automated DIB fitting. Similar to Fig. 3, but for 5780Å DIB.

IOP Conf. Series: Journal of Physics: Conf. Series **1231** (2019) 012009 doi:10.1088/1742-6596/1231/1/012009

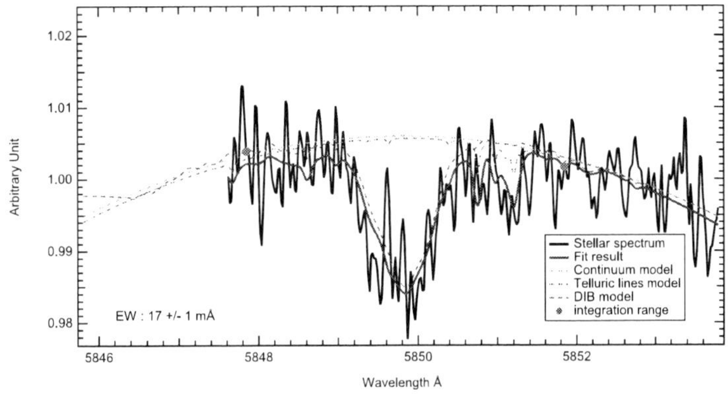


**Figure 5.** An example of automated DIB fitting. Similar to Fig. 3, but for weaker DIB, i.e., 5849.81Å DIB. Stellar line is masked when present.

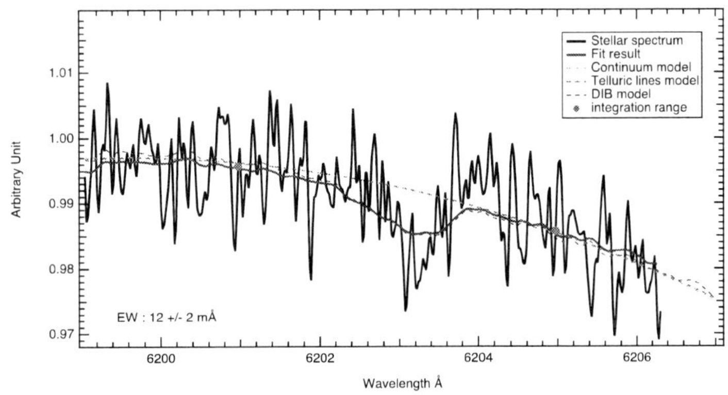


**Figure 6.** An example of automated DIB fitting. Similar to Fig. 3, but for weaker DIB, i.e., 6203.05 Å DIB. Stellar line is masked when present.

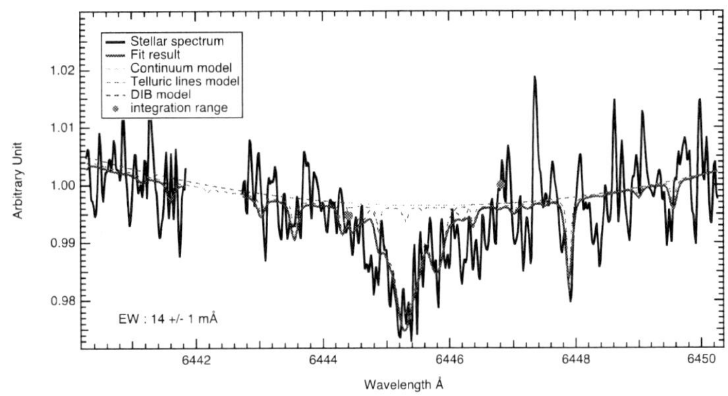


**Figure 7.** An example of automated DIB fitting. Similar to Fig. 3, but for weaker DIB, i.e., 6445Å DIB. Stellar line is masked when present.

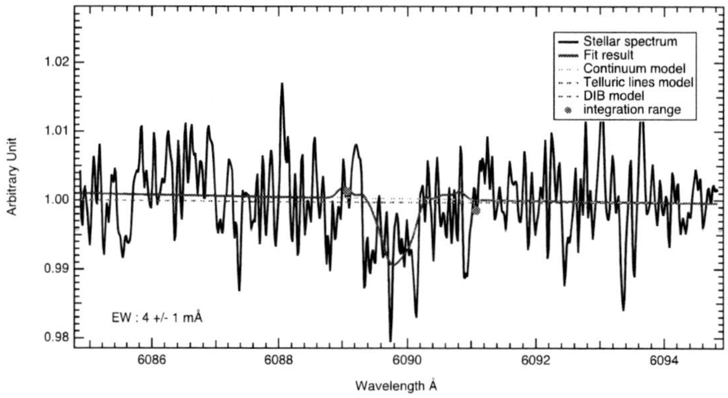


**Figure 8.** An example of automated DIB fitting. Similar to Fig. 3, but for weaker DIB, i.e., 6090Å DIB.

**Table 2.** DIB measurements from spectra of ε Aur and IQ Per.

| **DIB** | **ε Aur** | | **IQ Per (1)** | | **IQ Per (2)** | |
|---|---|---|---|---|---|---|
| **λcen (Å)** | **Wλ** | **ΔWλ** | **Wλ** | **ΔWλ** | **Wλ** | **ΔWλ** |
| **6283.84** | 422.94 | 26.49 | 187.59 | 8.83 | 224.60* | 28.68 |
| **5780.48** | 193.35 | 9.72 | 117.26 | 2.59 | 117.88 | 10.05 |
| **5797.06** | 63.76 | 5.17 | 36.65 | 1.93 | 30.94 | 5.59 |
| **6613.62** | 117.78 | 5.12 | 63.25 | 1.87 | 67.29 | 5.66 |
| **5849.81** | 15.83 | 4.14 | 16.88 | 1.47 | 20.12 | 4.59 |
| **6203.05** | 45.89 | 8.69 | 11.91 | 2.34 | 6.65 | 8.96 |
| **5705.08** | 42.33 | 13.36 | 17.11 | 3.08 | 15.55 | 13.50 |
| **6195.98** | 30.91 | 2.48 | 14.39 | 1.17 | 17.73 | 2.89 |
| **6445.28** | 14.17 | 4.60 | 13.93 | 1.49 | 11.15 | 5.02 |
| **6089.85** | 10.68 | 2.25 | 4.18 | 1.13 | 4.84 | 2.65 |

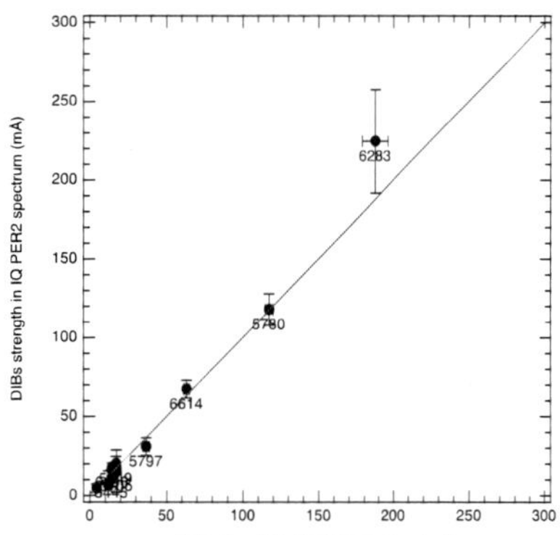


**Figure 9.** Comparison of DIB measurements from the IQ Per (1) spectrum and the IQ Per (2).

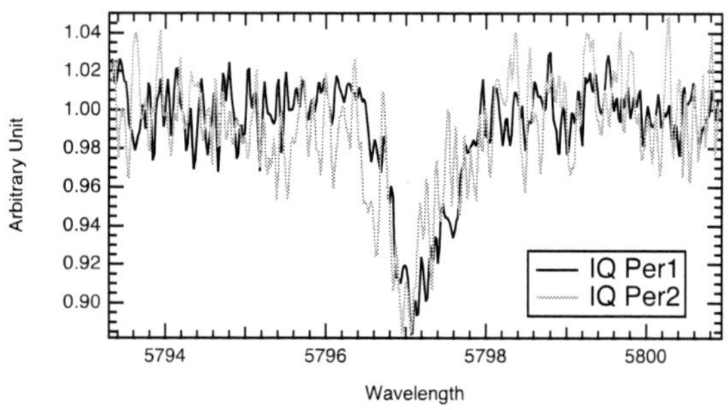


**Figure 10.** Potentially sub-structure variability in the DIB profile by comparing the spectrum of IQ Per (1) and IQ Per (2) (see the right wing of the profile).

## 5. Conclusions

From this study, DIBs have been correctly extracted from our spectroscopic data. We have performed spectral fitting of a combination of a smooth stellar continuum, empirical DIB model, and telluric lines model which is proper for the spectral type of the target stars, i.e., early-type star spectra. Comparing with 3D ISM map [14], the target stars (ε Aur and IQ Per) are located behind the interstellar clouds that have a distance of 100-300 pc, i.e. Perseus region. The high-resolution GAOES spectra allow us to study not only strong DIBs, but also weak and barely studied DIBs. In addition, it may allow us to detect sub-structure variability in DIB profile. For future prospect, massive spectroscopy data from the past and ongoing survey are now or will be available in archive facility. They may help us to improve our understanding about DIBs and to map the distribution of the ISM.

**Acknowledgments**
We thank the ITB for research fund (Riset Peningkatan Kapasitas 2018).

## 6. References

[1] Bertaux J L, Lallement R, Ferron S, Boone C, & Bodichon R 2014 *A&A* **564** A46
[2] Cami J, Sonnentrucker P, Ehrenfreund P, and Foing B H 1997 *A&A* **326** 822
[3] Campbell E K, Holz M, Gerlich D, and Maier J P 2015 *Nature* **523** 322
[4] Clough S A, Shephard M W, Mlawer E J et al 2005 *Journal of Quantitative Spectroscopy & Radiative Transfer* **91** 233
[5] Gaia Collaboration, Prusti T, de Bruijne J H J et al 2016 *A&A* **595** A1
[6] Gaia Collaboration, Brown A G A, Vallenari A et al 2018 *A&A* **616** A1
[7] Geballe T R 2016 *J. Phys.: Conf. Ser.* **728** 062005
[8] Hakkila J, Myers J M, Stidham B J, & Hartmann D H 1997 *ApJ* **114** 2043
[9] Hashimoto O, Kinugasa K, Nishihara E et al 2002 *8th Asian-Pacific Regional Meeting* **II** 7
[10] Hashimoto O, Malasan H L, Yamamuro T, et al 2003 *IAU Joint Discussion* 20
[11] Hashimoto O, Malasan H L, Taguchi H et al 2006 *The 9th Asian-Pacific Regional IAU Meeting* 295
[12] Herbig G H 1995 *Annual Review of Astronomy and Astrophysics* **33** 19
[13] Hobbs L M, York D G, Snow T P et al 2008 *ApJ* **680** 1256
[14] Lallement R, Vergely J L, Valette B et al 2014 *A&A* **561** A91
[15] Jenniskens P and Desert F X 1994 *Astronomy and Astrophysics Suppl.* **106** 39
[16] McCall B J and Griffin R E 2013 *Proc. R. Soc. A* **469** 2151
[17] Merrill P W 1934 *PASP* **46** 206
[18] Puspitarini L, Lallement R, Chen H-C 2013 *A&A* 555 A25
[19] Puspitarini L, Lallement R, Babusiaux C et al 2015 *A&A* 573 AA35
[20] Puspitarini L, Malasan H L, Aprilia et al 2018 *Journal of Physics Conference Series* **1011** 012082
[21] Rothman L S, Gordon I E, Barbe A et al 2009 *Journal of Quantitative Spectroscopy & Radiative Transfer* **110** 533
[22] Sarre P J 2006 *Journal of Molecular Spectroscopy* **238** 1

# Introducing the Zwicky Transient Facility and the Be star variability program: a progress report at the National Central University

**C-C Ngeow[1,*], C-D Lee[1], P-C Yu[1], F Masci[2], R Laher[2], T Kupfer[3,4] and V Z Golkhou[5,6,7] on behalf of the ZTF Collaboration**

[1]Graduate Institution of Astronomy, National Central University, Taoyuan 32001, Taiwan
[2]Infrared Processing and Analysis Center, California Institute of Technology, Pasadena, CA 91125, USA
[3]Kavli Institute for Theoretical Physics, University of California, Santa Barbara, CA 93106, USA
[4]Department of Physics, University of California, Santa Barbara, CA 93106, USA
[5]DIRAC Institute, Department of Astronomy, University of Washington, Seattle, WA 98195, USA
[6]The eScience Institute, University of Washington, Seattle, WA 98195, USA
[7]Moore-Sloan, WRF Innovation in Data Science, and DIRAC Fellow

Email: cngeow@astro.ncu.edu.tw

**Abstract**. The Zwicky Transient Facility (ZTF) is a modern-day wide-field optical survey to systematically explore the transient and variable sky. The ZTF utilizes the 48-inch Samuel Oschin Schmidt Telescope located at the Palomar Observatory. This telescope is equipped with a mosaic CCD camera that provides a field of view of 47 squared degrees. The allocated observing time of ZTF can be divided into partnership time (40%), public time (40%) and Caltech time (20%). The public time contains two surveys: a 3-day cadence for the Northern Sky Survey and a 1-day cadence for the Galactic Plane Survey. Astronomical communities in South East Asian countries are encouraged to explore the public ZTF data once it is released in March 2019. Taiwan's National Central University (NCU) is one of the partnered institutions, and a major ZTF-related project carried out at NCU is the ZTF Be stars variability (ZTF-BeV) program. The main goal of our program is to study the variability of Be stars in the range of ~13.5 to ~20.5 magnitudes.

***Keywords:*** *The Zwicky Transient Facility and Be star variability program*

## 1. Introduction

The 48-inch Samuel Oschin Telescope or P48 Telescope, located at the Palomar Observatory, has a preeminent history in sky surveys. These imaging surveys ranged from the Palomar Observatory Sky Survey (POSS) in mid-20th century, to the POSS II Survey in late 1980s, and to the Palomar Transient Factory (PTF, [1,2]) which commenced about a decade ago. A full list of the surveys that were carried out using the P48 can be found on the P48 website[1]. Since 2017, the P48 has been upgraded to

[1] http://www.astro.caltech.edu/palomar/about/telescopes/oschin.html

perform one of the largest imaging surveys before the Large Synoptic Survey Telescope (LSST) project – the Zwicky Transient Facility (ZTF), named after the Caltech (California Institution of Technology) astronomer Fritz Zwicky. In brief, ZTF is a dedicated very-wide-field synoptic survey for time domain astronomy, by repeatedly imaging the night sky accessible at the Palomar Observatory in three filters.

By design, the P48 is a wide-field telescope. The curved focal plane of the P48 can accommodate the plates used in POSS II Survey to reach a field of view (FOV) of 43.6 degrees$^2$. In the CCD era, the largest FOV achieved on P48 prior to ZTF is ~7.3 degrees$^2$ using the PTF mosaic CCD camera that consists of 11 active CCDs. The PTF camera was replaced with a more compact 600-MP (mega pixel) mosaic CCD camera for the ZTF – a camera with sixteen 6.1k x 6.1k CCD in a 4x4 configuration that provides a FOV of 47 degrees$^2$, while maintaining the same pixel scale of ~1 arc-second per pixel as the PTF camera. With a much improved readout time of ~8.2 seconds, ZTF can survey ~3750 degree$^2$ of the sky in an hour using a nominal exposure time of 30 seconds, reaching to a depth of ~20.5 magnitude. ZTF employs three custom-made filters, the ZTF-g, ZTF-r and ZTF-i filters (hereafter simply referred as g, r and i filters), their transmission curves are similar to the Sloan Digital Sky Survey (SDSS) and the Pan-STARRS1 (PS1) filters as shown in Figure 1.

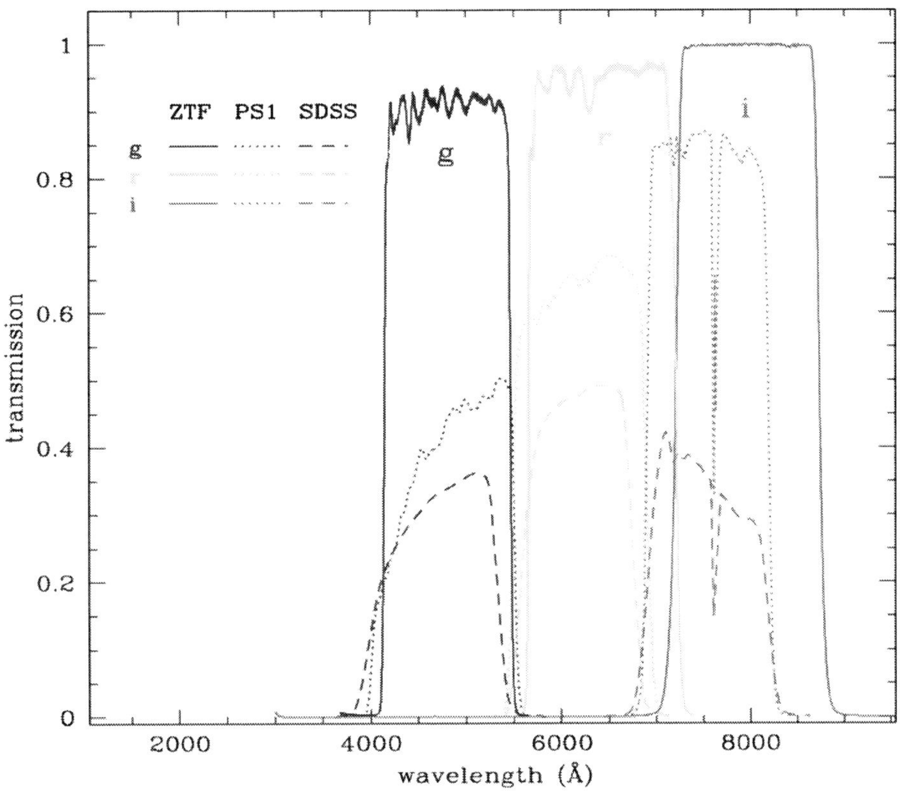


**Figure 1.** Comparison of the transmission curves for ZTF (solid curves), PS1 (dotted curves) and SDSS (dashed curves) in the g (blue curves), r (green curves) and i (red curves) band. Transmission curves for SDSS and PS1 filters are taken from the "Filter Profile Service" available at the Spanish Virtual Observatory (http://svo2.cab.inta-csic.es/theory/fps3/index.php?mode=browse). Note that the transmission curves for SDSS and PS1 include the CCD quantum efficiency, optics and atmospheric effects, while the transmission curves for ZTF are based on filters only.

The ZTF observing time is divided into three portions: partner time, public time and Caltech time. A list of ZTF partnered institutions can be found in the acknowledgment, of which the National Central University (NCU) joined ZTF through the TANGO Consortium of Taiwan. These partnered institutions share 40% of the ZTF observing time, which comprise a variety of observing programs – ranging from the search for various types of supernovae, high cadence observations of Galactic Plane variable stars, to the detection of streaking near-Earth asteroids. Since ZTF is also partly supported by the Mid Scale Innovation Program (MSIP) from the National Science Foundation, an equal amount of the 40% observing time was allocated to carry out two public surveys in g and r filters: (a) The Northern Sky Survey with a 3-day cadence; and (b) The Galactic Plane Survey with a 1-day cadence. The combination of both ZTF public surveys ensures the entire sky north of declination = -31° will be covered. Figure 2 shows the recent sky coverage from the public surveys. The first release of ZTF public-survey data is tentatively scheduled for March of 2019. The remaining 20% of the observing time is reserved for Caltech investigators.

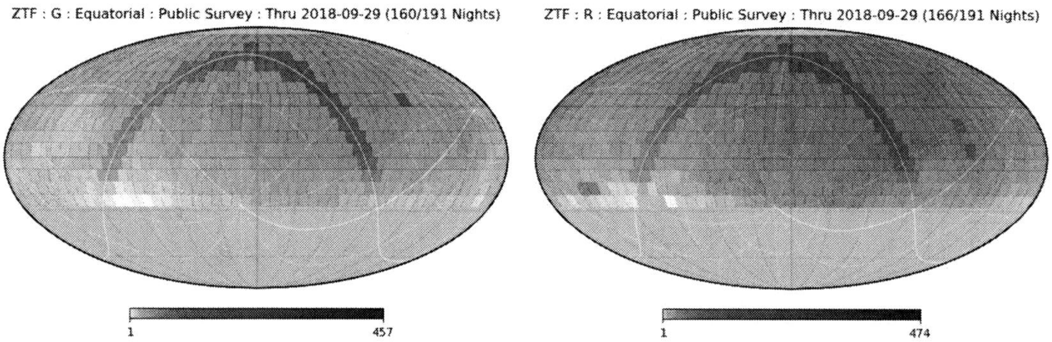


**Figure 2.** Sky coverage of the ZTF public surveys in g (left panel) and r (right panel) filters. The colour scales represent the number of accumulated observations on each ZTF fields (shown as tiling patterns on the map). These sky coverage maps are represented in equatorial coordinates.

## 2. The ZTF Be Star Variability Program

Be stars are B-type main-sequence, or near main-sequence, stars that exhibit emission lines, especially the Balmer series, in their spectra at least once. They are also fast rotators as their rotation velocity at equator can be near the critical velocity. Consequently, a "decretion" disk (instead of an accretion disk) was formed around the Be stars, which explains the observed emission lines. Most Be stars also display detectable variability, and sometimes they are referred as Gamma Cassiopeia (GCAS) type or Lambda Eridani (LERI) type variable stars, which exhibit a wide range of photometric variability (from outburst to periodic variation). Some reviews on Be stars, as well as their variability, can be found, for example, in [3,4,5] and references therein.

Based on KELT (kilodegree Extreme Little Telescope) time-series observations, [6] summarized the photometric variability of Be stars into the following classes: outburst variation, semi-regular outbursts (time scales of 10-200 days), long-term variations (time scales of years), non-radial pulsation (time scales of less than 2 days), intermediate periodicity (time scales of 2-200 days) and eclipsing binaries (time scale of 1-200 days). Given that ZTF will frequently visit the same part of the sky (due to its very large FOV and quick readout), time-series ZTF data can be used to probe the majority of the Be stars' variability, as demonstrated in Figure 3, except for the long-term variations that are longer than three years. We therefore initiated a program to investigate the photometric variability of Be stars using ZTF data at NCU, which we called the ZTF-BeV (ZTF Be stars Variability) program. The main goal of our program is to study the variability of those Be stars with brightness between ~13.5

magnitude and ~20.5 magnitude covered by the ZTF data. As an illustrative example, we show the light curves, based on the first 5 months of ZTF commissioning data downloaded to NCU (see Appendix), for a Be star in Figure 4. This Be star is identified in [7] but excluded from the KELT sample ([6]). Light curves from the CoRoT observations ([8]) for this Be star clearly displayed variation, and the ZTF light curves also show a similar variability. The ZTF-BeV program complements the studies from [6] using the KELT data, because KELT can only observe bright Be stars within the range of 7 to 13 magnitudes in V band. Combining both samples from bright to faint, an unbiased sample can be used to reveal the nature of variability seen in the Be stars.

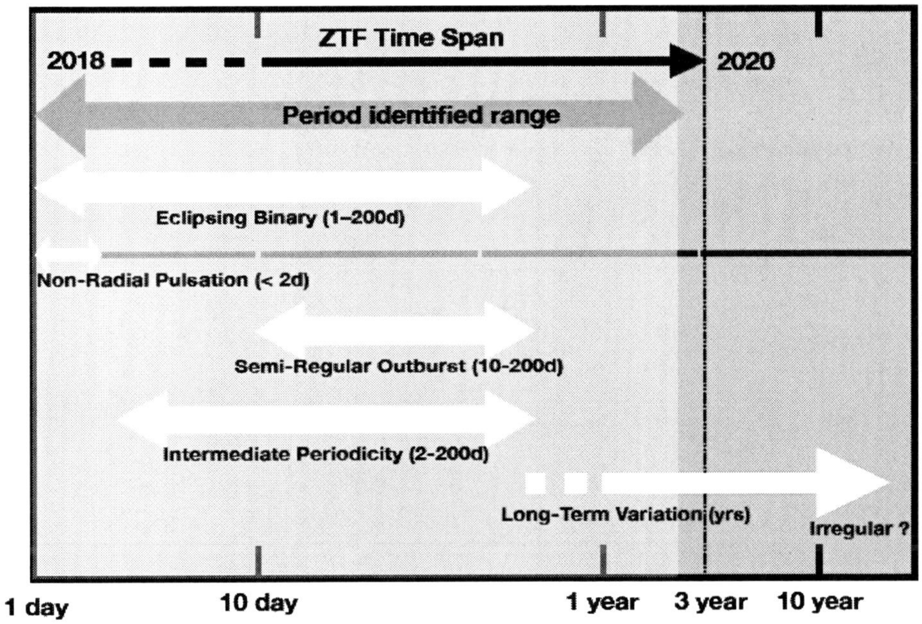


**Figure 3.** Illustration of time scales for Be stars' variability that can be probed by ZTF within its three years (2018-2020) of operation. The top axis represents the time span of ZTF, while the bottom axis shows the time scale for the different variability exhibited in Be stars. Clearly, ZTF cannot probe variability with time scales longer than three years (shown as the orange area in the right of the plot).

## 3. Conclusion

In this article, we briefly introduce the ZTF survey with an aim to promote ZTF to the broader astronomical community within South East Asian countries. ZTF contains two public surveys, the Northern Sky Survey and the Galactic Plane Survey, for which the data will be publicly available in March 2019 with sequential annual releases. Further information about ZTF can be found on the ZTF official website, available at https://www.ztf.caltech.edu/, and in [9,10]. The time-series imaging data in g and r filters will enable a wide range of scientific investigations, and an example study of the variability of Be stars is presented here. A number of technical papers that describe ZTF in more detail have been submitted to the Publication of the Astronomical Society of the Pacific at the time of this article was written; several noteworthy papers are:

a) Graham M *et al.*, "The Zwicky Transient Facility: Science Objectives" [11]
b) Bellm E *et al.*, "The Zwicky Transient Facility: System Overview, Performance, and First Results" [12]
c) Masci F *et al.*, "The Zwicky Transient Facility: Data Processing, Products and Archive" [13]
d) Dekany R *et al.*, "The Zwicky Transient Facility: Observing System" [14]
e) Mahabal A *et al.*, "Machine Learning for the Zwicky Transient Facility" [15]

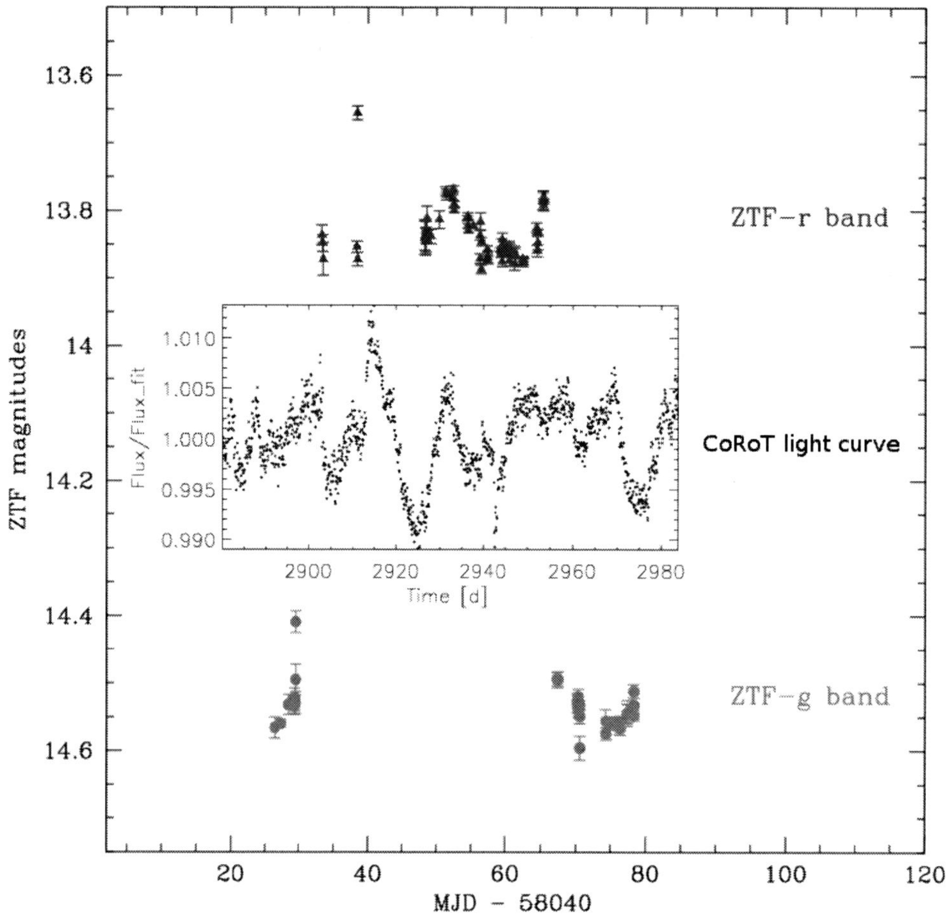


**Figure 4.** Comparison of the light curves for a Be star, CoRoT 102762536, from ZTF commission data and the CoRoT observations, shown as inserted plot in the middle (this plot is taken from [8]).

**Acknowledgments**
Based on observations obtained with the Samuel Oschin Telescope 48-inch and the 60-inch Telescope at the Palomar Observatory as part of the Zwicky Transient Facility project. Major funding has been provided by the U.S National Science Foundation under Grant No. AST-1440341 and by the ZTF partner institutions: the California Institute of Technology, the Oskar Klein Centre, the Weizmann Institute of Science, the University of Maryland, the University of Washington, Deutsches Elektronen-Synchrotron, the University of Wisconsin-Milwaukee, and the TANGO Program of the University System of Taiwan. CCN would also like to thank the financial support from the Ministry of Science and Technology grant 107-2119-M-008-014-MY2.

**Appendix: ZTF Data Download at NCU**
The ZTF imaging data is processed almost in real-time nightly at the Infrared Processing and Analysis Center (IPAC, located in the Caltech campus), hence in the morning data from previous night would be fully processed. At NCU, a Linux server was setup to run the `cronjob` program daily, which in

turn executes the `wget` command to automatically download the nightly ZTF data. Due to limit on bandwidth and storage space, we only download the ZTF catalog products. Figure A1 displayed the data volume downloaded in the first 12-months of ZTF observations, with a total data volume of 25.71 terabytes (TB) to our Linux server.

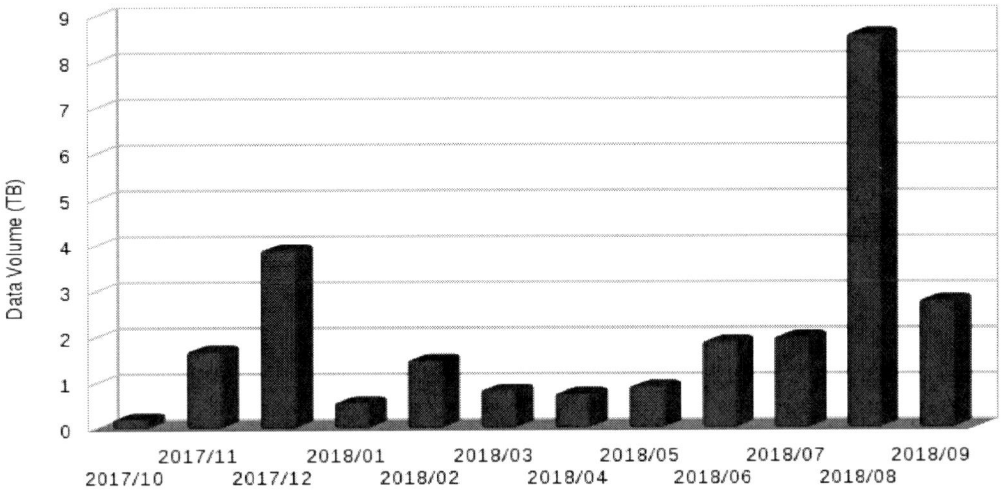


**Figure A1.** The monthly data volume of the downloaded ZTF catalogue products at NCU, expressed in terabyte (TB). Note that ZTF was under commissioning from October 2017 to 17 March 2018 – the official starting date of ZTF for both partnership and public surveys. We have downloaded the commissioning data as well as the ZTF partnership data after the official starting date. The ZTF public data can only be downloaded after the first data release, tentatively scheduled in March 2019.

**References**

[1] Law N *et al.* 2009 *Pub. of the Astronomical Society of the Pacific* **121** 1395
[2] Rau A *et al.* 2009 *Pub. of the Astronomical Society of the Pacific* **121** 1334
[3] Porter J M and Rivinous T 2003 *Pub. of the Astronomical Society of the Pacific* **115** 1153
[4] Saad S M, Hamdy M A and Abolazm M S 2012 *NRIAG J. of Astronomy and Geophysics* **1** 97
[5] Rivinius T, Carciofi A C and Martayan C 2013 *Astronomy & Astrophysics Rev.* **21** 19
[6] Labadie-Bartz J *et al.* 2017 *Astronomical J.* **153** 252
[7] Chojnowski S D *et al.* 2015 *Astronomical J.* **149** 7
[8] Affer L, Micela G, Favata F and Flaccomio E 2012 *Monthly Notices of the Royal Astronomical Society* **424** 11
[9] Laher R R *et al.* 2018 *Robotic Telescopes, Student Research and Education (RTSRE) 2017 Conference Proceedings* vol 1 ed Fitzgerald M, James C R, Buxner S and White S p 329
[10] Bellm E and Kulkarni S 2017 *Nature Astronomy* **1** 0071
[11] Graham M *et al.* 2019 *Pub. of the Astronomical Society of the Pacific* accepted
[12] Bellm E *et al.* 2019 *Pub. of the Astronomical Society of the Pacific* **131** 018002
[13] Masci F *et al.* 2019 *Pub. of the Astronomical Society of the Pacific* **131** 018003
[14] Dekany R *et al.* 2019 *Pub. of the Astronomical Society of the Pacific* submitted
[15] Mahabal A *et al.* 2019 *Pub. of the Astronomical Society of the Pacific* submitted

# Spectroscopic Study of Be Star β Lyrae

**S P Putra**[1,*], **I Imaduddin**[1], **Aprilia**[1], **D G Ramadhan**[1], **H L Malasan**[1,2], and **M I Arifyanto**[1]

[1] Department of Astronomy and Bosscha Observatory, Faculty of Mathematics and Natural Sciences, Institut Teknologi Bandung, Bandung, Indonesia, 40132
[2] ITERA Astronomical Observatory, Lampung, Indonesia

Email: singgihpranaputra@outlook.com

**Abstract.** We present the spectroscopic observation of β Lyrae, an interacting eclipsing binary, with orbital rotation period of 12.9 days, with one component known as the B-emission (Be) star. The primary star is a B6-B8II of 3 $M_⊙$ and the secondary is a 13 $M_⊙$ B0.5V star. The secondary star is embedded in the accretion disk produced by the infalling matter from the primary star, and this disk is estimated as the source of the emission lines. The observations were conducted at Bosscha Observatory, Lembang, Indonesia from May to August 2018, using 10" Meade LX-200R Telescope ($D$ = 254 $f/D$ = 9.84), equipped with a Littrow High Resolution Spectrograph (LHIRES) III, grating of 1200 grooves/mm yielding in the resolution of R~5900, and CCD SBIG ST-402 XME camera with backfocus at 17.5 mm. Seeing during observation was estimated to be 2 arc seconds. During this period, we have obtained 15 spectra in various wavelength coverage, 4 spectra covering around λλ = 5779 – 6046 ÅÅ (typical S/N = 0.02), 1 spectrum covering around λλ = 6430.257 – 6694.434 ÅÅ (typical S/N = 0.06), and 10 spectra covering around λλ = 6487 – 6752 ÅÅ (typical S/N = 0.06). The observed spectra show Hα-emission profiles with $V/R$ variation, He I 6678 Å with P-Cygni profiles, and He I 5876 Å with P-Cygni profiles. These variations are due to the fact that β Lyrae is a binary star system, and also it is suspected that another possible mechanism might be involved.

***Keywords:*** *Be Star β Lyrae and Spectroscopic*

## 1. Introduction

The second brightest star of the Lyrae constellation, β Lyr (HD 174638, HIP 92420, HR 7106) is one of the first known Be stars discovered by Angelo Socchi [7] in 1867 who also observed Hβ emission line in the spectrum. It is also an eclipsing binary, was first discovered by John Goodricke in 1785 with orbital period of 12.9 days [8] and inclination 86° [4]. Goodricke and Johann G. Palitsch, a German amateur astronomer, also found that β Lyr is an algol-type star. Β Lyr system has an early B-type mass-gaining star embedded in an optically thick disk. This disk is an accreting gas that flows from a less massive companion of B6-B8 II. Harmanec [5] and Harmanec [3] suggested there's more structures besides disk at β Lyr system. These structures are jet-like structure which is perpendicular to the orbit's orientation, and a hot-spot (the joint point between gas stream and the disk). It is suggested that these structures are the source where the emission lines of β Lyr came from, which makes this star as a Be-type star. Β Lyr shows some emission lines, eg in He I 5876 Å, Hα and He I 6678 Å. According to some papers [1][2][3][5], their emission lines originated from a jet-like structure. The artistic view of these structure of β Lyr had been drawn in figure 1 in Harmanec's paper [3] and we copy it in this paper as figure 1 with additional information about the position of these structures.

In this paper, we report part of our recent spectroscopic data, focused on He I 5876 Å, Hα and He I 6678 Å lines. The following section 2 describes our research methodology i.e. the observation and data reduction methods. In section 3, we do analysis of our observed spectral lines and we give our conclusion in section 4.

## 2. Methodology: Observation and Analysis

The spectroscopic observation of β Lyr was conducted at Bosscha Observatory, Lembang, Indonesia from May to August 2018. We used 10" Meade LX-200R Telescope ($D$ = 254 f/$D$ = 9.84), equipped with a Littrow High Resolution Spectrograph (LHIRES) III, grating of 1200 grooves/mm yielding in the resolution of R~5900, and CCD SBIG ST-402 XME camera with 1x1 on chip binning. We obtained 15 spectra in various wavelength coverage, 4 spectra around λλ = 5779 – 6046 ÅÅ, 1 spectrum around λλ = 6430.257 – 6694.434 ÅÅ, and 10 spectra around λλ = 6487 – 6752 ÅÅ. Table 1 shows observation date and orbital phase of He I 5876 Å, Hα and He I 6678 Å lines. The orbital phase is calculated using quadratic orbital ephemeris [2]. We also fit our spectra images to calculated orbital phase at figure 4 and 5. We use IRAF (http://iraf.noao.edu) for data reduction and to obtain spectral parameters such as FWHM, EW, V/R ratio and E/C ratio of He I 5876 Å, Hα and He I 6678 Å lines. De-blending technique in IRAF was used to estimate the EW and FWHM of He I 5876 Å, Hα and He I 6678 Å lines. De-blending technique with combination of gaussian and Voight functions for Hα and gaussian fitting for He I 5876 Å and He I 6678 Å lines

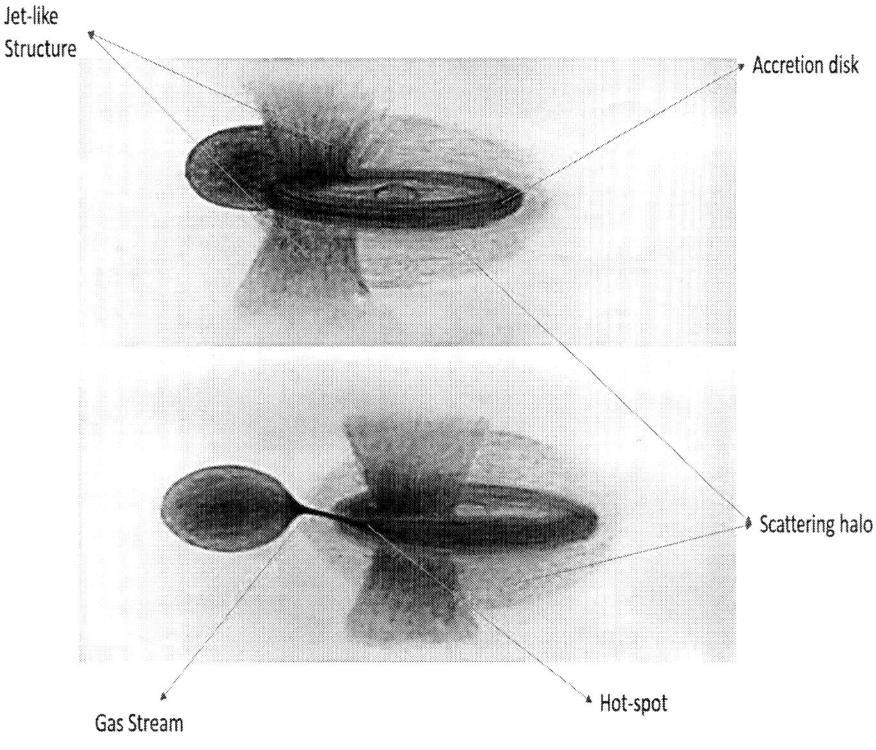


**Figure 1.** The artistic structure of the β Lyr's structure at different orbital phases. If we set the observer is at the right side, upper panel is around phase 0.9 and lower panel is at phase 0. This structure shows 5 different parts i.e. gas stream, hot-spot, accretion disk, jet-like structure, and halo/stellar wind above the polar region. [3]

IOP Conf. Series: Journal of Physics: Conf. Series **1231** (2019) 012011 doi:10.1088/1742-6596/1231/1/012011

**Table 1.** Observation date and Orbital Phase of He I 5876 Å, Hα and He I 6678 Å lines

| Observation Date | Orbital Phase | Lines |
|---|---|---|
| May 6 2018 | 0.061 | Hα and He I 6678 Å |
| May 7 2018 | 0.136 | Hα and He I 6678 Å |
| May 10 2018 | 0.369 | Hα and He I 6678 Å |
| June 28 2018 | 0.226 | Hα and He I 6678 Å |
| June 29 2018 | 0.234 | Hα and He I 6678 Å |
| July 10 2018 | 0.161 | Hα and He I 6678 Å |
| July 12 2018 | 0.236 | Hα and He I 6678 Å |
| July 27 2018 | 0.463 | Hα and He I 6678 Å |
| July 27 2018 | 0.469 | He I 5876 Å |
| July 28 2018 | 0.547 | Hα and He I 6678 Å |
| July 29 2018 | 0.551 | He I 5876 Å |
| August 2 2018 | 0.924 | He I 5876 Å |
| August 2 2018 | 0.929 | Hα and He I 6678 Å |
| August 3 2018 | 0.007 | He I 5876 Å |
| August 3 2018 | 0.008 | Hα and He I 6678 Å |

## 3. Results

### *3.1. He I 5876 Å line*

All the spectra of He I 5876 Å line show two unique profiles. Those profiles are He I 5876 Å line shows P-Cyg profile, that can be seen at figure 2, and a quite strong wing at R region (5880 Å - 5911 Å). Around R region, we assumed there is mixing of plenty of resonance lines such as of C I, C II, Fe I, Fe II, and F III. We checked the wavelength of each resonance lines at NIST Atomic Spectra Database Lines Form (https://physics.nist.gov/PhysRefData/ASD/lines_form.html). We estimate that there is mixture of those 5 resonance lines at between 5889 Å and 5896 Å and a mixture of only 2 resonance lines of Fe I, and Fe II from 5897 Å to 5900 Å. Because β Lyr is an algol-type star, and it is at the Roche-lobe overflow phase as a semidetached binary system with mass transfer, we presume that the metallic resonance lines detected as the absorption lines at β Lyr spectra are from the evolved primary component. This component transfers mass with heavy materials to secondary component which forms the accretion disk. The secondary component is also in the evolved phase so that some material from this secondary mix with the material from the primary. This is the reason why we detected very strong R-region wing which is closer to He I 5876 Å line.

### *3.2. Hα and He I 6678 Å lines*

As we see at figure 3, Hα and He I 6678 Å lines show different profile. Hα always shows double-peak profile with $V/R < 1$. We measured the line parameters of He I 5876 Å, Hα and He I 6678 Å, i.e. FWHM, EW, V/R and E/C, shown in the table 2. We plot these parameters to the observation date on figure 6-8 to show their variations.

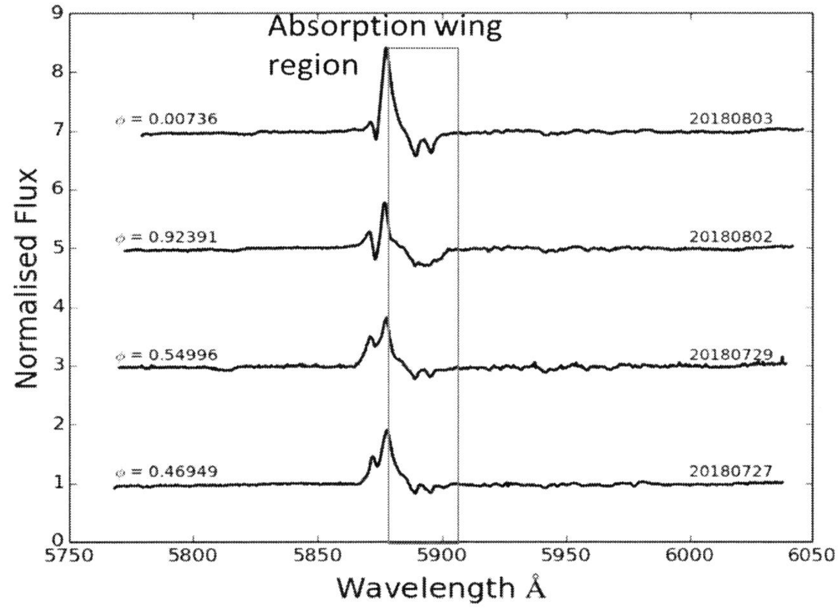


**Figure 2.** The β Lyr spectra at 5876 Å line ordered based on observation date (on the right) and orbital phase (on the left) with absorption wing region at the red side.

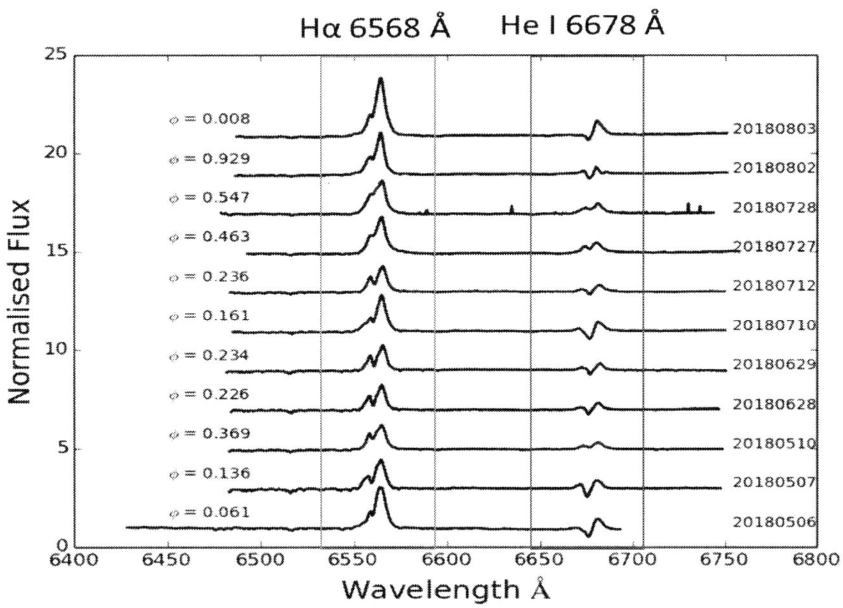


**Figure 3.** The β Lyr spectra at Hα and He I 6678 Å lines ordered based on observation date (on the right) and orbital phase (on the left).

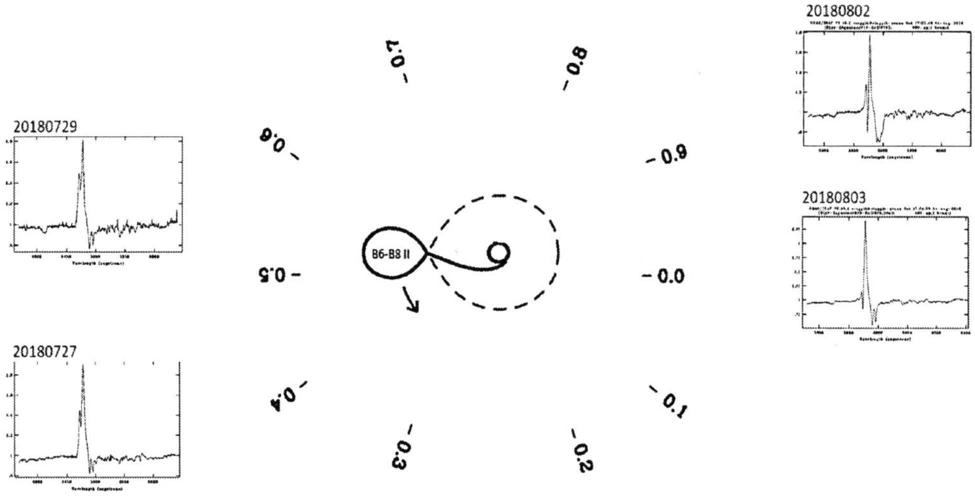


**Figure 4.** The illustration of spectra line of He I 5876 Å fitted with orbital phase.

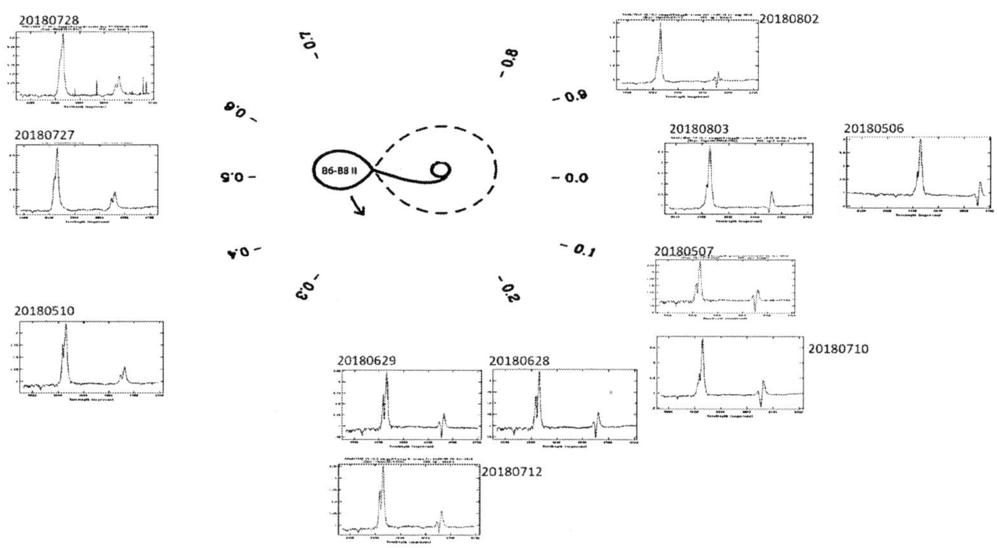


**Figure 5.** The illustration of spectra lines of Hα and He I 6678 Å fitted with orbital phase.

Our measurements show that the FWHM of Hα shows no trend with times and orbital phase its EW decreases with times however it increases with orbital phase. We still can not predict the pattern of Hα FWHM at this observational period, however it does not mean there is no periodicity of Hα FWHM, we can say it is monotonically increasing or decreasing trend [9]. Probably, we will see the periodicity of Hα FWHM at long-term observation. For the He I 6678 Å line, on the other hand, the FWHM of He I 6678 Å line shows no trend (figure 10A) however the EW of He I 6678 Å line increase with times (figure 10B). The FWHM of He I 6678 Å line shows no trend with orbital phases (figure 11A) but its EW decreases with orbital phases (figure 11B). For He I 5876 Å, its FWHM increases and EW shows no trend with times but its FWHM and EW both increase with orbital phases.

**Table 2.** The parameters of Hα and He I 6678 Å lines

| Observation Date | Lines | FWHM (Å) | EW (Å) | V/R | E/C |
|---|---|---|---|---|---|
| May 6 2018 | Hα | 5.5±0.2 | -8.67±0.2 | 0.26±0.1 | 1.36±0.2 |
| | He I 6678 Å | 5.35±0.03 | -0.72±0.03 | -1.04±0.5 | -0.01±0.03 |
| May 7 2018 | Hα | 5.04±0.12 | -6.09±0.12 | 0.43±0.19 | 1.06±0.12 |
| | He I 6678 Å | 3.57±0.015 | -0.08±0.015 | -1.46±1.14 | -0.08±0.015 |
| May 10 2018 | Hα | 5.42±0.1 | -5.34±0.1 | 0.53±0.3 | 0.93±0.1 |
| | He I 6678 Å | 5.43±0.005 | -1.48±0.005 | 0.50±0.4 | 0.24±0.005 |
| June 28 2018 | Hα | 4.27±0.06 | -4.31±0.06 | 0.53±0.37 | 0.9±0.06 |
| | He I 6678 Å | 3.11±0.004 | -0.24±0.004 | -1.02±0.5 | 0±0.004 |
| June 29 2018 | Hα | 4.21±0.08 | -4.37±0.08 | 0.55±0.4 | 0.92±0.08 |
| | He I 6678 Å | 3.01±0.005 | -0.46±0.005 | -0.83±0.3 | 0.03±0.005 |
| July 10 2018 | Hα | 5.92±0.11 | -7.39±0.11 | 0.30±0.13 | 1.2±0.11 |
| | He I 6678 Å | 7.35±0.01 | -0.63±0.01 | -1.56±0.85 | -0.17±0.01 |
| July 12 2018 | Hα | 4.63±0.09 | -5.30±0.09 | 0.55±0.34 | 1.01±0.09 |
| | He I 6678 Å | 7.49±0.007 | -1.10±0.007 | -1.19±0.26 | -0.04±0.007 |
| July 27 2018 | Hα | 4.6±0.17 | -5.14±0.17 | 0.22±0.23 | 1.09±0.17 |
| | He I 6678 Å | 7.2±0.02 | -1.74±0.02 | 0.51±0.56 | 0.32±0.02 |
| July 27 2018 | He I 5876 Å | 4.78±0.04 | -3.77±0.045 | 0.47±0.42 | 0.65±0.04 |
| July 29 2018 | He I 5876 Å | 5.2±0.04 | -3.76±0.04 | 0.63±0.52 | 0.67±0.04 |
| July 28 2018 | Hα | 6.09±0.17 | -7.91±0.17 | 0.55±0.3 | 1.24±0.17 |
| | He I 6678 Å | 5.8±0.02 | -2.28±0.02 | 0.56±0.48 | 0.16±0.02 |
| August 2 2018 | Hα | 5.1±0.22 | -7.78±0.22 | 0.40±0.27 | 1.47±0.22 |
| | He I 6678 Å | 2.23±0.008 | 0.09±0.008 | -1.07±2.1 | -0.01±0.008 |
| August 2 2018 | He I 5876 Å | 8.83±0.04 | -3.44±0.04 | -0.88±0.12 | 0.07±0.04 |
| August 3 2018 | Hα | 5±0.4 | -10.94±0.4 | 0.28±0.13 | 1.82±0.4 |
| | He I 6678 Å | 5.68±0.02 | 4.5±0.02 | -0.42±0.07 | 0.46±0.02 |
| August 3 2018 | He I 5876 Å | 6.1±0.04 | -5.67±0.04 | -1.17±0.02 | 0.25±0.04 |

The V/R ratios of Hα decrease with times but increase with orbital phase. Its E/C ratio increases with times and orbital phase. The V/R and E/C ratio of He I 6678 Å show no trend with orbital phases (figure 11C and 11D). The V/R and E/C ratio of He I 6678 Å line show no trend with times (figure 10C and 10D). For the He I 5876 Å line, on the other hand, the V/R ratio decreases with times (figure 6C) and E/C ratios shows no trend (figure 6D). Its V/R ratio increases with orbital phases but its E/C ratio decreases with orbital phases (figure 7C and 7D). These trends relate with the position of these spectra lines inside the system that is related to the orbital period at the time of observation. We predict that He I 5876 Å emission originates at around jet-like structure, Hα emission originates from inside the disk and He I 6678 Å emission line originates from gas stream or the hotspot. The reason why He I 6678 Å is at those positions is because He I 6678 Å is more far above the ground level to collisionally excite and must arise from helium recombination within hot medium [1]. Because of the rotation of the system, the position of these emission lines periodically changes. If those lines moved towards us, EW and V/R ratio will show increasing trend and vice versa. Because EW and V/R ratio of He I 5876 Å and Hα lines increase with orbital phases, we surely suggest that He I 5876 Å Hα and lines are moving towards observer. For He I 6678 Å line we cannot predict where it moves against observer due to V/R ratio and EW with orbital phase of He I 6678 Å show different trend. Due to the increases of E/C ratio and the decreases of the EW and V/R and value of Hα line with times, we confidently predict that Hα line region will be filled with much material which come from more massive star component. It is because Hα line will shrink towards to more massive star in the future. He I 6678 Å emission line region probably will be detached from the emission region of β Lyr system, and might be back to the less massive companion since its position is around the hot-spot area or at the gas stream due to its EW increase with times. He I 5876 Å region shrinks towards more massive star it is because the trends of E/C and V/R ratios with times of He I 5876 Å indicate so.

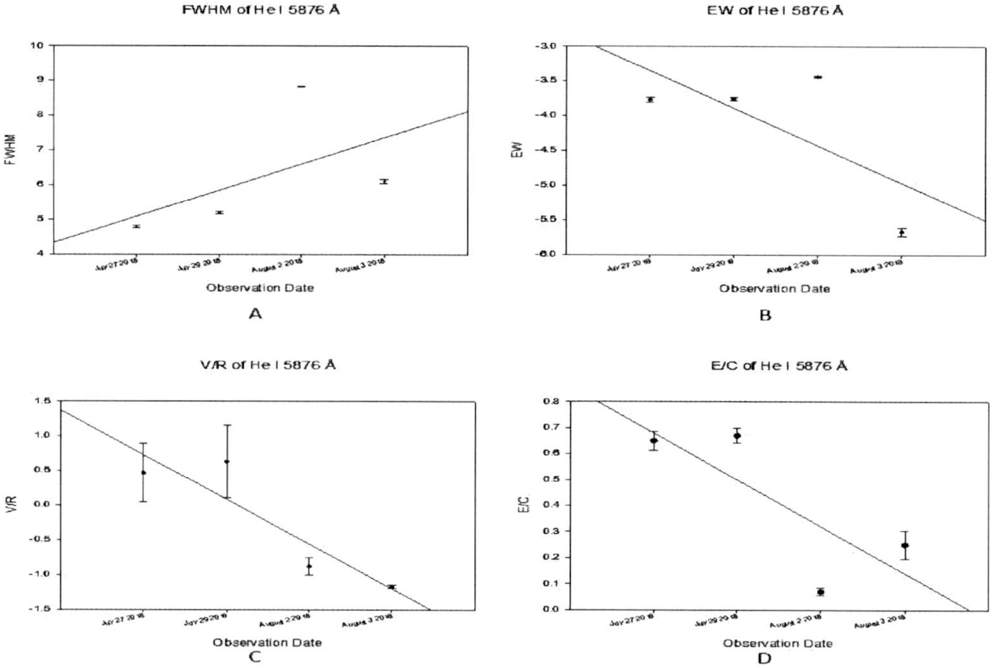


**Figure 6.** He I 5876 Å line plotted to observation date (x-axis) based on the data at table 3. (A) FWHM, (B) EW, (C) V/R ratio, and (D) E/C ratio.

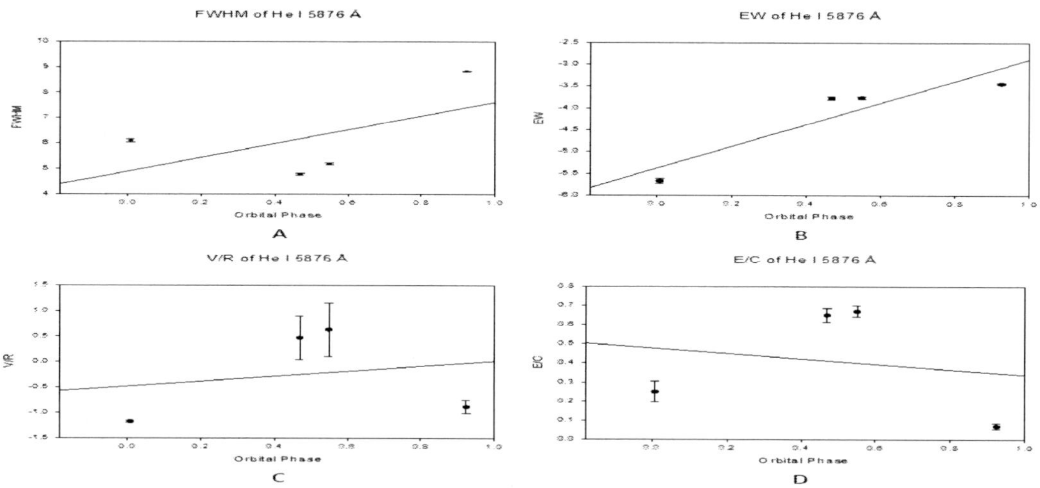


**Figure 7.** He I 5876 Å line plotted to orbital phase (x-axis) based on the data at table 1 and 3. (A) FWHM, (B) EW, (C) V/R ratio, and (D) E/C ratio.

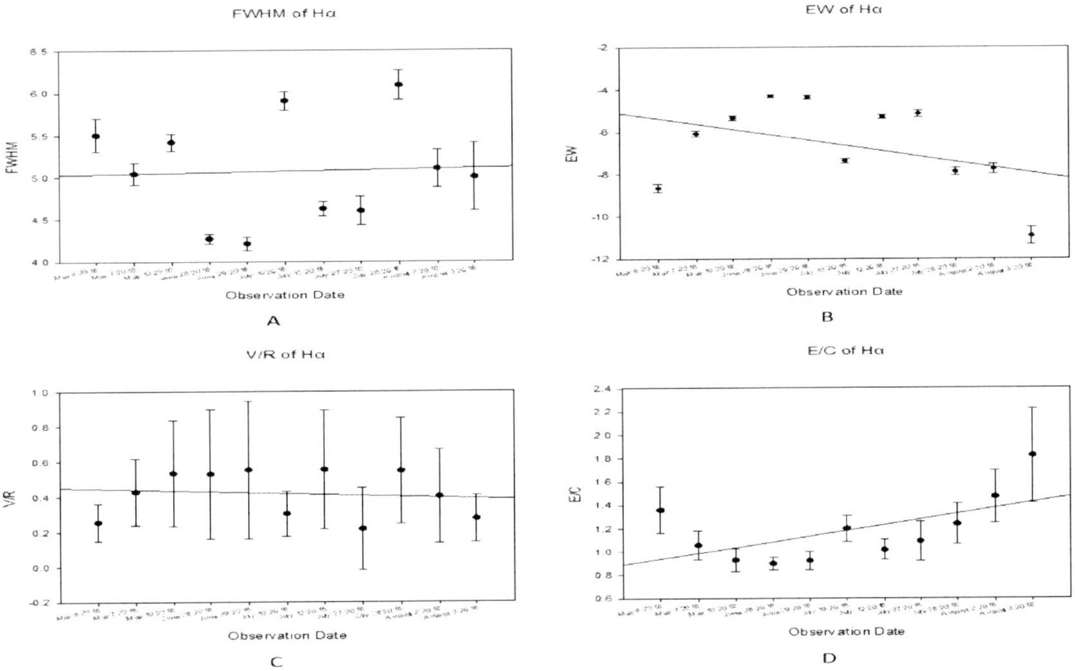


**Figure 8.** Hα line plotted to observation date (x-axis) based on the data at table 3. (A) FWHM, (B) EW, (C) V/R ratio, and (D) E/C ratio.

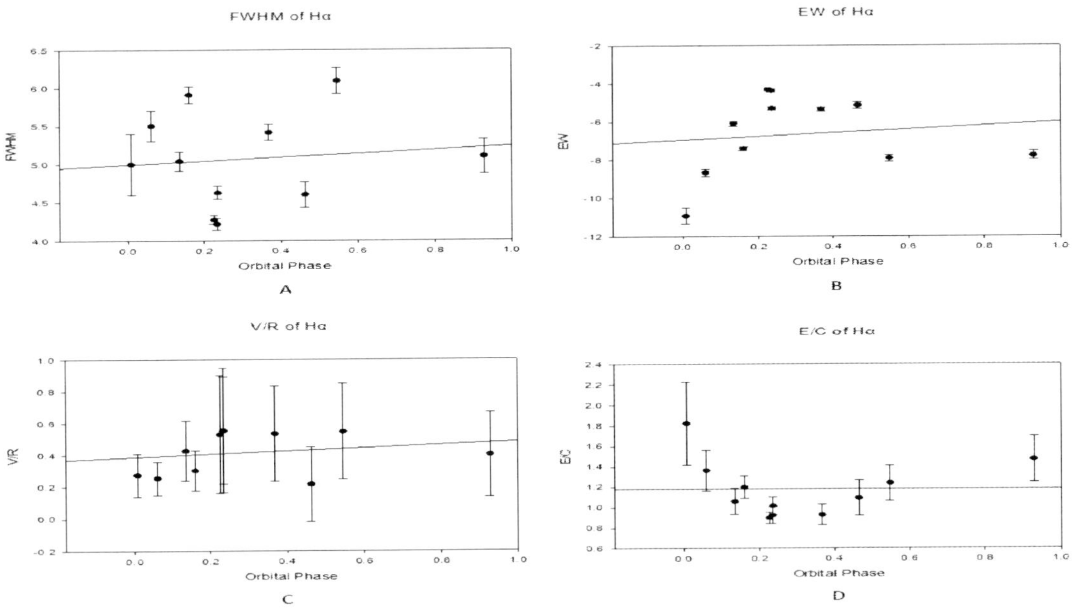


**Figure 9.** Hα line plotted to orbital phase (x-axis) based on the data at table 1 and 3. (A) FWHM, (B) EW, (C) V/R ratio, and (D) E/C ratio

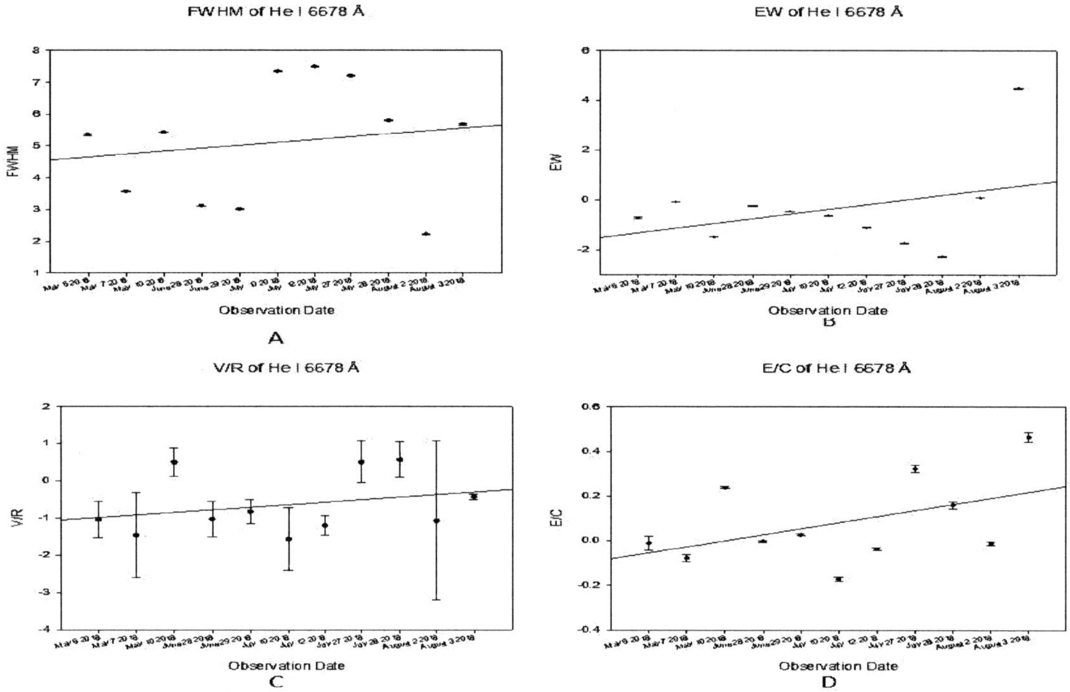


**Figure 10.** He I 6678 Å line plotted to observation date (x-axis) based on the data at table 3. (A) FWHM, (B) EW, (C) V/R ratio, and (D) E/C ratio.

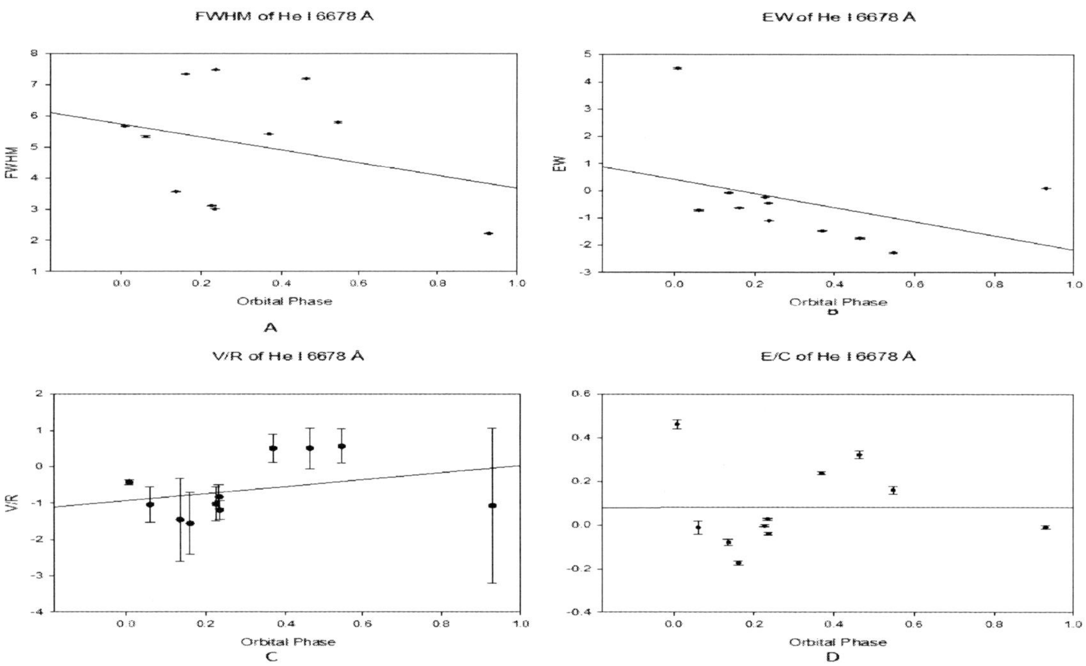


**Figure 11.** He I 6678 Å line plotted to orbital phase (x-axis) based on the data at table 1 and 3. (A) FWHM, (B) EW, (C) V/R ratio, and (D) E/C ratio

## 4. Conclusion

From this study, we conclude several points such as First He I 5876 Å emission originates at around jet-like structure, Hα emission originates from inside the disk and He I 6678 Å emission line originates from gas stream or the hotspot. Second, The FWHM of Hα line the E/C ratio of He I 5876 Å line and the V/R and E/C ratio of He I 6678 Å line do not show any trend with times. Third, The FWHM of He I 6678 Å line, and the V/R and E/C ratio of He I 6678 Å do not show any trend with orbital phases. Forth, Hα and He I 5876 Å lines are moving toward the observer due to the increases of EW Hα line. Fifth, He I 6678 Å emission line region probably will be detached from the system and might be back to the less massive companion since its position is around the hot-spot area or at the gas stream. Sixth, Hα and He I 5876 Å emission lines region is shrinking toward more massive star in the future due to the decreases of the E/C and V/R ratios of He I 5876 Å and due to the decreases of EW and V/R ratio of Hα and the increases of E/C ratio of Hα.

**Acknowledgment**
The authors would like to thank to ITB for providing research fund from P3MI.

## 5. References

[1] Bonneau, D., Chesneau, O., Mourard, D., Bério, Ph., Clausse, J. M., Delaa, O., Marcotto, A., Perraut, K., Roussel, A., Spang, A., Stee, Ph., Tallon-Bosc, I., McAlister, H., ten Brummelaar, T., Sturmann, J., Sturmann, L., Turner, N., Farrington, C., and Goldfinger, P. J., *Astronomy & Astrophysics,* **532** (2011) A148
[2] Ak, H., Chadima, P., Harmanec, P., Demircan, O., Yang, S., Koubský, P., Škoda, P., Šlechta, M., Wolf, M., Boži´ c, H., Ruždjak, D., and Sudar, D., *Astronomy & Astrophysics,* **463**, (2007) 233
[3] Harmanec, P., *Astronomische Nachrichten*, **323**, 2 (2002) 87
[4] Linnell, A. P., *Monthly Notices of the Royal Astronomical Society*, **319**, 1 (2000) 255-266
[5] Harmanec, P., Morand, F., Bonneau, D., Jiang, Y., Yang, S., Guinan, E. F., Hall, D. S., Mourard, D., Hadrava, P., Bozic, H., Sterken, C., Tallon-Bosc, I., Walker, G. A. H., McCook, G. P., Vakili, F., Stee, P., and Le Contel, J. M. *Astronomy and Astrophysics*, **312**, (1996) 879-896
[6] O. R. Pols, J. Cote, L. B. F. M. Waters, J. and Heise, *Astronomy and Astrophysics*, **241**, (1991) 419-438
[7] Socchi, A., *Sugli spettri prismatici delle stelle fisse : memoria del A[ngelo] Secchi, by Secchi, Angelo*, **1**, (1867) 67
[8] Goodricke, J., Pols, O. R., Cote, J., L. Waters, B. F. M., J. and Heise, *Philosophical Transactions of the Royal Society of Londons*, **75**, (1785) 153-164
[9] Jones, C.E., Tycner, C., and Smith, A.D, *The Astronomical Journal*, **141**, Issue 5 (2011) 150
[10] Wilson, R. E., and van Hamme, W., *Monthly Notices of the Royal Astronomical Society*, **303** (1999) 736-754.

# Search for blue straggler stars in open clusters from the VVV survey (preliminary result)

**S N Amalina[1,*] and M I Arifyanto[2]**

[1] Undergraduate Program of Astronomy, Faculty of Mathematics and Natural Sciences, Institut Teknologi Bandung, Jalan Ganesha 10, Bandung 40132, Indonesia
[2] Department of Astronomy, Faculty of Mathematics and Natural Sciences, Institut Teknologi Bandung, Jalan Ganesha 10, Bandung 40132, Indonesia

Email: amalina.shinta@gmail.com

**Abstract.** Blue straggler stars (BSS) are a group of stars that are found within some stellar clusters through their color-magnitude diagrams (CMDs). The BSS are located on the upper-part of the CMD's turn-off (TO) point along the zero-age main sequence (ZAMS) line. BSS are commonly found within intermediate-age to old stellar clusters, typically $10^8$ years. Based on the catalogue of BSS in open clusters (OCs), the OCs toward the Galactic Center (GC) that host BSS are less than the ones toward the anti-center. Using the VVV survey data in near-infrared, we plan to investigate the presence of BSS in OCs toward the GC. To reach our objectives, we generally divide the works into: (1) determine cluster membership probability, (2) fit the decontaminated CMDs with its respectable isochrones, (3) detect any stars that fall into BSS area in CMD and recheck their position in spatial diagram. Out of the 4 OCs we planned to investigate, only 1 OC had been analyzed up to its membership probability; NGC 5999. However, we found 1541 stars as the most probable members, exceeded way too many compared to the literature. We still need to refine our method in the future work to clean more field stars.

***Keywords:*** *Blue straggler stars, Open clusters, and VVV survey*

## 1. Introduction

The discovery of blue straggler stars (BSS) back in 1953 by Allan Sandage [1] started long years of quest to understand their nature and stellar evolution in general. BSS are found through the color-magnitude diagrams (CMDs) of stellar clusters (SCs) as a sequence that resides on the upper-part of the CMDs' turn-off (TO) point along the zero-age main sequence (ZAMS) line. They are commonly found within intermediate-age to old SCs, typically around $10^8$ to $10^{10}$ years. BSS' location on the CMD is defying the stellar evolution theory in SCs where there should be no main sequence stars that are bluer and more massive than the ones on the TO point. This implies they seem to stay longer than they should on the MS [2], raising a question towards our idea that members of SCs start their evolution together. The BSS sequence is more prominent on the CMD of old SCs such as globular clusters. It is interesting to investigate the presence of BSS in younger stellar population such as intermediate-age open clusters (OCs). In 2007, Ahumada and Lapasset published 'New Catalogue of Blue Stragglers in Open Clusters' (hereafter as AL07) based on their studies on large numbers of OCs in Milky Way [2]. The catalogue itself is an extended version and a revision of their previous work, AL95 [3]. However, in AL07, the OCs which are located towards the Galactic Center (GC) are less than the ones on the anti-center direction. This may be due to thick obscuration of ISM and the dense field. In this work, we focus on the searching for the BSS presence in OCs located towards the GC.

## 2. Objects and Data

### *2.1. The VVV Survey*

Our choice to investigate the OCs toward the GC led us to make use the public data from the VISTA Variables in the Vía Láctea (VVV) Survey by European Southern Observatory (ESO). The survey was conducted in near-infrared wavelength, divided into 5 passbands: Z (0.87 μm), Y (1.02 μm), J (1.25 μm), H (1.64 μm), and Ks (2.14 μm). The survey lasted for about five years from 2010, covering 520 $deg^2$ of bulge and disc region that are divided into hundred tiles [4]. We used the data from ZYJHKs Catalogue in the Vía Láctea Version 2 (2015) where it is based from the VVV DR 2.

### *2.2. Objects*

The objects were chosen from AL07 and 'New Galactic star clusters discovered in the VVV survey' [5]. Our objectives are rechecking the number of BSS in some OCs in AL07 with our own method and also analyzing OCs with unknown number of BSS. They were then cross-matched with the criteria of OCs that would most likely to host BSS [6] which are shown in table 1.

**Table 1.** Comparison of properties of OCs in AL07 which host BSS and the ones without

| Parameter | OCs with BSS | OCs without BSS |
|---|---|---|
| Location from Galactic Disc | Further | Closer |
| Average of log(age) | 8.58 | 7.52 |
| Average of (J-H) | 0.43 ± 0.170 | 0.19 ± 0.185 |
| Average of (H-Ks) | 0.11 ± 0.074 | 0.09 ± 0.084 |
| Average of (J-Ks) | 0.54 ± 0.235 | 0.23 ± 0.270 |
| Average of J | -5.29 ± 1.061 | -4.59 ± 1.072 |
| Average of H | -5.71 ± 0.984 | -4.78 ± 1.019 |
| Average of Ks | -5.83 ± 0.098 | -4.77 ± 1.020 |
| Ratio of number density of the core to the total | 1.47 ± 0.201 | 1.39 ± 0.193 |

The next part is matching the OCs that fall into criteria with the VVV data. This is because not every OCs inside VVV survey region was observed. We decided to analyze four OCs; NGC 5999, Pismis 18, VVV CL008, and VVV CL070.

**Table 2.** The chosen OCs' properties

| Parameter | NGC 5999 | Pismis 18 | VVV CL008 | VVV CL070 |
|---|---|---|---|---|
| RA(J2000.0) | 15h 52m 8.64s | 13h 36m 55.92s | 11h 55m 28.99s | 16h 21m 48.00s |
| DE(J2000.0) | -56° 28′ 22.12″ | -62° 03′ 54.00″ | -63° 56′ 24.00″ | -51° 44′ 11.00″ |
| Galactic longitude ($l$) | 326.016° | 308.233° | 296.833° | 332.406° |
| Galactic latitude ($b$) | -1.923° | 0.337° | -1.733° | -1.329° |
| Log(age) | 8.4 | 8.6 | 8.6 | 8.8 |
| Apparent diameter | 3′ | 4′ | 1′ 14″ | 52″ |
| $N_{members}$ | 65 | 150 | 25 | 30 |
| Distance (kpc) | 1.629 | 2.309 | 1.400 | 2.000 |
| $N_{BSS}$ | 4 | 3 | unknown | unknown |

We also added some constrains before requesting the data for each objects. We only picked data points with the flag "stars" and gave range of values for each magnitudes to make sure there is no empty value within the data set requested. The chosen data points are plotted spatially and the OCs which have cropped or incomplete spatial distributions were singled out.

## 3. Cluster Membership Analysis

Before going further to isochrones fitting and finding the BSS, we need to separate the stars that are members of the clusters and the ones belong to the back- and foreground. This step is very important because the objects lie toward a crowded region. The objects, too, are located near the Galactic disc where star formation activity is high. For the cluster membership analysis, we followed the algorithm explained by Froebrich, et al [7].

The first step is to define the cluster region ($A_{cl}$) and a control field region ($A_{con}$) around the cluster's center coordinate. For this step, we made use the value of core radius in Milky Way Star Cluster (MWSC) Catalogue by Kharchenko, et al [8]. Stars that are located nearer than 3 times the cluster core radius belong to the cluster region $A_{cl}$. All stars farther away than 5 times the cluster core radius up to the limit of the frame requested belong to control field $A_{con}$. This step is illustrated in figure 1.

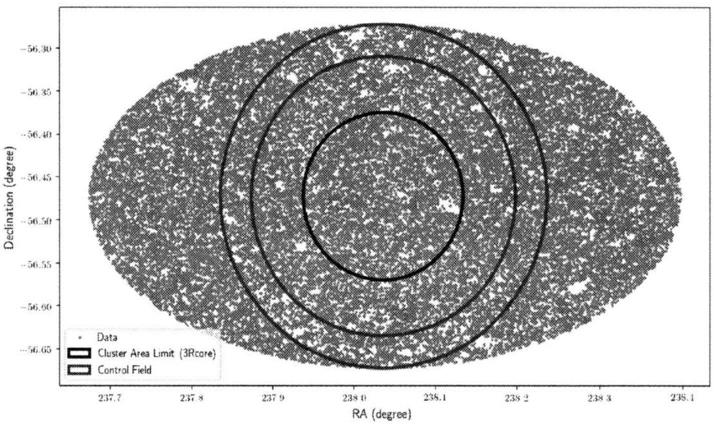


**Figure 1.** The spatial distribution of NGC 5999 with the projected cluster region $A_{cl}$ (inner circle, black in digital version) and control field $A_{con}$ (region between two outer circles, blue in digital version). The second circle is 5Rcore while the outermost is 12 arcmin. The dots represent the stars.

We then applied the decontamination procedure that makes full use of the photometric data. For each stars within each OCs' respective frames, we only requested the magnitudes in J, H, and Ks, hence each stars have their own CMD coordinates in color-color-magnitude (CCM). For each stars $i$ with the apparent magnitude $J^i$, $H^i$, and $Ks^i$, color $H^i - Ks^i = HKs^i$, and color $J^i - Ks^i = JKs^i$, we calculate the CCM distance, $r_{ccm}$, to every other stars where $j \neq i$ in the following way:

$$r_{ccm} = \left[\frac{1}{2}\left(Ks^i - Ks^j\right)^2 + \left(JK^i - JK^j\right)^2 + \left(JH^i - JH^j\right)^2\right]^{\frac{1}{2}} \qquad (1)$$

The factor of ½ in front of the difference in Ks-band accounts for generally larger spread of values compared to colors. For each star $i$ in $A_{cl}$, we then listed all the tenth smallest CCM distance ($r_{ccm10}$) values over all stars where $j \neq i$. We then compared the stars in $A_{con}$ which have similar CCM coordinates with the stars in $A_{cl}$ then we computed the number of stars in $A_{con}$ within the CCM distance $r_{ccm10}$ computed before ($N_{ccm}$). Then we computed probability of star $i$, $P_{ccm}$, to be a member of the cluster in the following way:

$$P_{ccm} = 1.0 - \left(\frac{N_{ccm}}{10}\right)\left(\frac{A_{cl}}{A_{con}}\right) \qquad (2)$$

If $P_{ccm}$ of a particular star is negative, then its membership probability is decided to be zero. We accept stars with $P_{ccm}$ more than or equal to 0.8 as most probable members of the cluster.

After selecting the stars which have the highest chance to be members of the cluster, we construct a decontaminated CMD for the cluster. The decontaminated CMD will then be used to fit with the isochrones. The fitted CMD will be used to determine the area where BSS should reside. We then will detect star(s) which fall into the BSS area and recheck their spatial distribution. If there are any BSS detected, we will recheck whether they are really located in cluster radius and towards the cluster area or they appear further away, closer to the field [9].

## 4. Preliminary Result

Out of four OCs we planned to analyze, we already partially analyzed NGC 5999 up to counting its membership probabilities $P_{ccm}$. We detected 5441 stars in its $A_{cl}$ and 7162 stars in $A_{con}$. The CMDs of Acl, Acon, and the decontaminated CMD are shown below in figure 2.

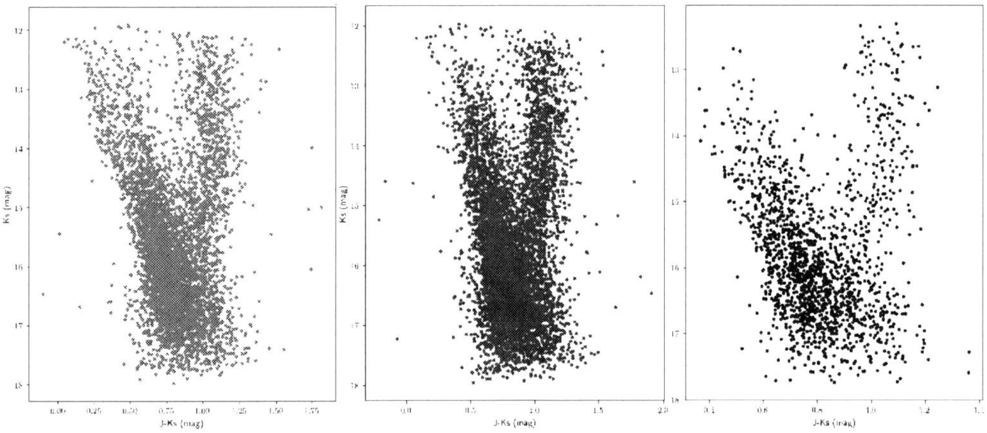


**Figure 2.** Contaminated CMD of NGC 5999 in $A_{cl}$ (left, in red dots), CMD of $A_{con}$ of NGC 5999 (middle, in blue dots), decontaminated CMD of $A_{cl}$ (right, in black dots). All axes of the CMDs are in J-Ks versus Ks.

The result of the membership analysis we conducted on NGC 5999 is we get 1541 stars as the most probable members of NGC 5999. This result draws the number of members way too many compared to AL07 (65 stars). We assume there is still contamination of field stars that made us unable to proceed in searching for the BSS. We will recheck and refine our membership determination method to clean more field stars in future work.

## 5. Conclusion

Blue stragglers (BSS) are unique objects that are challenging to study. BSS can be found in some stellar clusters, typically the old ones. A catalogue of OCs with BSS had been published in 2007 by Ahumada & Lapasset (AL07) where the OCs toward the Galactic Center are less studied. In this work, we focus our study in searching for BSS in OCs towards the GC. We made use the public data provided by the VVV Survey by ESO in near-infrared wavelength. The membership analysis is an important step in studying OCs, especially for OCs that are located toward GC due to the high density of projected stars on the field. We followed the algorithm by Froebrich, et al (2010) to determine cluster membership probability. We have planned to do the search for BSS up to the number of BSS determination, however, the work in this paper are still unfinished. We have analyzed only NGC 5999 up to its membership probability determination. We found 1541 stars as the most probable members out of 5441 we detected in its cluster region. This result exceeded way too many compared to AL07. We decided to refine our method of membership analysis in the future work.

**Acknowledgements**

We express our sincere gratitude for the travel and accommodation support given by P3MI who made it possible for us to contribute for the $10^{th}$ SEAAN Meeting 2018 in Bandar Lampung, Indonesia. We also thank the VISTA Variables in the Vía Láctea (VVV) Survey by European Southern Observatory (ESO) that gave us the opportunity to use the public data service.

**References**

[1] Sandage A 1953 *The Astronomical Journal* **58** 1207 p 61-75
[2] Ahumada J A and Lapasset E 2007 *A&A* **463** p 789 (AL07)
[3] Ahumada J A and Lapasset E 1995 *A&A* **109** p 375 (AL95)
[4] Saito R K et al 2012 *A&A* **537** A107
[5] Boffin H M J (ed), Carraro G (ed), and Beccari G (ed) 2014 *Ecology of Blue Straggler Stars* (Heidelberg: Springer-Verlag Berlin Heidelberg)
[6] Lee H-U and Chang H-Y 2017 *Journal of the Korean Astronomical Society* **50** p 51-59
[7] Froebrich D, Schmeja S, Samuel D, and Lucas P W 2010 *MNRAS* **409** p 1281-128
[8] Kharchenko N V et al 2013 *A&A* **558** A53
[9] Carraro G, Vázquez R A, and Moitinho A 2008 *A&A* **48**
[10] Borissova J et al 2011 *A&A* **532** A131

# Applying Deep Neural Networks (DNN) for Measuring Photometric Redshifts from Galaxy Images: Preliminary Study

**M R I Syarifudin[1*], M I Hakim[1] and M I Arifyanto[1]**

[1]Department of Astronomy, Institut Teknologi Bandung, Bandung, West Java, Indonesia

E-mail: ilmi.mris@gmail.com

**Abstract.** In the cosmological and extragalactic study, distance to a galaxy is an important parameter, by knowing the distance, we can find any other physical parameter such as mass, luminosity, star formation rate, and metallicity. By applying the specific cosmological model, we can measure a distance from the redshift. The exact redshift can only measure by using the spectroscopic technique (Doppler effect), but spectroscopic observation limited to brighter objects and numbers of objects in a single field of view (FoV). While photometric observation can capture fainter objects and more objects in a single FoV. Measurements of photometric redshift could have done by comparing the SED curves of the elliptical galaxy with known spectroscopic redshifts from other elliptical galaxies which we want to find the photometric redshift. Another method is to do linear or non-linear regression, by assuming the redshift is a function of magnitude in each band-pass filter. Therefore, we propose a technique that using full galaxy images in each measured bands and machine learning method for measuring photometric redshift. We pass entire multi-band galaxy images into the machine learning architecture to get an estimated redshift. In this work, we use galaxies images at $0 \leq z \leq 1$ from SDSS DR 10 as the datasets and we use DenseNet, one of the Deep Neural Networks (DNN) architecture.

***Keywords**: Photometric-redshifts, Galaxies, Machine Learning, Deep Neural Networks*

## 1. Introduction

The exact redshift only could measure by using the spectroscopic technique (Doppler effect), but spectroscopic observation limited to brighter objects and numbers of objects in a single field of view (FoV). While photometric observation can capture fainter objects and more objects in a single FoV [1,2]. We used data from telescope survey Sloan Digital Sky Survey (SDSS). SDSS Data Release 10$^{th}$ had 859,322 galaxies spectrums and 208,478,448 galaxies images [3]. Photometric redshift first measured by [4]. He used SED pattern of the elliptical galaxy which known the redshift as a template to measure the direct shifting in the x-axis (wavelength) of another SED from an elliptical galaxy, thus we get the redshift of that galaxy relative to the template. [5] created a new method that inspired by [4], they made SED templates from the galaxy that known redshift each galaxy morphology. [6] used linear and non-linear regression method to estimate the redshift, using assumption redshift was the function of magnitude. We called those methods as template-fitting methods. [7] used Artificial Neural Networks (ANN) for pattern recognition and as an estimator. They used ANN to find a pattern between magnitudes and redshifts. [8] used other machine learning algorithms (MLAs) called k-nearest neighbors (k-NN) and [9] used the random forest. These methods have similar or better performances when a large spectroscopic training set available. The accuracy of the output photometric redshifts from these methods is limited by the photometric measurements [10].

Nowadays, DNN techniques have revolutionized the field of image recognition. The advantages of DNN is no longer required manual features extraction and feature selection. [11] showed that Deep Convolutional Neural Networks (CNN) were able to provide accurate photometric redshifts from multichannel images, instead of extracted features, taking advantage of all the information contained in the pixels, such as galaxy surface brightness and size, disk inclination, or the presence of color gradients and neighbors. In this paper, we present a DNN model for estimation of photometric redshifts and their associated PDF using the deep learning framework Keras[1] and TensorFlow[2]. In contrast to previous studies, our input consists of the following magnitude in *g, r, i* band-pass and we use Densely Connected Convolutional Networks (hereafter DenseNet) designed by [12]. The paper is organized as follows. In Section 2, we describe the data used in this study and the pre-processing steps to prepare the images for the DNN. In Section 3, we introduce the CNN concepts and the DenseNet architecture. In Section 4, we present the analysis and the results for the estimation of photometric redshifts and associated PDFs. In Section 5, we conclude and discuss.

## 2. Methodology

The galaxy data in this study are drawn from the SDSS Data Release 10 [3]. We retrieved 1,918,221 sources classified as galaxy, with the photometric selection criteria following [11]. In detail, we run the MySQL query as shown in the appendix in the CasJobs server. We took ~60000 random sample from the sources before with the spectroscopic redshift below 2 and only got 59861 galaxies downloaded successfully in the following three photometric bands; *g,r,i,* using python module Astroquery[3]. Each pixel in the FITS images file has a resolution of 0.396 arc seconds and contain the measured flux which has been calibrated with flat fielding and sky subtraction. For every entry, we do image registration for the three SDSS filters to resample to a common pixel grid and stack all the available image data using python module Astroalign[4]. We rely on the WCS parameters in the input image headers for the astrometry. All pixel fluxes are converted to pixel magnitudes following [13]. We apply a further extinction correction to account for galactic dust using the maps of [14] which is available from the `photoObjAll` table in the CasJobs server. The extinction corrections are subtracted from magnitude in each pixel in the corresponding FITS image file. We choose to use images of size 72x72 pixels, corresponding to 28.5 arcseconds on a side. The result is 72x72x3 pixel data cube in a gnomonic projection centered on the galaxy coordinates. We convert data cube into RGB images following [15] using python module Astropy. In the term classification problem in supervised learning, the data have a label. We group the value of the spectroscopic redshift from 0.00 to 1.00 into 100 class with bins 0.01 as the labels (e.g. the label from a galaxy with z = 0.54123 is 54). Then, we randomly shuffle and subdivide the sample into training, and test sample by the ratio 70:30. Before we train the MLA on the training sample, we apply k-folds cross-validation with k=7. This approach randomly dividing the set of training sample into k groups, or folds, of approximately equal size. The first fold is treated as a validation set, and the MLA train on the remaining k -1 folds then yield a model. To obtain the predictions of the photometric redshift, we pass the test sample through the model. Fig. 1. show the distribution of the spectroscopic redshift from train and test sample, with these redshift distributions produce a fair estimate of the ability of the MLA to predict redshifts for other galaxies which are representative of the training sample.

---

[1] https://keras.io
[2] https://www.tensorflow.org
[3] https://github.com/astropy/astroquery
[4] https://github.com/toros-astro/astroalign

### *2.1. Deep neural networks*

Deep neural networks (hereafter DNN) are based on standard neural networks. DNN depart from these simple neural networks by constructing many hidden layers, with many multiples connected neurons per layers. The power of DNN comes from advances in how the many connections between the many millions of neurons are trained. Previously the many millions of connections would quickly over fit even large training sample, and thereby lose the predictive power [11]. The ways to avoid overfitting model were use data augmentation and add regularization like Dropout and L1/L2 regularization [16].

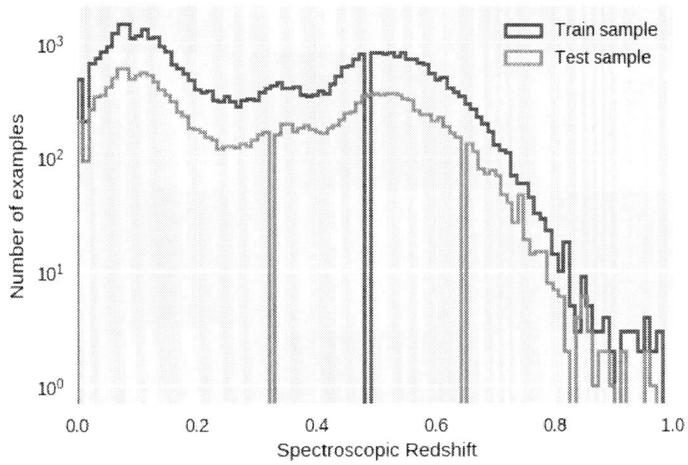


**Fig. 1.** The redshift number distribution of training (blue line) and test (green line) galaxies used in this work. The stepped lines represent the classification bins which are of width 0.01.

### *2.2. Convolutional neural networks*

DNN architecture which can process images as input called Convolutional Neural Networks (hereafter CNN). CNN are typically composed of a number of convolutional and pooling layers followed by fully connected layers. A convolutional layer operates on a data cube, and computes one or several feature maps, also stored as a data cube. For the first convolution layer input data cube is typically a multichannel or multispectral image. Subsequent layers operate of feature maps from previous layers. The pooling layers have function to progressively reduce the spatial size of the representation to reduce the amount of parameters and computation in the network, and hence to also control overfitting. Neurons in fully connected (FC) layer have full connections to all activations in the previous layer, as seen in regular NN. The FC layers are in charge of further processing the features extracted by convolutional layers before, for classification or regression. In this work, we choose to handle the estimation of photometric redshift as a classification problem. The galaxy at the center of the input multichannel image belong to a single class. We applied *softmax* activation function [17] to the output layer, then the sum of the probability must be 1.

### *2.3. Densely Connected Convolutional Networks (DenseNet)*

DenseNet is a network architecture where each layer is directly connected to every other layer in a feed-forward fashion (within each dense block). For each layer, the feature maps of all preceding layers are treated as separate inputs whereas its own feature maps are passed on as inputs to all subsequent layers [12]. We choose DenseNet with hyperparameter as follow; growth rate = 12, depth = 40, number of dense block = 3. We also use regularization the dropout technique with rate 0.4. In total the number of trainable parameter is 1,100,428.

IOP Conf. Series: Journal of Physics: Conf. Series **1231** (2019) 012013 doi:10.1088/1742-6596/1231/1/012013

### *2.4. Photometric redshift estimation*

We estimate the photometric redshift of a given galaxy by computing its expected value from its PDF $P(z_k)$:

$$z_{phot} = \Sigma_{k=1}^{N_c} z_k P(z_k) \tag{1}$$

where $z_k$ is the midpoint of value of the *k*-th redshift bin. Assuming that $P(z_k)$ is reliable over the whole redshift range.

## 3. Results and Discussion

We train the MLA with two scenarios. First, we train the MLA only on half of total sample dataset and second we train the MLA on all sample dataset. We want to know how the difference accuracy between the model trained only with around 21 K galaxies and the model trained with around 42 K galaxies. Each scenario, we trained model use 7-folds cross validation and we saved the model every validation accuracy has improvement, we called the early stopping model or checkpoint model and model which saved at the latest iteration of cross-validation we called as CV model.

### *3.1. Metrics*

To assess the quality of the photometric redshift, we adopt the statistics following [11]; $\Delta_z = z_{photo} - z_{spec}$ as residual vector, μ, $\sigma_{68}$, $\sigma_{95}$, $\eta$, respectively correspond to the median value of $\Delta_z$, the 68% spread of $\Delta_z$, the 95% spread of $\Delta_z$, and the outlier fraction $\left|\frac{\Delta_z}{1+z_{spec}}\right| > 0.15$. The statistics for the performance of the models doing prediction on the testing sample reported in Table 1, we include Hoyle's work at the first row as comparison. See CV model of 50% sample and CV model of 100% sample from Table 1, we know that the model trained with more sample is better than model trained with less sample. If we look only at the model of 50% sample, checkpoint model performs better than CV model. But, if we look at the model of 100% sample, checkpoint model does not perform better than CV model even if checkpoint model had the best validation accuracy.

**Table 1.** Metric from the estimated photometric redshift using DenseNet.

| Model | N training sample | N testing sample | μ | $\sigma_{68}$ | $\sigma_{95}$ | η |
|---|---|---|---|---|---|---|
| DNN [11] | 33,167 | 27,433 | 0.000 | 0.030 | 0.100 | 1.71% |
| CV model of 50% sample | 20,952 | 8,979 | 0.008 | 0.102 | 0.204 | 6.72% |
| Checkpoint model of 50% sample | 20,952 | 8,979 | 0.013 | 0.086 | 0.171 | 3.35% |
| CV model of 100% sample | 41,902 | 17,959 | 0.004 | 0.089 | 0.177 | 3.68% |
| Checkpoint model of 100% sample | 41,902 | 17,959 | 0.013 | 0.089 | 0.179 | 4.14% |

### *3.2. Histogram of the estimated photometric redshift*

We plot the distribution of spectroscopic redshift from testing sample over the distribution of estimated photometric redshift, see Fig. 2. The shape of the plot distribution of $z_{photo}$ and $z_{spec}$ look similar. However, at the upper, there are missing value at range $z_{photo}$ 0.6 – 1.0, and imply over estimate at range $z_{photo}$ 0.5 – 0.6. That happened because lack of training sample at range 0.6 – 1.0. Meanwhile, at the bottom, the estimated photometric redshifts distributed more likely the spectroscopic redshifts distribution and $z_{photo}$ at range 0.6 – 1.0 are more filled. With this visualization, we could say that model trained with more training sample will performed better than model trained with less training sample. Also, at the bottom, we could see that Checkpoint model of 100% sample more filled and distributed likely the spectroscopic distribution than CV model. So, we also could say that Checkpoint model is better than CV model.

IOP Conf. Series: Journal of Physics: Conf. Series **1231** (2019) 012013 doi:10.1088/1742-6596/1231/1/012013

### *3.3. Plot $z_{photo}$ against $z_{spec}$*

We use plot $z_{photo}$ against $z_{spec}$ to visualize accuracy of the model. More closer data point ($z_{spec}$,$z_{photo}$) to the diagonal line (which mean $z_{photo} = z_{spec}$) means more accurate the estimation comes from the model. From Fig. 3. at the upper either at the bottom, we could say that the Checkpoint models have better accuracy than the CV models from the indication of density near at the diagonal line, blue show the low density and red show the high density.

### *3.4. Histogram of the residual vector*

We also use histogram to visualize the distribution of the residual vector. From Fig. 4. we could judge the best model from how high the bar at the median point or at $\Delta_z$ = 0.0 and also from the spread of $\Delta_z$. The model which have the highest point at $\Delta_z$ = 0.0 is the Checkpoint model of 100% sample. The spread of $\Delta_z$ between the Checkpoint model of 50% sample, the CV model of 100% sample, and the Checkpoint model of 100% sample are more likely equal. So, from this visualization we could say the best model is the Checkpoint model of 100% sample.

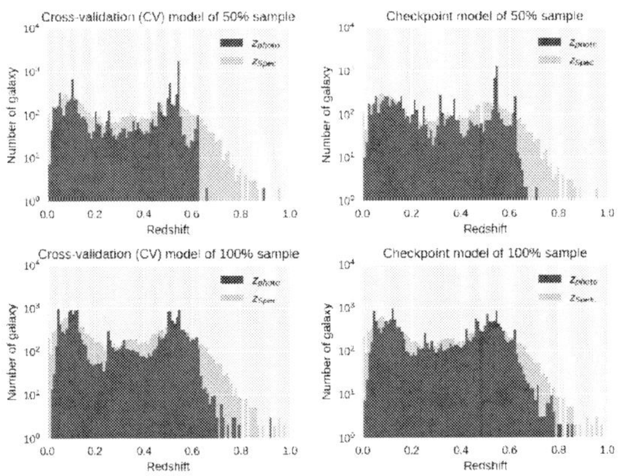


**Fig. 2.** The upper shows the histogram results from testing the 50% model data set and the bottom for the 100% data set model. The left part is a CV model and the right part is a Checkpoint model.

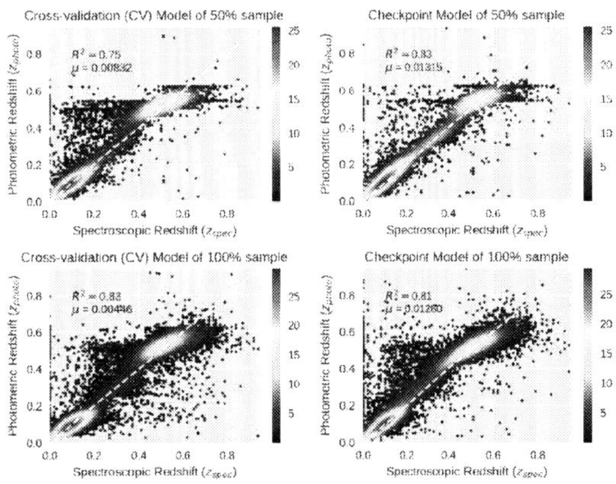


**Fig. 3.** The upper shows the plot the estimated photometric redshift against the spectroscopic redshift from testing the 50% model data set and the bottom for the 100% data set model. The left part is a CV model and the right part is a Checkpoint model.

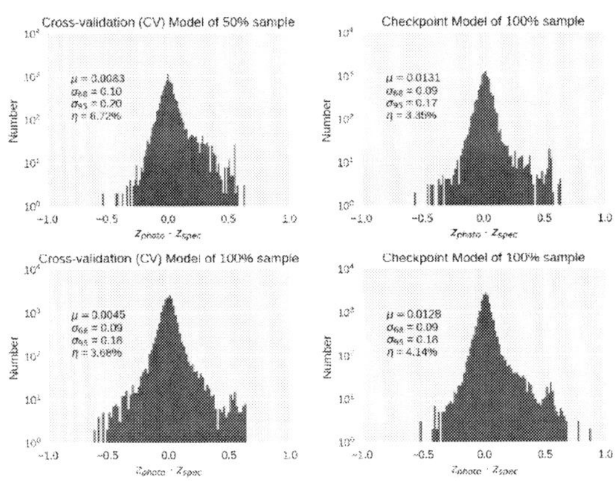


**Fig. 4.** The upper shows the plot the distribution of residual vector $\Delta_z$ from testing the 50% model data set and the bottom for the 100% data set model. Meanwhile, the left part is a CV model and the right part is a Checkpoint model.

In this paper we have presented DenseNet, one of the DNN architecture. Because of DenseNet using dense block, we could reduce the numbers of trainable parameter, so it can avoid the risk of overfitting. We used DenseNet with the following hyperparameters; growth rate = 12, depth = 40, number of dense block = 3, and we use the dropout with rate 0.4, thus we only had 1,100,428 parameters. We compiled DenseNet model using Keras and TensorFlow. We had 59,861 sample of the galaxy images and we would to test the DenseNet performance when trained with all of the sample and with half of the sample. We group the value of the spectroscopic redshift from 0.00 to 1.00 into 100 class as the labels. We subdivide the sample into 70% for training sample and 30% for testing sample. We used 7-folds cross-validation, which means we use 6/7 of the training sample for training and 1/7 of the training sample for validating the model. We trained the model using workstation with Intel Xeon E5 8 cores, 8GB of RAM, and GPU Quadro M2000 4GB. We evaluate the model using the testing sample. We pass the testing sample through the trained DenseNet models and thus yield each sample had the PDF. We estimate the photometric redshift from the expected value of PDF. We calculated the metrics to measure the performance of DenseNet models and we visualized the estimated photometric redshift in the following several plots; histogram of the estimated photometric redshift, plot versus, and histogram of residual vector. From the metrics and the visualizations, we concluded more data sample will give the model with better accuracy or better performance and the model saved when had the best validation accuracy is the highly recommended model. Our models do not have much better performance than Hoyle's model. In the future work, we will try to vary the hyperparameter of DenseNet, add more sample data, and do cross-validation to find the best model.

**Acknowledgments**
I would like to thank Hakim L. Malasan, Hesti T. R. Wulandari, and Yayan Sugianto from Departement of Astronomy, Institut Teknologi Bandung for helpful conversations and advices, particularly when this work presented for thesis project. Then, we acknowledged Program Penelitian, Pengabdian kepada Masyarakat, dan Inovasi (P3MI) for a funding travel grant to SEAAN 2018.

**Appendix A. MySQL data Query**
We select data from the SDSS CasJobs website by running the following MySQL query in the Data Release 10:

```
select p.objid, s.specobjid, s.ra, s.dec, s.z as spec_z, s.zerr as err_spec_z,
p.dered_u,p.dered_g,p.dered_r, p.dered_i,p.dered_z,p.PETRORAD_R, p.extinction_g, p.extinction_r,
p.extinction_i,p.extinction_z into mydb.DR10_DNN from Specobjall s join photoPrimary p on (s.bestobjid
=p.objid) and p.deVRad_r >0 and p.deVRad_r<30 and p.dered_r>0 and p.dered_r < 22 and s.z>0 and s.z<2
and s.zerr>0 and s.zerr<0.1 and p.expRad_r>0 and p.expRad_r <30 and p.type=3
```

**References**
[1] Smail, I, Hogg, D W, Yan, L, & Cohen, J G 1995 *ApJ* **449**
[2] Newman, J A, Cooper, M C, & Davis, M 2013 *ApJS* **208** 5
[3] Ahn, C P, Alexandro, R, Prieto, et al. 2014 *ApJS* **211** 17
[4] Baum, W A 1962 *Problems of extra-galactic research* **15** 390
[5] Loh, E D, & Spillar, E J 1986 *ApJ* **303** 154
[6] Connolly, A J, Csabai, I, Szalay, A S, et al. 1995 *ApJ* **110** 2655
[7] Collister, A. A., & Lahav, O. (2004). *PASP* **116** 345-351
[8] Csabai, I, Dobos, L, Trencséni, M, et al. 2007 *Astronomische Nachrichten* **328** 852
[9] Carliles, S, Budavári, T, Heinis, S, Priebe, C, &, A S 2010 *ApJ* **712** 511
[10] Pasquet J, Bertin E, Treyer M, et al. 2018 *A&A* **621**
[11] Hoyle, B 2016 *Astronomy and Computing* **16** 34-40
[12] Huang, G, Liu, Z, Maaten, L V D, et al. 2017 Densely connected convolutional networks *IEEE Conference on Computer Vision and Pattern Recognition (CVPR)*
[13] Lupton, R H, Gunn, J E, & Szalay, A S 1999 *AJ* **118** 1406-1410
[14] Schlegel, D J, Finkbeiner, D P, Davis, M, 1998 *ApJ* **500** 525–553
[15] Lupton, R, Blanton, M R, Fekete, G, et al. 2004 *PASP* **116** 133-137
[16] Nielsen, M A, 2015 Neural Networks and Deep Learning *Determination Press*
[17] Bridle, J S 1990 *Neurocomputing, ed. F. F. Soulié & J. Hérault (Berlin, Heidelberg: Springer Berlin Heidelberg)* 227–236

# Radial Fourier transforms in exoplanetary imaging and potential uses at Timau National Observatory

**H Albanna[1,*], D Doelman[1] and F Mumtahana[2]**

[1] Leiden Observatory, Leiden University, P.O. Box 9513, 2300 RA, Leiden, The Netherlands
[2] National Institute of Aeronautics and Space (LAPAN), Indonesia

Email: albanna@strw.leidenuniv.nl

**Abstract.** Due to Fourier transforms nature between the field detected on the image and its corresponding input, astronomical imaging can be modelled mathematically. In exoplanetary imaging, we aim to detect exoplanets whose typical contrast are approximately one over a million times dimmer compared to their parent stars. Among the possible approaches to accomplish that is optical apodization, a technique to purposely modify the input signal profile such that the 'Airy rings' on the resulting image are suppressed while keeping the central brightness high. In the paper, we pedagogically describe this technique applying Fourier transforms of radially-symmetric functions; and investigate potential future uses at Timau National Observatory.

***Keyword :*** *Imaging, observatory, exoplanet, fourier transforms*

## 1. Introduction

The existence of worlds other than the earth has always been a great interest for humankind. Beginning from Galileo Galilei's observation of the phase of Venus in 1610, confirming that the planets orbit around the Sun, it then makes sense that other stars must also have companion planets. In fact, the discovery of exoplanets, planets around stars other than the Sun, has already started more than two decades ago. In 1992, Wolszczan and Frail detected the first exoplanet(s) orbiting not a solar-type star, but a pulsar [1]. They measured the timing of pulses of the pulsar which seemed to be irregular, indicating the existence of two planets. The first detection of an exoplanet around a Sun-like star follows not so long after, using radial velocity method. Mayor and Queloz observed a periodic variation on the radial velocity of 51 Pegasi. Another detection method is called transit method, first performed by Charbonneau *et al.*, which measures periodic dimming of the starlight during the planetary transit across the host star [2]. The exoplanet itself, HD 209458b, was already discovered with radial velocity method [3]. Other important methods are astrometry and microlensing, which makes use of the variation of the position of the star and the gravitational perturbation due to the foreground planet, respectively [4, 5, 6].

The techniques mentioned above are indirect methods to probe exoplanets. Together, those allow for determination of the exoplanet properties, such as period, radius, mass and temperature. For further characterization of the exoplanet, however, direct imaging is required. It gives ways to a more complete story than the indirect counterparts. Yet, it comes with several challenges, the biggest of which is the extremely dim flux of the planet, overwhelmed by the glare of its host star. To date, only 3 percent out of over 3500 exoplanet detections are detected directly, as shown in Figure 1. Only the largest telescopes and the most advanced techniques have been able to carry out such difficult tasks.

This paper is aimed to lessen those gaps, introducing a simple yet powerful principle behind exoplanetary imaging: optical apodization. Throughout the paper, we mostly use pedagogical approach in describing the technique; how and why it works, but we leave out the practical and/or manufacturing aspects, as those would require a different discussion and expertise.

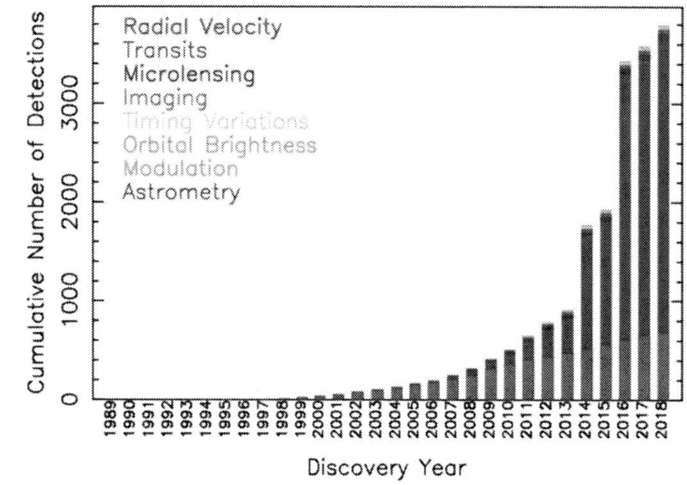


**Figure 1.** Cumulative number of exoplanet detections. Radial velocity and transits methods contribute the most with ~3600 detections combined. Direct imaging, being the most difficult technique, only shares less than 100 detections. Source: *https://exoplanetarchive.ipac.caltech.edu/*

## 2. Fourier transforms

### *2.1. Physical backgrounds*

Using the Huygens-Fresnel principle which states that each point on a wave front can be regarded as a source of spherical secondary wavelets and the amplitude of the field at any point beyond is the superposition of these wavelets, one can arrive to the following result [7, 8, 9]

$$E(u,v) = \frac{e^{ik\left[z+\frac{u^2+v^2}{2z}\right]}}{i\lambda z} \iint_{-\infty}^{\infty} \tilde{E}(x,y) e^{-\frac{i2\pi(ux+vy)}{\lambda z}} dxdy, \tag{2.1}$$

where the field density, the aperture and the induced phase are embedded inside $\tilde{E}(x,y)$. This equation exactly resembles Fourier transform in two-dimension, with a multiplicative factor before the integral. Fixing $\lambda = z = 1$ to be unity and expressing the resulted lengths in terms of $\lambda$, the corresponding Fourier transform is then

$$F(u,v) = \iint_{-\infty}^{\infty} f(x,y) e^{-i2\pi(ux+vy)} dxdy. \tag{2.2}$$

After the normalization, the electric field pair $\tilde{E}(x,y) - E(u,v)$ and the Fourier transform pair $f(x,y) - F(u,v)$ are completely interchangeable, with $(x,y)$ and $(u,v)$ represent the coordinate systems in the pupil plane and the focal plane, respectively. In other words, from this point on it is possible to switch completely to mathematics without losing any physical meaning.

### *2.2. Fourier transforms in polar coordinates*

Due to circular shape of telescope apertures, it is naturally more convenient to work in polar coordinate system instead of Cartesian. Moreover, when a function in the pupil plane has a circular symmetry, so does its resulting image, since they are related through Fourier transforms and Fourier transforms preserve the symmetry. If this is true, suppressing 360 degrees of starlight halo would also require patterns with circular symmetry, which suggests the use of polar coordinates. It will be more evident once the expressions are derived. Cartesian to polar coordinate system conversion is performed by introducing the following coordinates

$$x = r\cos\theta\,,\ y = r\sin\theta\,,\ u = \rho\cos\psi\,,\ v = \rho\sin\psi. \quad (2.3)$$

Substituting them into equation (2.2), we obtain

$$F(\rho,\psi) = \int_0^\infty \int_0^{2\pi} r f(r,\theta) e^{-i2\pi\rho r\cos(\theta-\psi)} dr d\theta \quad (2.4)$$

Any function $f(r,\theta)$ is always periodic in angular direction, so rewriting it in terms of Fourier series is possible

$$f(r,\theta) = \sum_{n=-\infty}^{\infty} a_n e^{i2\pi\left(\frac{n}{T}\right)\theta} \text{ with } a_n = \frac{1}{T}\int_{-\frac{T}{2}}^{\frac{T}{2}} f(r,\theta)\, e^{-i2\pi\left(\frac{n}{T}\right)\theta} d\theta, \quad (2.5)$$

to which we can plug a period of $T = 2\pi$, since it is periodic every full circle, and obtain

$$f(r,\theta) = \sum_{n=-\infty}^{\infty} f_n(r) e^{in\theta} \text{ with } f_n(r) = \frac{1}{2\pi}\int_{-\pi}^{\pi} f(r,\theta)\, e^{-in\theta} d\theta. \quad (2.6)$$

Hence, the overall Fourier transform becomes

$$F(\rho,\psi) = \sum_{n=-\infty}^{\infty} \int_0^\infty \int_0^{2\pi} r f_n(r) e^{in\theta - i2\pi\rho r\cos(\theta-\psi)} dr d\theta. \quad (2.7)$$

Part of the exponents can be expanded with the following relation which contains Bessel function of the first kind $J_n$ of order $n$ [10]

$$e^{-2\pi\vec{\rho}.\vec{r}} = e^{-2\pi\rho r\cos(\theta-\psi)} = \sum_{n=-\infty}^{\infty} (-i)^n 2\pi J_n(2\pi\rho r) e^{-in\theta} e^{in\psi}. \quad (2.8)$$

Plugging equation (2.8) into (2.7), we get the general expression of Fourier transforms in polar coordinates

$$F(\rho,\psi) = \sum_{n=-\infty}^{\infty} (-i)^n e^{in\psi} 2\pi \int_0^\infty r f_n(r) J_n(2\pi\rho r) dr, \quad (2.9)$$

where the azimuthal variation is stored in $f_n(r)$, taking the form

$$f_n(r) = \frac{1}{2\pi}\int_{-\pi}^{\pi} f(r,\theta)\, e^{-in\theta} d\theta. \quad (2.10)$$

If we examine the last two equations, it is made clear that the function $f(r,\theta)$ can consist of both radial and azimuthal modes. In this paper, we restrict ourselves only to the radial mode due to some considerations. First, working in radial mode would simplify the problem significantly. Removing angular mode in polar coordinates basically leaves one dimension to work with while keeping the two-dimensional shape. Second, Fourier transform symmetry-preserving nature suggests radially symmetric patterns for 360 degrees dark zone. However, when there is a need for further investigation in azimuthal functions, such as examining the effect of secondary mirror spiders, it is suggested to begin with equation (2.9) and (2.10).

### 2.3. *Radial Fourier transforms*

For radially symmetric functions, it is assumed that $f(r,\theta) = f(r)$ only depends on the radius. Hence, the Fourier series coefficients in equation (10) can be calculated as

$$f_n(r) = \frac{f(r)}{2\pi}\int_{-\pi}^{\pi} e^{-in\theta} d\theta = \frac{f(r)}{2\pi}\left[\frac{e^{-in\theta}}{-in}\right]_{-\pi}^{\pi} = \begin{cases} 0 & n \neq 0, \\ f(r) & n = 0. \end{cases} \quad (2.11)$$

IOP Conf. Series: Journal of Physics: Conf. Series **1231** (2019) 012014 doi:10.1088/1742-6596/1231/1/012014

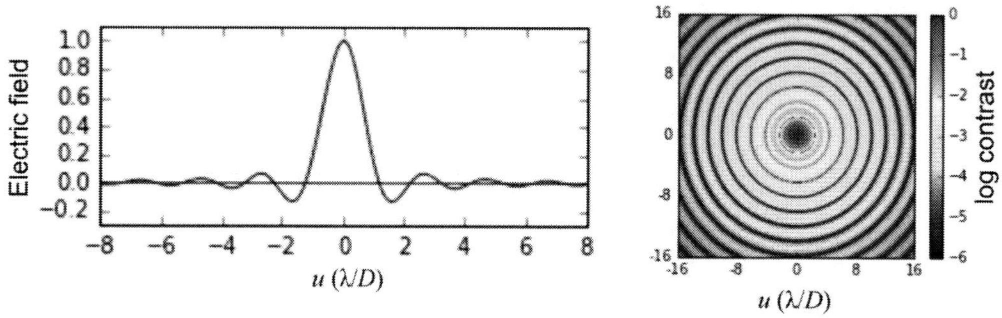


**Figure 2.** Airy function, a pattern resulted from a circular aperture. It closely resembles damped sinusoidal function. If there are more than one source, each source through the aperture would result in the same pattern with different amplitude based on its brightness.

This result significantly reduces the summation in equation (9) to only remains one non-zero term, which is for n = 0

$$F(\rho) = 2\pi \int_0^\infty r f(r) J_0(2\pi\rho r) dr. \qquad (2.12)$$

The expression above makes it clearer that radial patterns at the pupil plane would surely generate radially symmetric structure at the focal plane. Hereafter, we refer Fourier transforms for radially symmetric functions as Radial Fourier Transforms (RFT).

## 3. Applying RFT

### *3.1. Apodization*

Using equation (2.12), we can, for instance, calculate the image detected at the focal plane for a circular aperture by putting circular function

$$f(r) = \mathrm{circ}\left(\frac{r}{a}\right) = \begin{cases} 1 & r < a, \\ 0 & r > a, \end{cases} \qquad (3.1)$$

where $a$ is the radius of the aperture. Then, making use of the following integral property of Bessel function

$$\int_0^\alpha x J_0(x) dx = \alpha J_1(\alpha), \qquad (3.2)$$

we arrive at the historical Airy function

$$F(\rho) = a \frac{J_1(2\pi\rho a)}{\rho}. \qquad (3.3)$$

The function is visually illustrated in Figure 3. Typical hot Jupiters would be located at around $2 - 5\lambda/D$, covered completely by the second or third Airy rings. This makes it extremely difficult to directly image exoplanets without any additional optical components. Two different kinds of modifications can be applied at the pupil plane: amplitude- and phase-apodization.[11][12] Both methods are principally similar, deciding the function $f(r)$ in equation (2.12)

$$f(r) = \begin{cases} A(r) & \text{amplitude-apodization,} \\ A(r).\, e^{i\phi(r)} & \text{phase-apodization.} \end{cases} \qquad (3.4)$$

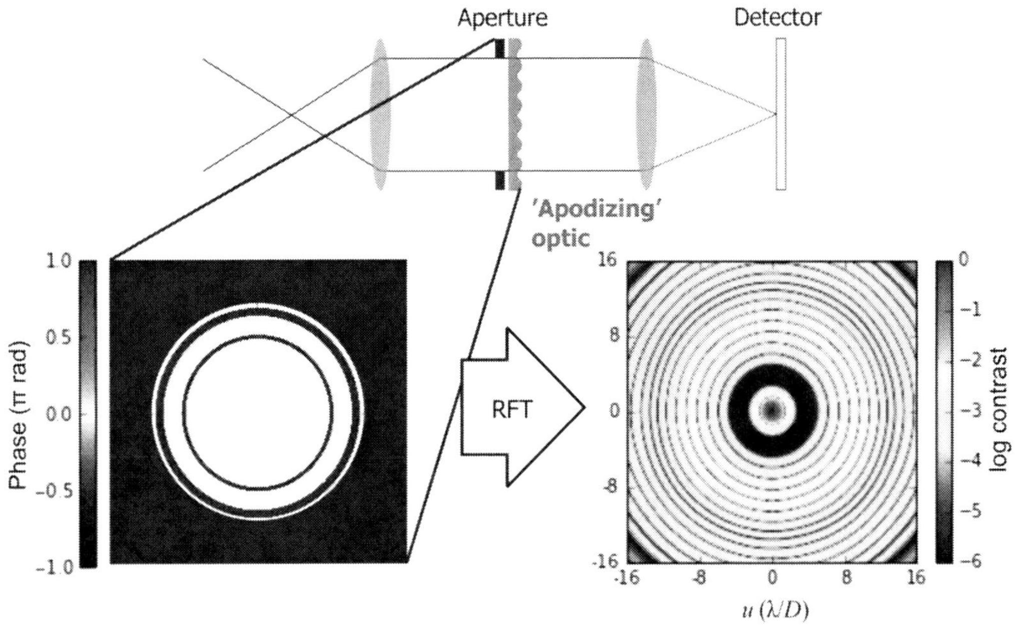


**Figure 3.** Configuration of phase-apodization principle. An 'apodizing' optic is applied at the pupil plane to form an extended dark region around the peak of the resulting image. At the bottom panel is a RFT pair of a working sample design of phase-apodizing coronagraph.

In amplitude-apodization, $A(r)$ consists of the original aperture shape and an amplitude mask to modify the signal; while in phase-apodization, $A(r)$ is the original aperture shape and $\phi(r)$ is a phase-inducing optics. The configuration is illustrated at the top panel of Figure 3. The first phase inducing-optics were made from ZnSe diamond-turned plates with varying height. Nowadays, devices to induce phases to liquid crystals have been created [13, 14]. Despite the different manufacturing methods, the desired outputs of those two methods are the same: suppressing the Airy rings within certain radius (hereafter working angle), while keeping the central brightness to be as high as possible. Mathematically, the first requirement translates to

$$\text{minimize}\left(\frac{1}{N}\sum_{i=0}^{N}\left|\frac{F_{\text{mod}}(\rho_i)}{F_{\text{mod}}(0)}\right|^2 - 10^{-6}\right), \tag{3.5}$$

with $N = (\rho_f - \rho_i)/\Delta\rho$ is the number of points to be evaluated and $\Delta\rho$ is small enough value. The assumed required contrast of $10^{-6}$ can be adapted accordingly. The *mod* subscripts stand for *modified*. Similarly, the second requirement translates to

$$\text{maximize}\left|\frac{F_{\text{mod}}(0)}{F_{\text{init}}(0)}\right|^2, \tag{3.6}$$

where *mod* represents the modified function with the mask/optic applied and *init* represents the initial function without one. These two requirements are generally sufficient to generate working coronagraph designs. In Figure 3, we present an example of phase-apodization with a working angle of $(\rho_i, \rho_f) = (3, 5)\ \lambda/D$.

### *3.2. Potential uses*

There are at least three criteria to enable direct imaging of exoplanets: an extremely high contrast, a very high angular resolution, and a total field of view that includes the working angle.[15] Based on these criteria, we did a case study towards a future 3.8-meter optical telescope at Timau National Observatory, Indonesia,[16] coupled with a RFT-generated phase-apodizing coronagraph. With simulation, we managed to reach a high contrast of $10^{-6}$; a typical exoplanet has a relative brightness of $10^{-5} - 10^{-10}$ compared to its parent star. For the angular resolution, we found a value of 0.04 arcsec at 700 nm, which can be considered to be high. Finally, an inner working angle of 3.1 $\lambda/D$ from the simulation approximately coincides with 0.11 arcsec at 700 nm. As a reference, a 'twin Jupiter' at 10 pc from us would have a separation of 0.5 arcsec, which would be within our simulated working angle. We positively conclude that the configuration can potentially detect exoplanets directly.

## 4. Conclusion

The basic principle behind one of exoplanetary imaging methods, optical apodization, has been discussed, starting from Huygens-Fresnel principle to a specific application of Fourier transforms for radially symmetric functions. With this technique, the future 3.8-meter telescope at Timau National Observatory should be able to directly image Jupiter-sized exoplanets at 10 pc distance.

## Acknowledgements

We warmly thank Christoph Keller who have contributed the most to our understandings about the topic. We thank the referees for their useful comments on this paper. The study was partly funded by Indonesia Endowment Fund for Education (LPDP).

## References

[1] Wolszczan A and Frail D A 1992 *Nature* **355.6356** p 145
[2] Mayor M and Queloz D 1995 *Nature* **378.6555** p 355
[3] Charbonneau D *et al* 1999 *Astrophys. J. Lett.* **529.1** p L45
[4] Mazeh T *et al* 2000 *Astrophys. J. Lett.* **532.1** p L55
[5] Perryman M *et al* 2014 *Astrophys. J.* **797.1** p 14
[6] Gaudi B S 2012 *Annu. Rev. Astron. Astrophys.* **50** pp 411–453
[7] Huygens C 2012 *Treatise on Light* tredition
[8] Goodman J W 1996 *Introduction to Fourier Optics* McGraw-Hill Book Co New York
[9] Hecht E 1998 *Optics 4th edition* Addison Wesley Longman Inc
[10] Kyatkin A B and Chirikjian G S 2000 *Engineering Applications of Noncommutative Harmonic Analysis: with Emphasis on Rotation and Motion Groups* CRC press
[11] Codona J L *et al* 2006 *Proc. SPIE* **6269** 62691N
[12] Kenworthy M A *et al* 2007 *Astrophys. J.* **660.1** p 762
[13] Snik *et al* 2012 *SPIE* **8450** p 84500M
[14] Otten G P *et al* 2014 *SPIE* **9151** p 91511R
[15] Burrows A 2005 *Nature* **433** pp 261–268
[16] Mumpuni E S *et al* 2017 *Selayang Pandang Observatorium Nasional Timau* PT Gramedia Jakarta

# Twenty years of the Gunma Astronomical Observatory, experimental trials of how to use a 1.5-m telescope at a public observatory

**O Hashimoto[1,*] and H L Malasan[2,3]**

[1] Gunma Astronomical Observatory, 6860-86 Nakayama, Takayama, Agatsuma, Gunma 377-0702, Japan
[2] Institut Teknologi Sumatera (ITERA), 35365, South Lampung, Sumatera, Indonesia
[3] Astronomy division and Bosscha Observatory, FMIPA, Institut Teknologi Bandung, Jl. Ganesa 10 Bandung 40132, Indonesia

Email: osamu@astron.pref.gunma.jp

**Abstract.** Gunma Astronomical Observatory (GAO) is regarded as a pioneer of the public observatories equipped with a large telescope of far beyond one meter class. It was established in 1999 by Gunma prefecture local government. Its main telescope is a 150-cm reflector which has an eyepiece system for the star gazing by public people in addition to advanced measuring instruments such as a very powerful high resolution spectrograph for the scientific research. Using this telescope we have carried out a number of activities and studies in the fields of scientific research, education and public outreach. Many of them were kinds of experimental trials of how to use such a telescope at a public observatory. We present some examples of our experiences in these two decades.

***Keywords:*** *1.5-m telescope and Gunma Astronomical Observatory*

## 1. Introduction

In 1999 Gunm Astronomical Observatory ( hereafter GAO ) was established in Takayama village in Gunma prefecture which is about 120 km north-west of Tokyo ( N 36° 35' 47", E 138° 58' 22" ) as a public observatory aiming at effective education and outreach of astronomy and astrophysics for all kinds of people from young children to professional researchers [1,2]. It was build by the local government of Gunma prefecture. GAO is operated by the board of education of the Gunma prefecture without belonging to any universities and other institutions. There are a number of telescopes with aperture sizes from 10-cm to 150-cm at GAO. All of them are mainly prepared for the public activities as GAO is a public observatory. Even its main telescope with a primary mirror as large as 150-cm in diameter is equipped with an eyepiece system for the star gazing as seen in Fig.1. GAO can be regarded as a pioneer of the public observatories equipped with such a large telescop that is sufficiently powerful also for the scietific observation by the professional researchers.

The 150-cm reflector of the GAO main telescope was the second largest telescope in Japan when it was built, following the largest one of the 188-cm reflector at the Okayama Astrophysical Observatory, National Astronomical Observatory of Japan (OAO, NAOJ) which was designed and used only for the scientific studies by professional researchers. The GAO 150-cm reflector was the world largest telescope at that time far beyond the second one with which people can watch stars by their own eye through its large aperture. In order to utilize the full capability of the telescope for the scientific observations of high standard in addition to the public services, some powerful measuring instruments have been designed and made for the GAO 150-cm reflector as seen in Fig.2. The largest and the most powerful instrument is the Gunma Astronomical Observatory Echélle Spectrograph (GAOES) [3,4], that is a high resolution echélle spectrograph set at the Nasmyth platform. It makes us possible to carry out very detailed studies of stellar astrophysics through the spectra of extremely high spectral resolution up to $R = \lambda/\delta\lambda \sim 10^5$. The Gunma Infra-Red Camera and Spectrograph (GIRCS) at the Cassegrain focus [5] and the Gunma LOW resolution Spectrograph (GLOWS) at the bent-Cassegrain focus are very powerful as well. Using those instruments on the 150-cm telescope we have carried out a lot of scientific observations. Thanks to the high sensitivity of modern detector devices, telescopes with an aperture size about two meters can be very useful and powerful tools for the studies of astrophysics. In particular, there are a lot of interesting stars which are bright and suitable enough for the detailed studies using the telescopes of such a scale.

**Figure 1.** The 150-cm telescope of GAO [1,2]. It has a star gazing optic at a Nasmyth focus.

## 2. Strategy for scientific observations

In order to make full use of the observational performance given to the GAO 150-cm telescope, it is essentially important to choose the most suitable observation methods focusing on the most suitable scientific subjects. Spectroscopic study of stellar objects is one of the most suitable ways in practice. There are a lot of stars which are bright enough to be observed very precisely by the 150-cm telescope, while it is much smaller than the world largest telescopes of 8 – 10-m class.

In addition, we must take into account the observation condition at the site where the telescope is located. Usually the world largest telescopes are located in the sites ideal for the observations, where the sky is dark and clear, and the atmosphere is extremely stable with a number of clear nights, such as at the top of Mt. Mauna Kea in Hawaii. It is somewhat unfortunate that GAO may not be located in such an ideal place. However, spectroscopic observations with a sufficiently high spectral resolution can overcome the less ideal observation condition of us. Since the photons coming into a spectrograph are divided into a number of pixels on the detector of different wavelengths, the number of photons from the city lights becomes very small at each pixel. Thus, the light pollution does not affect the astronomical measurements so seriously in the spectroscopic observations.

Poor seeing which makes the stellar image larger at GAO does not matter extremely much either for the spectroscopic observations as long as a slit width can be made wide enough against the size of stellar image. Generally speaking, any spectrographs for a big telescope at a poor seeing condition necessarily become very large in order to realize good observation efficiency because the size of a spectrograph is proportional to the aperture size of telescope, spectral resolution and the size of stellar image caused by the seeing condition in principle. On the other hand, it also means that it is much easier to realize a spectrograph of a reasonable size with sufficiently good observation efficiency and high spectral resolution for the smaller telescopes even if somewhat poorer seeing condition is taken into account.

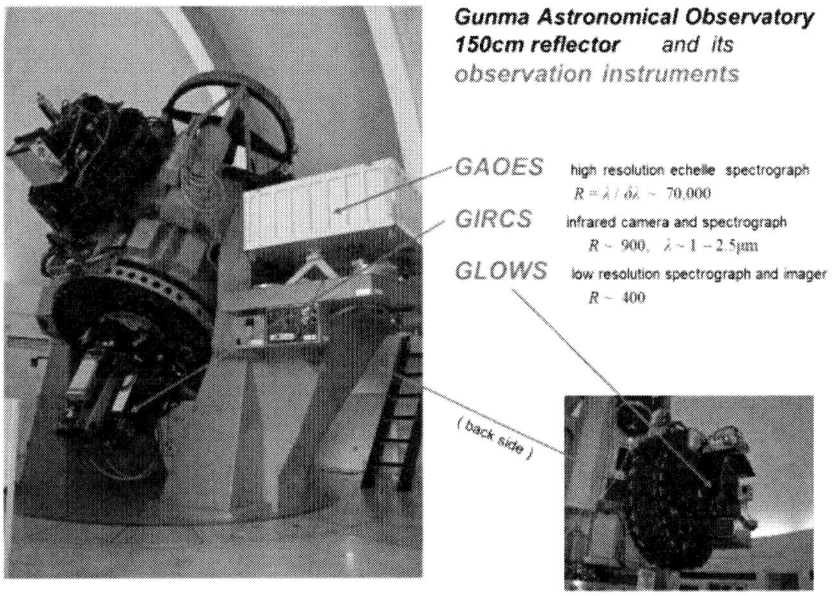


**Figure 2.** Observation instruments on the GAO 150-cm telescopes [2].

It is not always easy to get much observation time of the world largest telescopes. It is very rare in practice that observations for a long period are carried out by some of the world largest gigantic telescopes. On the other hand, we can easily expect more observation time for smaller telescopes. With a number of observations for longer periods by middle class telescopes like the GAO 150-cm telescope, we can obtain statistically valuable data of large samples, or time dependent information of a number of variable events for long periods. It is no doubt that the spectroscopic observations with a middle class telescope can provide unique and promising opportunities for the scientific studies.

It is noted that GAO is operated independently. We can use all the observation times of our telescopes flexibly based on our own decision. It allows us to make ourselves focus on a limited number of specific scientific subjects. GAO has definite advantages in making the best use of our telescopes for the scientific researches of astrophysics. A number of unique observational studies can be carried out with the use of our 150-cm telescope and its measuring instruments. In practice this telescope has published a number of papers based on its own observations.

## 3. Scientific researches by the GAO 150-cm telescope

Spectroscopic study of the chemical abundances in the Pleiades open cluster M45 [6] is a good example of the scientific research which was carried out based on a large sample obtained by the middle class telescopes. We could ensure the homogeneity of chemical abundance in M45 by the high resolution spectroscopic observations of a number of stars in the cluster using GAOES on the 150-cm telescope. There were few observational evidences definitive enough to confirm the homogeneous metalicity in the clusters, while such a result had been expected from the view point of cluster evolution since it was believed that all the stars in a cluster were born from the same interstellar matter of homogeneous chemical composition. Chemical abundances measured for a number of stars in a large sample also provide some important suggestions about stellar evolution in the Galaxy. We found that chemical abundances of carbon and oxygen of field stars nearby the Sun are significantly different from those of the stars in some open clusters. It indicates their evolutions are not the same and they have evolved in different manner at the different places in the Galaxy [7].

Our observation system of the GAO 150-cm telescope is more effective and useful in the observational studies of time dependent phenomena of variable targets. For instance, some details of the dynamic structure of an X-ray emitting binary system consists of a hot star and an evolved compact neutron star have been made clear by the continuous repetition of high resolution spectroscopic observations by GAOES for a long period. We detected direct evidences of the non-axisymmetric features of an eccentric disc around one star in the system [8,9,10]. Also, we have made a significant contribution by the GAOES observations for several years to solve some enigmas of a strange eclipsing binary system of ε Aurigae which is well known by its very long variation period of about 27 years. Our observation results indicate a complex structure of the rotating disc around the visibly unseen secondary star [11]. As the most recent study of us, we have been observing a unique binary system of α Aurigae ( Capella ) which consists of two G-type giant stars [12]. One is a normal G-type giant star while the secondary is a rapid rotating G giant. Using the high resolution spectra obtained by the GAOES for a period of several months, individual spectra of those two stars were clearly separated by the spectrum disentangling technique [13]. It is found that more careful treatment of chromospheric effect is necessary for the abundance analysis of the fast rotating giant stars of high activity as the conventional model atmosphere analysis does not work well any longer.

Transient phenomena such as novae, supernovae, gamma-ray bursters are also very suitable targets for the middle size telescopes since rapid responses by the flexible operation and time dependent continuous follow up observations for certain periods are essentially important for the studies of them. We have made a number of observations of such transient targets using the GAO 150-cm telescope. Many observation reports have been published in addition to the usual papers in journals. It is noted in particular that cooperative observations with various telescopes and observatories are essentially effective for the studies of such time varying transient objects. It is not only useful to avoid bad weather condition at each individual observation site but makes us possible to get more observation points in time for a longer period with higher density. In practice, most of our studies of the transient and variable targets have been carried out as collaboration works with various telescopes at various observatories in the world. While many of them are international collaborations, GAO has been intensively contributing to the Japanese collaboration network the Optical and Infrared Synergetic for Education and Research (OISTER) [14] in the observations of the transient events. After the first detection of the gravitation waves from stars [15], quick response and follow up of the very transient phenomena caused in the sources of gravitation waves have become more and more important in order to solve a number of essential enigmas of universe such as the origin of heavy elements and the universe itself.

## 4. Education and outreach

Basically, GAO is a public observatory which has a mission of education and outreach for various kinds of public people while its main telescope is very powerful for the scientific research. Star gazing events to the public people using the telescopes at the observatory are the most important activities of us in the aspects of education and the public outreach.

However, the telescopes with an aperture size much larger than one meter are not always very useful and suitable for providing beautiful clear images of high spatial resolution for the human eyes in the case of star gazing since the seeing effect by the atmosphere of the earth works. As the larger aperture gathers lights coming through the more distant air in the sky, the image quality is affected much more seriously by the seeing condition. For example, Saturn may not be seen more beautiful than by a smaller telescope in many cases.

On the other hand, light collecting power of the telescopes always works well almost regardless of the seeing condition. We can always expect brighter images by the larger telescope. A lot of photons gathered by the telescope with a sufficiently large aperture can provide good opportunities to see the interesting aspects of the astrophysics in the stellar light even by a human eye which is not very sensitive to the faint light. In addition to the differences of colors from star to star, it is quite impressive that various features in the stellar spectra can also be directly seen through the large telescopes if a suitable spectroscope is used.

Based on such experiences of us at GAO as a public observatory we have proposed and developed a unique spectroscopic eyepiece system for the GAO 150-cm telescope [1,16]. It provides a direct image and a spectrum of target star at two separate viewing points simultaneously. Observers using this newly developed instrument should be able to understand the physical meanings of the spectrum and the color of the observing target much easier by direct comparison between them.

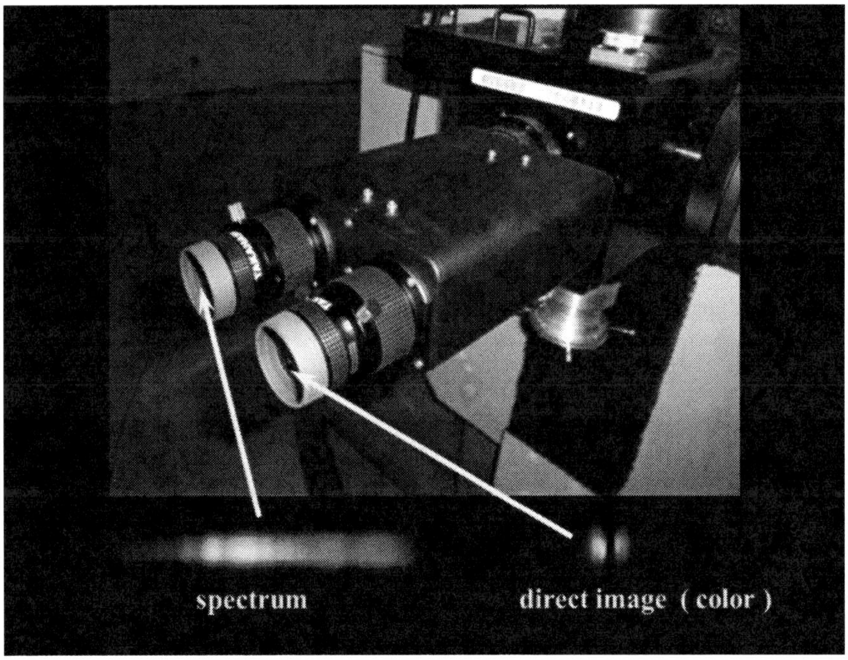


**Figure 3.** The spectroscopic eyepiece system set on the star gazing optics at the Nasmyth focus of the GAO 150-cm reflector [2.4]. Its right viewing point provides a direct image of the target star on the slit of the spectroscope as the observers can see the color of the target. The vertical gap in the stellar image is made by the slit. The left one shows a spectral image of the target star at the same time.

Figure 3 shows the spectroscopic eyepiece system set at the star gazing focus of the GAO 150-cm telescope. It is made as small and light as it can be easily used instead of a usual 50.8-mm eyepiece used for the star gazing. It has two viewing points. One provides a direct image on the slit of the spectroscope where the stellar image is divided into two parts by a gap caused by the slit as seen in the right-bottom panel in Fig.3. It makes observers easy to see the color of the target because all the photons gathered by the telescope are focused at the direct image on the slit. The light coming into the slit is processed to show the spectrum of the target at the other viewing point. Observers can compare the color and spectrum directly at the same time. We have been investigating some applications of this unique instrument in the education programs of astronomy and astrophysics using the 150-cm telescope at GAO. It is found that the direct comparison of the stellar spectrum and color is quite effective in practice for understanding the physical meanings of them intuitively.

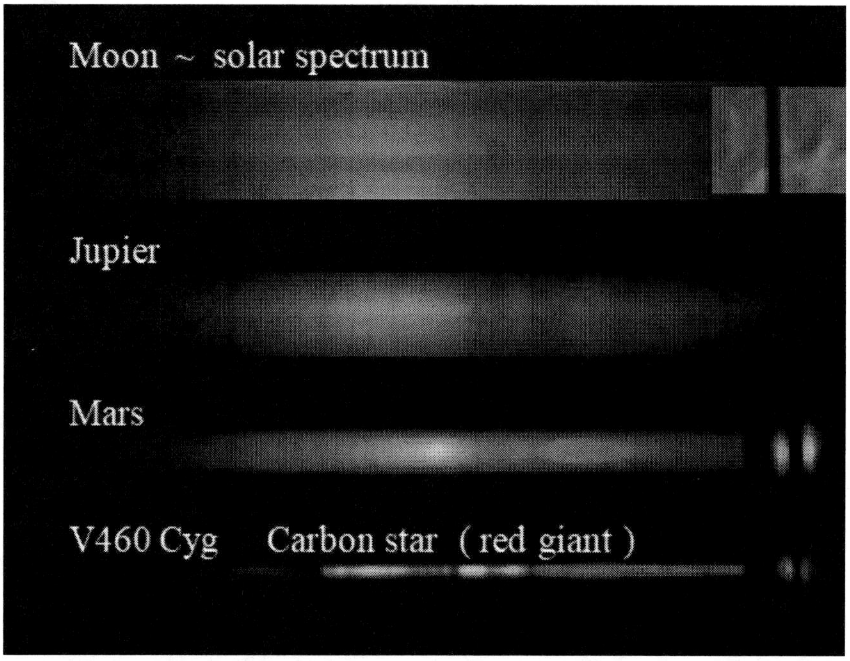


**Figure 4.** Spectral images of the Moon, Jupiter, Mars and a carbon star V460 Cygni taken by the GAO 150-cm telescope.

For example, we can easily see the difference of spectra between Mars and the red giant stars, even though their red colors in the direct images seem very similar for human eyes. Figure 4 shows images of spectra of Moon, Jupiter and Mars in addition to the spectrum of a carbon star V466 Cygni. All the spectra of Moon and planets Jupiter and Mars seems almost the same as the solar spectrum which is often observed as a rainbow. It is quite natural since those solar system bodies are not shining by themselves but just reflecting the lights coming from the Sun. We can easily understand such a basic fact directly from the comparison of those spectra.

On the other hand, the spectrum of the carbon star, which is a kind of red giant star, in the bottom panel of Fig. 4 is completely different from that of Mars. It indicates that the carbon star is not reflecting solar light but shining by itself. Further, we can estimate its surface temperature as about 2500 K from its spectrum. Some structures in the spectrum such as absorption features indicate the chemical characteristics of the stellar atmosphere as well. We can see the existence of a lot of carbon molecules in this star even though we can never touch the stellar body directly. Through our experimental trials of the education and outreach using our 150-cm telescope we have learned an interesting fact. Most students and peoples are impressed and activated very much by knowing that the telescope and instruments in front of them are really used and active in the real scientific research on the cutting edge. It is essentially important for the effective education and outreach to make full use of the telescopes also for the scientific research in practice.

## 5. Summary

Gunma Astronomical Observatory (GAO) can be regarded as a pioneer of public observatory equipped with a telescope of 2-m class, which is sufficiently powerful for the scientific research on the cutting edge by professional researchers. With the use of the 150-cm telescope, GAO has been carrying out a number of scientific studies. Many of them are studies of stellar astrophysics based on the large samples of observation data or focusing on the time variation of the targets for long periods or quick response to the transient phenomena. Such observations are the most suitable in general for the middle size telescopes, since it is not usually easy to occupy some of the world largest telescopes for long periods focusing on a limited number of scientific subjects. Collaborations with various observatories make it more effective to get more data and cover more observation points in time for longer periods. It is not only limited in the optical wavelengths but collaborative observations covering various wavelengths from gamma rays to radio wavelengths make the studies more fruitful. Moreover, since the first detection of the gravitation waves from the merging black holes [15], cooperative observations of transient phenomena using various media other than electromagnetic waves are getting more and more important. The middle size telescopes under a sufficiently flexible operation are really suitable for such collaboration studies so called as "multi-messenger observations".

It is proved through our experimental trials at GAO using the 150-cm telescope that the middle size telescopes are also useful for the education and outreach activities both for the students and public people. It can provide some unique opportunities of effective studies of astronomy and astrophysics by taking the advantage of the strong light gathering power of the telescopes. It is noted that the students and all the people are much impressed and activated through their own experiences using the real facilities and instruments which are really used for the active research of science as well. It is essentially important for the effective education and outreach to make full use of the telescopes also for the scientific research in practice. All the telescopes with an aperture size around 2-m must be the very powerful tools for any of the astronomical research, education and outreach. We expect Lampung Astronomical Observatory in Sumatera which will be equipped with a telescope of such a scale should work well.

**Acknowledgments**

This paper is based on our keynote presentation at the 10th SouthEast Asia Astronomy Network (SEAAN) 2018 meeting held in Lampung Indonesia in October 2018. Osamu Hashimoto was supported by JSPS KAKENHI Grant Number JP16K00996 for the development of the spectroscopic eyepiece system. A substantial grant on the period 2002-2008, has been provided by the Hitachi Global Foundation to whom Hakim L. Malasan would like to record his deepest gratitude.

## References

[1] O. Hashimoto, H. L. Malasan, Y. Kozai, *J. Tech. Soc. Sci. 2,* **1** 1-10
[2] O. Hashimoto, S. Yamano, A. Igarashi, *International Symposium on the NAOJ Museum*, 58-64
[3] O. Hashimoto, H. L. Malasan, H. Taguchi, T. Kurata, T. Yamamurao, N. Takeyama, M. Shimizu, *Asian-Pacific Reagional Meeting 2005 (ARIM2005)* 295-296
[4] O. Hashimoto, H. L. Malasan, H. Taguchi, S. Honda, *ITB-GAO Joint Workshop in Astronomy and Science Education* 45-49
[5] H. Takahashi, E. Nishihara, and GAO 150cm telescope working group, *ITB-GAO Joint Workshop in Astronomy and Science Education*, 50-56
[6] H. Funayama, Y. Itoh, Y. Oasa, E. Toyota, O. Hashimoto, T. Mukai, *Publ. Astron. Soc. Japan* **61**, 930-939
[7] Y. Takeda, O. Hashimoto, S. Honda, *Publ. Astron. Soc. Japan* **69** 1-17
[8] Y. Moritani, D. Nogami, A. T. Okazaki, A. Imada, E. Kambe, S. Honda, O. Hashimoto, Y. Ishino, Y. Suzuki, J. Tanaka, *Mon, Roy. Astron. Soc.* **405** 467-476
[9] Y. Moritani, D. Nogami, A. T. Okazaki, A. Imada, E. Kambe, S. Honda, O. Hashimoto, K. Ichikawa, *Publ. Astron. Soc. Japan* **63** L25-L29
[10] Y, Moritani, D. Nogami, A. T. Okazaki, A. Imada, E. Kambe, S. Honda, O. Hashimoto, S. Mizoguchi, Y. Kanda, K. Sadakane, K. Ichikawa, *Publ. Astron. Soc. Japan* **65** 83
[11] K. Sadakane, E. Kambe, O. Hashimoto, S. Honda, B. Sato, *Publ. Astron. Soc. Japan* **65** L1-L4
[12] Y. Takeda, O. Hashimoto, S. Honda, *Astrophys. J.* **57** 862
[13] S. Ilijić, *ASP Conf. Ser.* **318** 107
[14] M. Yamanaka et al., *Publ. Astron. Soc. Japan* **68** 68
[15] B. P. Abbott et al. *Phys. Rev. Lett.* **116** 061102
[16] O. Hashimoto, H. Taghuchi, *Communicating Astronomy with the Public 2018 (CAP2018)* 56-57

IOP Conf. Series: Journal of Physics: Conf. Series **1231** (2019) 012016 doi:10.1088/1742-6596/1231/1/012016

# Diffuse Interstellar Bands (DIBs) as a tracer of warp in the Third Galactic Quadrant: Preliminary Results

**A N Istiqomah[1*], L Puspitarini[1,2] and M I Arifyanto[1,2]**

[1]Undergraduate Program, Astronomy Study Program, Institut Teknologi Bandung, Bandung, Indonesia
[2]Astronomy Research Division and Bosscha Observatory, Faculty of Mathematics and Natural Sciences, Institut Teknologi Bandung, Bandung, Indonesia

E-mail: istiqomah.adenur@gmail.com

**Abstract.** We measured 6613.6 Å diffuse interstellar band (DIB) recorded in the 318 stellar spectra of stars located in the Third Galactic Quadrant. In agreement with previous studies, we found that the DIB strength has a positive correlation with the interstellar extinction, even though significant deviations were found. The correlation study between the DIB strength and the extinction is limited by the fact that the 3D extinction survey in the Galaxy is incomplete. The DIB spatial distribution toward the South Galactic Pole indicates that the DIB carrier(s) are distributed similarly as other Galactic warp tracer. More data, especially DIB measurements from target stars above Galactic principal plane, are needed to demonstrate that DIB can be used as a tracer of Galactic warp.

***Keywords**: Galactic Warp, Interstellar Matter, Diffuse Interstellar Bands*

## 1. Introduction
The discovery of Galactic warp based on HI observations is groundbreaking revolution and is an important step to understand the structure of our Galaxy. From HI observation, it was found that the gas layer in outer disk is warped away from the Galactic principal plane ($b = 0°$) [1]. It becomes warped towards North Galactic Pole in the First and Second Galactic Quadrants; and towards South Galactic Pole in the Third and Fourth Quadrants [2]. Over decades, large-scale surveys have been conducted in order to reveal the mysterious warp of our Galacy. Interestingly, the Galactic warp is not only seen in Galactic gas, but also in other Galactic disk's tracer such as stars, molecular clouds, and interstellar extnction [3,4,5]. This indicates that the warp affects the distribution of both stellar and interstellar components in the Galaxy. A longstanding mystery in astrophysical spectroscopy associated with interstellar materials is diffuse interstellar bands (DIBs). The absorption appears to be weak and diffuse in spectra of objects that reside behind interstellar clouds. Almost 500 DIBs were cataloged along optical and infrared wavelength, but only two DIBs in NIR have known carrier. Even though the carriers of DIBs are still under investigation, DIBs can be used as they have interesting properties, e.g., have positive correlation with interstellar extinction, HI and $H_2$ column density [6,7]. Thus, DIBs can be used as alternative tracer of the interstellar medium.

## 2. Data Analysis
### *2.1 Data*
This work used high SNR, medium resolution (R ~ 17,000) spectra from Gaia-ESO Spectroscopic Survey Data Release 3.1 supplement [8,9] observed by using GIRAFFE instrument (with HR15N setting) of targets located in the Third Galactic Quadrant. The HR15N setting yields wavelength coverage from 6470 Å to 6790 Å, allowing us to measure strong DIB with a wavelength center of 6613.6 Å [10] (hereafter 6614 DIB). Most of the Gaia-ESO Spectroscopic Survey targets are members of both open and globular

clusters, as the GIRAFFE instruments was designed to obtain several objects at one exposure (multi-object spectrograph). The targets are dwarf to sub-giant late-type stars, with effective temperature ranging from 3800 K to 8000 K and wide range overall metallicity (-0.5 dex to 1.0 dex). Distance estimation towards each target was obtained from [11].

*2.2 DIB extraction and error estimation*

As mentioned earlier, the dataset consist of late-type stars. In such low stellar temperature, many atomic and molecular lines present. It is challenging to distinguish between the DIB feature and the stellar features. We, therefore, compared each observed spectrum to a stellar synthetic spectrum based on atomic line list in order to distinguish stellar lines from DIB absorption. Predetermined stellar parameters for each target were obtained from Gaia-ESO Catalogue (Version 2). Based on these parameters, we performed stellar modeling and spectrum synthetizing using MARCS [12] and SPECTRUM [13]. In the stellar modeling, we assumed that geometry of stellar atmosphere to be plan-parallel and obey Local Thermodynamical Equilibrium (LTE). Since microturbulence velocity can contribute to the line broadening, we adopted 2 km/s as a default value of microturbulence velocity. Stellar radial velocity was determined by using cross-correlation method, with initial value taken also from Gaia-ESO Catalogue. In this work, we didn't are negligible [14]. The equivalent width of DIB 6614 (corresponding to DIB strength) towards each target was measured by performing single Gaussian profile fitting with Levenberg-Marquardt algorithm to minimize chi-squared value [15]. The error of equivalent width measurement were estimated by taking the fitting error and continuum placement into account, thus $\sigma^2 = \sigma^2_{fit} + \sigma^2_{cont}$.

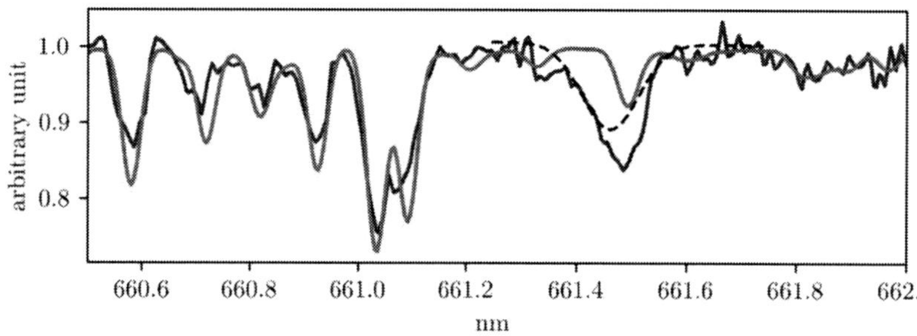


**Figure 1.** Observed spectra, synthetic spectra and fitted DIB profile in blue, red, and dashed black lines, respectively. The DIB profile is not only shifted from its wavelength center but also blends with stellar lines. The fitting region are showed by the dashed black line.

## 3. Results and Discussions

We have measured 318 DIB out of 696 spectra of GES DR3.1 (HR15N). However, we faced some problems regarding the spatial distribution of out target stars and the stellar model. Unfortunately, in this work, the available target stars are not widely and regularly distributed in the Third Galactic Quadrant. The measured targets distribution, leaving only field in 212.6° $< l <$ 213.0°, 251.2° $< l <$ 253.6° , and 262.2° $< b <$ 269.9° with Galactic latitude spans from -1.8° $< b <$ -8.9° as shown in Figure 2. Furthermore, we can see from Figure 1., there were overestimated stellar lines in the synthetic. As blended DIB extraction relies much on the synthetic spectra (see Figure 1.), we gave caution flag to mark blended DIB with stellar lines.

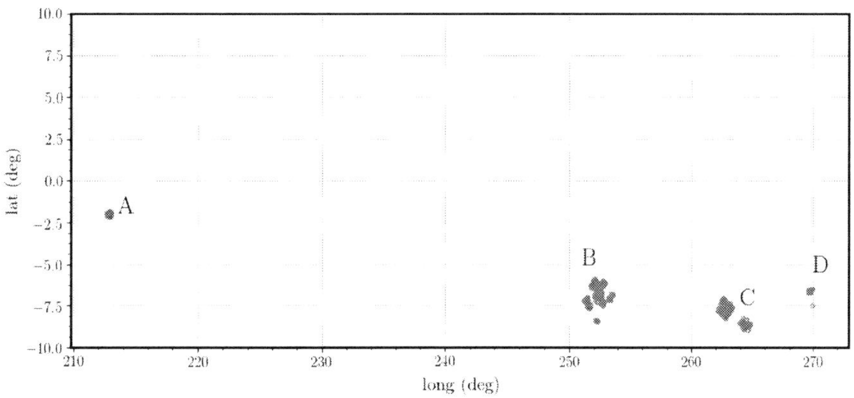


**Figure 2.** Distribution of targets with measured DIB. A, B, C, and D denote target fields.

Incomplete 3D interstellar extinction survey has limited our DIB correlation study. However, Figure 2. shows correlation of DIB and extinction toward late-type stars in Field A, which was the only field with available interstellar extinction value. Lines resulted from [7] and [17] were derived from nearby, early-type stars in all of their fields.

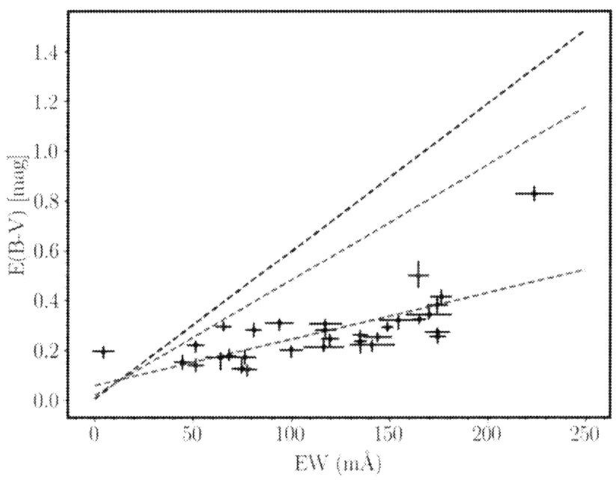


**Figure 3.** Equivalent width (EW) of 6614 DIB vs. E(B-V) toward Field A. Interstellar extinction value is adopted from [16]. Blended DIB was marked with blue points. The red, blue, and green lines respectively show correlation from this work, [17], and [7].

In this work, Galactic warp is inspected toward each target field by converting DIB abundance into interstellar extinction. This conversion makes use DIB correlation obtained by [14]. Interstellar extinction towards Field A, as shown in Figure 3., is in agreement with [16] work even though some scatters are present. Increasing in abundance in < 2 kpc and > 4 kpc denote dense regions toward this field. These regions are Local (Orion) and Perseus Arm as revealed by [18]. Plateau in 2 – 4 kpc indicates a less dense region, probably interarm region.

IOP Conf. Series: Journal of Physics: Conf. Series **1231** (2019) 012016 doi:10.1088/1742-6596/1231/1/012016

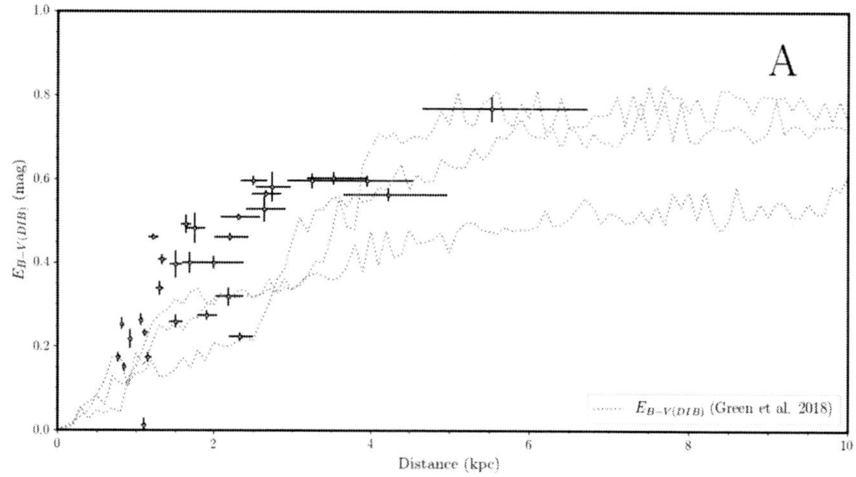


**Figure 4.** Interstellar extinction towards Field A. Top, middle, and bottom dashed gray lines are interstellar extinction curves toward $b \sim -3°$, $0°$, and $+3°$, respectively.

Due to incomplete interstellar extinction information in [16], we used interstellar extinction derived from stellar clusters studies by [19] instead. As we can see in Figure 4., DIB abundance has similar trend with those from stellar cluster studies. Despite of this fact, we can not draw conclusion on interstellar environment towards Field B and C for it shows noticeable scatter in > 2 kpc. This illustrates the complex structure of the interstellar matter itself as denotes by Galactic CO emission in [20]. Interestingly, interstellar extinction from stellar clusters studies fill in the gaps where there are lack of data from this work in < 2 kpc. Furthermore, DIB abundance in these fields also show steep increasing in < 2 kpc, mainly due to Local (Orion) Arm.

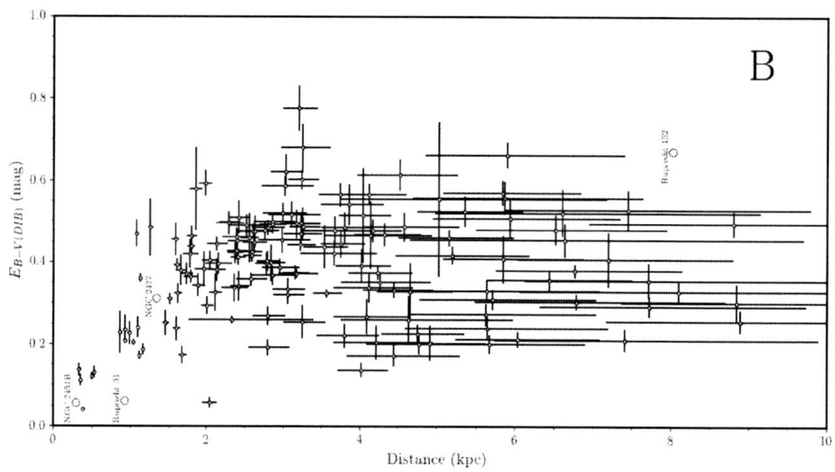

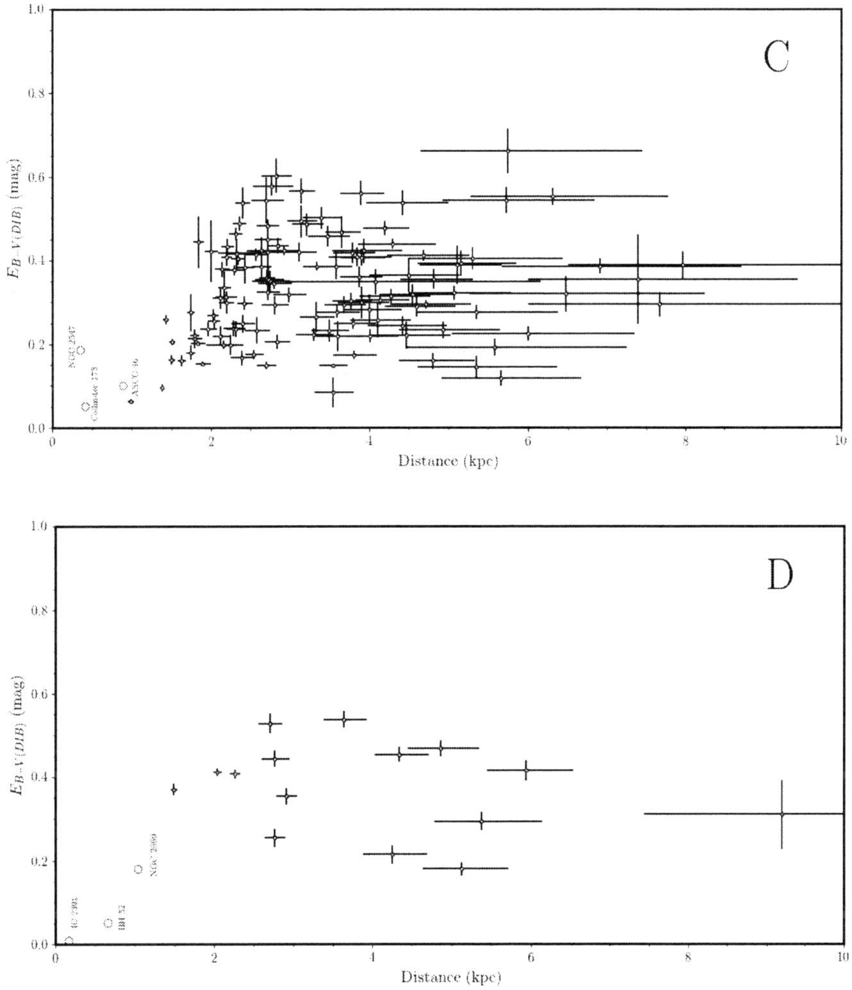


**Figure 5.** Similar to Figure 4., but for Field B (top panel), C (middle panel), and D (lower panel). Red circle is interstellar extinction from stellar cluster studies [18].

## 4. Conclusions

With the current data, although there was a subtle indication of increasing DIB abundance towards SGP and DIB strength is in agreement with interstellar extinction, it is still too early to conclude that we detect the warp by using DIB as ISM tracer. Therefore, we need more targets that are well-distributed along the galactic longitude and latitude also distances to be able to draw a conclusion about the Galactic warp traced by the interstellar matter.

**Acknowledgement**
ANI would like to thank P3MI for providing travel and accommodation support to attend SEAAN Meeting 2018.

IOP Conf. Series: Journal of Physics: Conf. Series **1231** (2019) 012016 doi:10.1088/1742-6596/1231/1/012016

**References**

[1] Burke B F 1957 *Astron. J.* **62**, 90
[2] Kerr F J 1957 *Astron. J.* **62**, 93
[3] Momany Y, Zaggia S, Gilmore G, Piotto G, Carraro G, Bedin L R and de Angeli F 2006 *A&A* **451(2)** 515 - 38
[4] May 1999
[5] Reyle C, Marshall D J, Robin A C and Scultheis M 2009 *A&A* **495(3)** 819-26
[6] Moutou C, Krelowski J, D'Hendecourt L and Jamroszczak 1999 *A&A* **351** 680-88
[7] Friedman S D, York D G, McCall B J, Dahlstrom J, Sonnentrucker P, Welty D E, Drosback M M, Hobbs L M, Rachfors B L and Snow T P 2011 *AJ* **727(1)** 13
[8] Gilmore G *et al* 2012 *The Messenger* **147** 25-31
[9] Randich S, Gilmore G and Gaia-ESO Consortium *The Messenger* **154** 47-49
[10] Hobbs L M *et al* 2009 *AJ* **705(1)** 32-45
[11] Bailer-Jones C A L, Rybizki J, Fouesneau M, Mantelet G and Andrae R 2018 *AJ* **156(2)** 58
[12] Gustafsson B, Edvardsson B, Erksson K, Jorgensen U G, Nordlund A and Plez B 2008 *A&A* **486(3)** 951
[13] Gray R O 1999 ascl:9910.002
[14] Puspitarini L *et al* 2015 *A&A* **573** 16
[15] Newville M, Stensitzki T, Allen D B, Rawlik M, Ingargiola A and Nelson A 2016 ascl:1606.014
[16] Green G *et al* 2018 *MNRAS* **478(1)** 651-66
[17] Puspitarini L, Lallement R and Chen H C 2013 *A&A* **555** 25
[18] Churchwell E, *et al* 2009 *PASP* **121** 213
[19] Dias W S, *et al* 2002 *A&A* **389** 871-73
[20] Dame T M, *et al* 2001 *ApJ* **547** 792-813

# Preliminary Report of Light Pollution in Indonesia Based on Sky Quality Observation

**A. G. Admiranto[1,*], R. Priyatikanto[1], S. Maryam[1], Ellyyani[1], and N. Suryana[1]**

[1]Space Science Center, Indonesian National Institute of Aeronautics and Space, Jl. Dr Djundjunan 133 Bandung

Email: gunawan.admiranto@lapan.go.id

**Abstract.** We observed night sky quality in several LAPAN stations (Agam, Bandung, Pontianak, Sumedang, Garut, Pasuruan, and Biak) which were conducted from April until July 2018 using Unihedron Sky Quality Meter LU-DL type. Observational data from all of the observational points were then sent regularly to a centralized database for further use. Although most of the measurements were done in overcast conditions, we were able to determine the representative clear sky brightness statistically. The results showed that the light pollution level of the most of the stations are moderate (the values at Biak, Agam, Sumedang, and Pontianak are 20.0, 19.5, 19.6, and 17.7 mpsas respectively) and the stations which are located near or in cities are high (Bandung and Pasuruan with 17.1 and 18.0 mpsas, respectively). In a particular station (Garut) the light pollution is low (20.6 mpsas), so it is good to make this spot to be a location of astrotourism.

***Keywords:*** *night sky observations; Unihedron Sky Quality Meter; light pollution; astrotourism*

## 1. Introduction

There are many challenges faced by astronomical communities, professionals and amateurs alike, related to the observing activities. There are some issues in this respect, e. g. location, equipment, and human resources, but the most pressing issue is the light pollution. Light pollution can disturb astronomical observations especially to the faint object observations. Other problems are that people can't see the beauty of night sky, the disruption of the circadian rhythm in human and animals, and the efficiency of energy in street lamps in cities. The investigated the light pollution and its impacts [1]. They analyzed the pollution which occurred in Europe, Asia, North America, and they concluded that light pollution is a global issue. They observed that light pollution affects 88% of world population and 99% of United States population. The discussion same problem and observed that this problem also affect many regions like Central Africa and Central Asia which was previously thought were free from this problem [2]. Nevertheless, they observed also that Venice is not affected by light pollution due to the policy of the local government related to the light pollution. This problem of light pollution is directly related to the sky quality. The better the sky quality, the lesser is the light pollution. The sky quality can be measured using Sky Quality Meter device. This device records the light emitted by the sky and calculates the incident radiation in various wavelength, and the pollution in each wavelength can be measured. The measurement sky quality of a city in Holland in blue, green, red, and infra-red wavelengths has been succesful [3]. His results indicated that the level of light pollution in a certain location depends on the dominant wavelength of the sky brightness of the location. The propagation of light in a polluted atmosphere was analyzed to obtain sky brigtness [4]. They used a radiation propagation model which applied to various atmospheric conditions including various types of aerosols, i.e. volcanic and meteor dust and atmospheric gases like ozone. Their calculation was then applied to make a map of air pollution in several locations and they found that the sky quality and air pollution depends on several parameters, including atmospheric composition and ground reflectance. The sky quality observation in 12 locations in Italy for 12 years [5]. The time of observations and locations are varied, from height of 15 m until 1366 m above sea level and during winter and non-winter season. The observations conducted after astronomical twilight. They found that the existence of ground snow has no effect to the sky quality. Sky quality observation of Postdam-Babelsberg which is located about 23 km from Berlin was conducted [6]. The observations were conducted in two conditions, during moonlit sky and during overcast. They found that the sky quality was more influenced by the cloud cover than by the changing lunar radiation due to the changing lunar phases. From this analysis we see that the sky quality observations are more extensively conducted in the more developed countries. So, there is some need to conduct similar observations in other regions which have no record of sky quality observations, especially Indonesia. It is hoped that from these observations we get some insights about the extent of light pollution in Indonesia and if there is any location in Indonesia which are best candidate to be a dark sky park to promote astrotourism.

This paper is organized as follows. Section 2 discussed data and methodology used in this investigation. Section 3 discussed the analysis of the data obtained, while section 4 presented the results of this analysis. Discussion and conclusion of this investigation is presented in section 5.

## 2. Methodology

Data of this investigation were obtained from several LAPAN stations, i.e. Agam, Pontianak, Sumedang, Garut, Biak, Pasuruan, and Bandung stations in which the locations are shown in Table 1 and Figure 1.

Table 1. Geographical coordinates and conditions of 7 observation stations

| Stations | Longitude | Latitude | Remarks |
|---|---|---|---|
| Agam (AGM) | 100 19' 10.00" E | 0 13' 48.00" S | Rural |
| Pontianak (PTK) | 109 20' 23.00" E | 0 02' 48.00" N | Peri urban |
| Bandung (BDG) | 107 40' 40.21" E | 6 55' 32.03" S | Urban |
| Sumedang (SMD) | 107 50' 13.97" E | 6 54' 47.08" S | Rural |
| Garut (GRT) | 107 41' 31.97" E | 7 39' 00.22" S | Rural |
| Pasuruan (PSR) | 112 40' 00.00" E | 7 34' 00.00" S | Peri urban |
| Biak (BIK) | 136 01' 00.00" E | 1 17' 00.00" S | Urban |

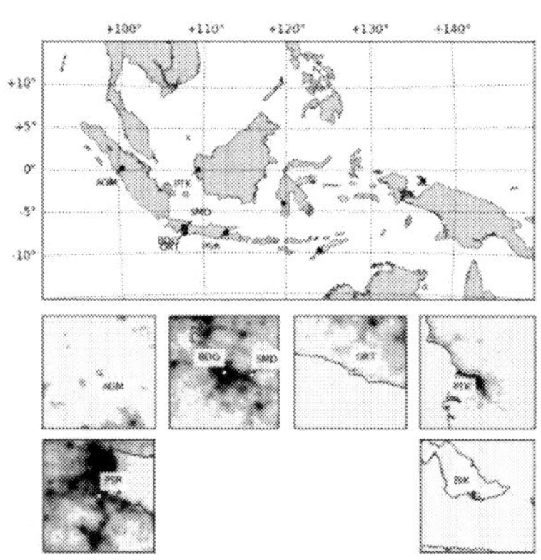

Figure1. Locations of sky quality observations using Unihedron Sky Quality Meter across Indonesia.

The Unihedron Sky Quality Meter are SQM-LU types in which the equipment can be set to make continuous observations in predetermined intervals. These equipment were put in a housing to protect them from humidity and direct sunlight. This type of photodiode-based instruments has effective field of view of ~20° and produce sky brightness data in mag/arcsec$^2$. All SQM was set to record zenithal sky brightness magnitude from 5 pm to 7 am with sampling rate of one minute, regardless weather conditions and moon phases. Figure 2 shows the equipment we use to record the sky quality.

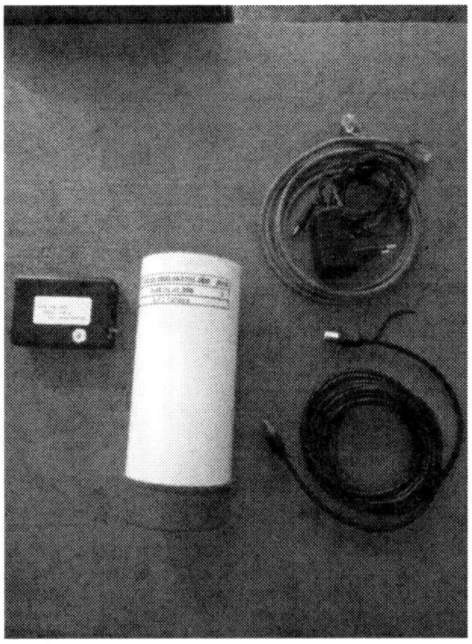

Figure 2. Unihedron SQM LU-DL equipments with its accessories.
This equipment was designed to fetch the data automatically in a certain period of time.

These data were obtained from April until July 2018. These data were obtained under different sky conditions, during dark and full Moon, low and high cloud cover, and clear and overcast condition. The data then sent to Bandung and stored in a repository and organized according to data source (7 stations) to be analyzed by the researchers. The results of the observations can be summarized in Table 2 and Figure 3 below. We can see that the sky conditions in Indonesia depended upon the proximities of those various locations to the source of night sky. There are places which are very dark, especially in Sumedang and Garut, but there are some places that are very bright like in Bandung and Pasuruan which are urban and peri urban in nature, respectively. From the data obtained, we then made some plot in which we want to see the correlation of the sky brightness with the time of observations.

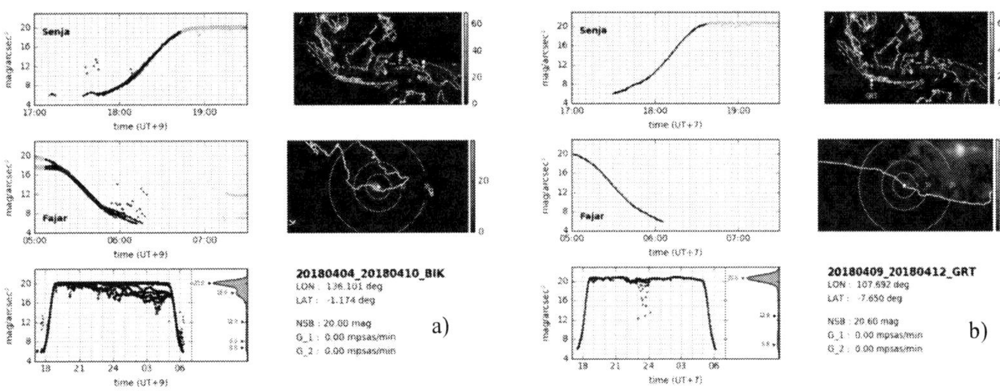

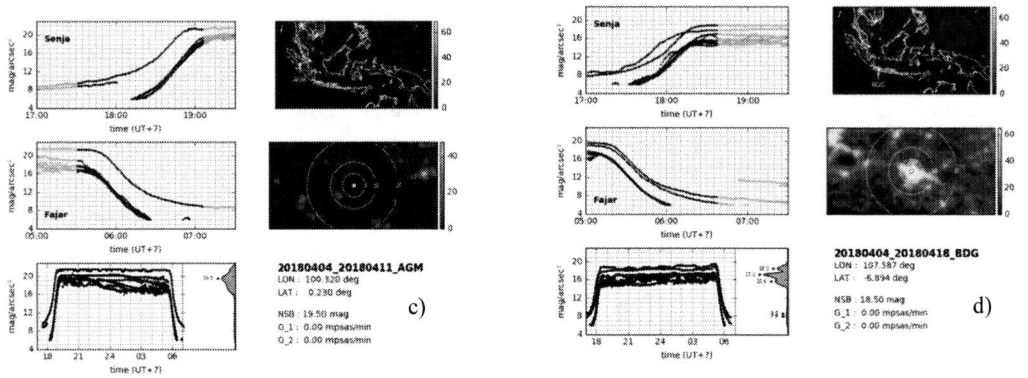


Figure 3. Data plot of several observational points, i.e. (a) Biak, (b) Garut, (c) Agam, (d) Bandung. These plots showed various conditions of sky qualities. These observations showed also the sky conditions above the observational points which are mainly in cloudy conditions.

Table 2. Results of observations from LAPAN stations. These results showed the nature of light pollution in those places.

| Code | Data Amount [nights] | Location | Brightness [mpsas] | Remarks |
|---|---|---|---|---|
| AGM | 29 | Agam | 19.8 | Rural |
| BDG | 50 | Bandung | 17.3 | Urban |
| BIK | 43 | Biak | 19.7 | Rural |
| GRT | 61 | Garut | 20.8 | Rural |
| PSR | 69 | Pasuruan | 17.6 | Peri-urban |
| PTK | 39 | Pontianak | 18.0 | Peri-urban |
| SMD | 67 | Sumedang | 20.1 | Peri-urban |

## 3. Results of Discussion

We then conducted other observations, in which the data was acquired in a different manner compared with the data obtained from the stations. Whereas the station's data were obtained from a fixed location in an extended period of time, the other method is taking the data in a moving fashion but in a shorter period of time compared with that of the fixed locations. The purpose of this observation is to get the night sky quality maps in a certain area so we can get some insight about the light pollution map in the area. This time we conducted three modes of observations. The first one is the same mode of observations as in the stations. The second one is the same as the first one, but we used a kind of baffle to prevent stray light from entering the equipment. The third mode used a handheld sky quality meter. The measurements were conducted in an area which surrounds Bandung and Sumedang which can be seen in the Figure 4.

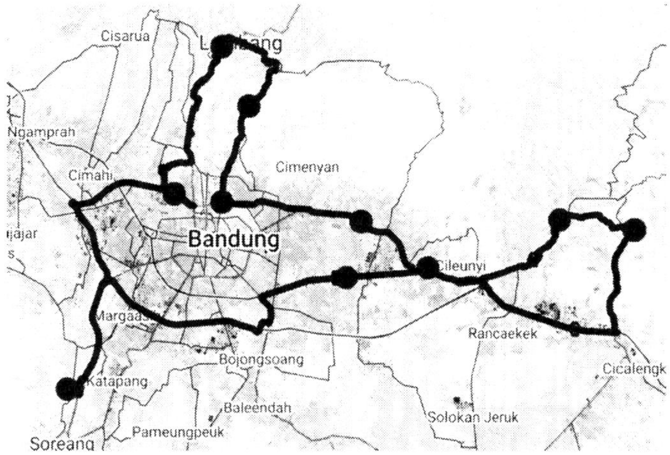


Figure 4. Area traversed during the measurements in moving mode.

In this mode of observations the data was recorded every 1 minute using the SQM and using the handheld equipment the data recorded every 1 km. Using this mode of observation we got about 3200 data in which the distance traversed is about 120 km. From the data obtained in moving mode we got values of night sky brightness along the distance traversed during the observations. We can see several bright points, but also some points which are very dark, i.e. the value of the sky brightness is more than 19. From this mode of observations, we can see that in general the night sky brightness of Bandung and its surrounding area are not very good due to the light pollution. Figure 6 summarized the observations and the data obtained in these modes. The NSB observations which had been conducted in Bandung-Sumedang area above had shown us the map of sky brightness/light pollution in these area. We can see some regions which are very polluted, and some other regions which are not polluted at all. These observations showed also that moving mode of NSB observations is a good way to get a detailed picture of sky brightness/light pollution in a certain region.

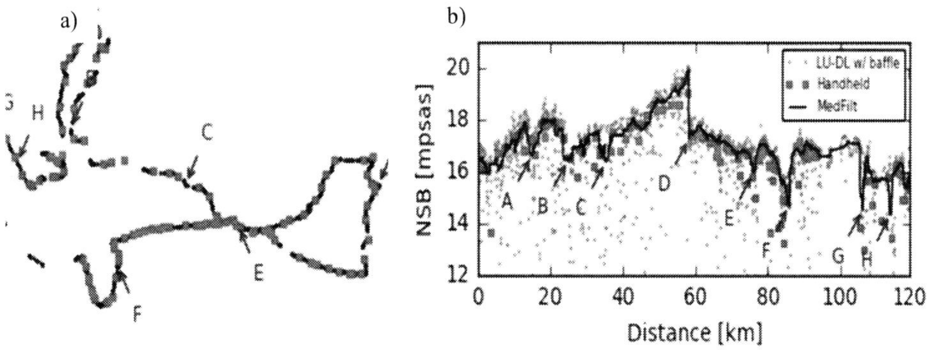


Figure 5. Results of NSB observations with some locations which were highlighted due to its brightness. (a) Highlighted points (b) the brightness of these points.

As a control mode, handheld observations had proved it to be a good measurement of the reliability of this mode. We made a correlation of this observation upon one another as shown in Fig 6. These pictures showed us that these three modes of observations have good correlation among one another.

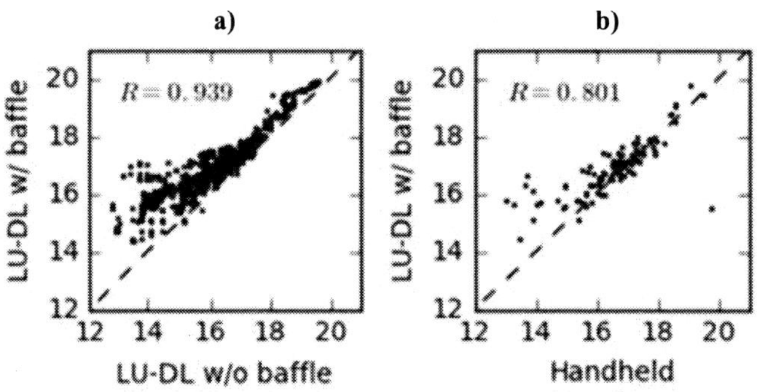


Figure 6. Correlation of three modes of observations, (a) correlation of LU-DL with and without baffle, and (b) LU-DL with baffle with handheld observations.

## 4. Discussion and Conclusion

We have observed night sky brightness (NSB) in several locations in Indonesia for some extended period of time. These observations had shown us the importance of the distributed observations and should be expanded, both spatially and temporally. Furthermore, moving NSB observations which conducted in some limited area can be used to measure the level of light pollution in this area. From the results of our investigation, we can propose some regions which can be used as a place of astrotourism because of its pristine environment with respect to the NSB observed. This could also be used as some example to other regions in preserving the beauty of the night sky so the light pollution in other areas can be reduced. We are planning to continue the NSB observations we have conducted in some other places so the map of light pollution over Indonesia can be completed and we can make further contribution to the international community about the level of light pollution in Indonesia.

## Acknowledgments

We would like express our gratitude to the Head of the Space Science Center of LAPAN the financial support we got in attending the South East Asia Astronomy Network 2018 in Lampung, October 19-21, 2018.

## References

[1] Cinzano, P., Falchi, F. Elvidge, C. D. 2001 *Mon. Not. R. Astron. Soc.* **328** 689
[2] Cinzano, P., Falchi, F. (2014), Journal of Quantitative Spectroscopy & Radiative Transfer **139** 13
[3] Falchi, F. 2010 *Mon. Not. R. Astron. Soc.* **412** 33.
[4] Falchi, F., Cinzano, P. Duriscoe, D., Kyba, C.C.M., Elvidge, C.D., Baugh, K.,. Portnov,B.A., Rybnikova, N.A., Furgoni, R. 2016 *Science Advances* **2** 6
[5] Puschnig, J., Schwope, A., Posch, T., Schwarz, R., 2014, *Journal of Quantitative Spectroscopy & Radiative Transfer* **139** 76

# Cosmic Evolution of Nearby Radio Active Galactic Nuclei

**Irham Taufik Andika[1], Mochamad Ikbal Arifyanto[2], Ade Nur Istiqomah[2], Shinta Nur Amalina[2], Aprilia[2], Muhamad Irfan Hakim[2]**

[1]Max Planck Institut für Astronomie, Königstuhl 17, D-69117, Heidelberg, Germany
[2]Department of Astronomy, Faculty of Mathematics and Natural Sciences, Institut Teknologi Bandung, Jl. Ganesha No. 10, Bandung 40116, Indonesia

E-mail: andika@mpia.de

**Abstract.** The exact formation mechanism of massive galaxy in the universe still become an open question in modern astrophysics. Radio emission from active galactic nucleus (AGN) is known to be suppressing stellar growth in the most massive galaxies, thus becoming significant ingredient in modeling galaxy formation process. Radio luminosity function across cosmic redshifts ($z$) is an important tool to constrain the co-evolutionary models of radio AGN and its host galaxy. Therefore, we aim to calculate radio luminosity function at 1.4 GHz frequency for radio AGN in the range of $0.1 \leq z \leq 1.3$ in this work. The radio data was taken from deep VLA observation of Stripe 82 field at angular resolution of 1.8" and 52 $\mu$Jy sensitivity. On the other hand, the optical/near-infrared data was taken from Dark Energy Survey DR1 observation in $g$, $r$, $i$, $z$, $Y$ bands with co-added catalog limiting magnitude of $i = 23.44$.

We estimated the photometric redshift (photo-$z$) of each sources by using combined two machine learning algorithms: neural networks and boosted regression trees. We successfully performed accurate photo-$z$ measurement with average bias $\langle\delta\rangle = -3.5\times10^{-3}$, scatter $\langle\sigma\rangle = 0.15$ and outlier fraction $\langle f(3\sigma)\rangle = 0.06$. By using $1/V_{\rm max}$ method, we calculated the luminosity function, then constrained their evolution with pure density evolution (PDE) or pure luminosity evolution (PLE) model. At median $z = \{0.31, 0.59, 0.88, 1.10\}$, we found the power-law index of PDE is $\alpha_D = \{1.29, 1.43, 1.73, 0.94\}$ while for PLE is $\alpha_L = \{2.28, 2.59, 3.19, 1.63\}$. Our result is consistent with previous studies and gives better constraint to radio AGN luminosity and density evolution power-law indexes due to larger number of sources (6900) and wider covered sky fraction (92 deg$^2$).

## 1. Introduction

The exact mechanism of massive galaxy formation still leaves big question in modern astrophysics. Radio emission from active galactic nucleus (AGN) is known to be suppressing star formation in the most massive galaxies, thus becoming significant ingredient in modeling galaxy formation process. By now, radio jets feedback from AGN has become a primary prescription in semi-analytic models and is able to well-reproduce the observed properties [1]. The produced outflows heats the massive galaxy halo gas and quenches its star formation, suppress the growth and thus set certain limit on the assembled mass.

Past studies have shown that there are different evolutionary schemes of radio AGN which is consistent with "cosmic downsizing" phenomena. This means faint radio-luminosity objects evolve more weakly compared to the luminous ones. The number density peak located at higher redshift for higher radio luminosity. Radio luminosity as a function of cosmic redshifts ($z$) is very useful to constrain the co-evolution of radio AGN and its host galaxy models [1]. In this work, we will determine the radio luminosity function at 1.4 GHz frequency for radio AGN at $0.1 \leq z \leq 1.3$.

## 2. Data

### *2.1. Radio catalog*

The radio sources were obtained from high resolution 1.4 GHz Very Large Array survey that was conducted by [2]. This observation has an angular resolution of 1.8" and 52 $\mu$Jy beam$^{-1}$ sensitivity over 92 deg$^2$ in Stripe 82 field. There are 17969 isolated radio sources from this catalog that can be used.

### *2.2. Optical photometry data*

We use the data from Dark Energy Survey Data Release 1 [3] to search the optical counterparts of our radio sources. DES is an optical/near-infrared imaging survey conducted by the Dark Energy Camera, attached on the 4-m Blanco telescope of CTIO in Chile. Current data release covers 5000 deg$^2$ in five broad photometric bands of the southern Galactic cap. The catalog depth for median co-added images with 1.95" aperture diameter at SNR = 10 is $\{g, r, i, z, Y\} = \{24.33, 24.08, 23.44, 22.69, 21.44\}$ mag. We focus on searching galaxy candidates in Stripe 82 field by using this SQL query:

```
SELECT *

FROM dr1_main main
JOIN dr1_magnitude mag ON main.COADD_OBJECT_ID = mag.COADD_OBJECT_ID

WHERE
(main.ra BETWEEN 315 AND 360 OR main.ra BETWEEN 0 AND 45)
AND main.dec BETWEEN -14 AND 5
AND main.imaflags_iso_z = 0
AND main.imaflags_iso_i = 0
AND main.imaflags_iso_r = 0
AND main.imaflags_iso_g = 0
AND main.flags_z < 4
AND main.flags_i < 4
AND main.flags_r < 4
AND main.flags_g < 4
AND main.spread_model_i BETWEEN -0.05 AND 0.05
AND (
(CASE WHEN spread_model_i + 1. * spreaderr_model_i > 0.003 THEN 1 ELSE 0 END) +
(CASE WHEN spread_model_i + 3. * spreaderr_model_i > 0.005 THEN 1 ELSE 0 END) +
(CASE WHEN spread_model_i - 1. * spreaderr_model_i > 0.003 THEN 1 ELSE 0 END)
) > 2
```

The obtained objects in this catalog is named as main optical catalog.

### *2.3. Spectroscopic catalog*

We assemble public spectroscopic catalogs which will be used later to determine photometric redshifts (photo-$z$) of objects in our main optical catalog. For all of these catalogs, we limit our search to include only objects within $0.01 < z < 9$, not stars or artifacts, and redshift error cut of $\sigma_z < 0.005\ (1+z)$. Additional criteria were applied for each survey to ensure high quality spec-$z$ measurements and summarized in Table 1. Then, we match our main sample with this

**Table 1.** The Best-fit parameters of LF evolution in each redshift bins.

| Survey | Criteria | Reference |
|---|---|---|
| SDSS/BOSS | zWarning = 0 | Abolfathi et al. (2018) |
| DEEP2 | qFlag = 4 (> 99.5% confidence) | Davis et al. (2003), Newman et al. (2013) |
| VIPERS | qFlag = 3 – 4 (> 95% confidence) | Garili et al. (2014) |
| VVDS | qFlag = 3 – 4 (> 95% confidence) | Le Févre et al. (2013) |
| GAMA | qFlag = ≥ 4 (highly confident) | Liske et al. (2015) |
| WiggleZ | qFlag = ≥ 4 (highly confident) | Drinkwater et al. (2010) |
| UDSz | qFlag = ≥ 4 (> 95% confidence) | Bradshaw et al. (2013), McLure et al. (2013) |

compiled spectroscopic catalogs by using 1" matching radius.

## 3. Photo$z$ Determination
We utilize ANNz2 [12], a new version of neural network photo-$z$ code ANNz [13]. This version incorporates additional machine learning algorithms more than Artificial Neural Networks (ANNs), such as k-Nearest Neighbors (KNN) and Boosted Decision Trees (BDTs). These methods are contained the TMVA package [14]. To train the software, we run ANNs 120 times and randomly varied the parameters such as training cycles count, number of nodes in each layer, the type of activation function, the utility of Bayesian regulator that minimizes the over-fitting risk, the type of variable transformation performed before training (e.g. PCA transformation and normalization), the number of convergence test, and initial random seed. The performance test is then measured through optimization scheme which leads to a photo-$z$ value. The full set of solutions is then used to make the probability distribution function (PDF) of photo-$z$. For the first step, each solution is folded with its error, that was derived before with KNN error calculation as in [15]. Then, the ensemble solution is constructed after the optimization is done and weighting scheme applied. This will allows us to calculate both the intrinsic uncertainties on the input parameters and the errors on the method itself. The aforementioned methodology is called "randomized regression". In addition, a useful feature in ANNz2 is the weighting scheme for training data [16]. This technique were applied to make the training data distributions becoming more representative of the test data.

## 4. Result and Analysis
In this section, we will discuss methods of determining the LF from our data, then show the fitting procedure, and present the LFs for our radio AGN. Derived parameters will give us constraint on how radio AGN and its host galaxy are connected, which is valuable ingredient in modeling this type of system.

### *4.1. Estimating the luminosity function*
First, the assumption that radio sources are dominant in radiating synchrotron emission in its radio spectrum and can be described with power-law $S_v \propto \nu^{\alpha}$ is made. Here $\alpha$ is the spectral index and $S_\nu$ is monochromatic flux density at specific frequency $\nu$. In cosmological context, standard radio $K$-correction is defined as $K(z) = (1+z)^{-(1+\alpha)}$. Hence, the rest-frame radio

luminosity $L_{\nu_1}$ at specific frequency $\nu_1$ can be derived from the observed redshift $z$, luminosity distance $D_L$, and flux density $S_{\nu_2}$ at frequency $\nu_2$ as

$$L_{\nu_1} = \frac{4\pi D_L^2(z)}{(1+z)^{1+\alpha}} \left(\frac{\nu_1}{\nu_2}\right)^{\alpha} S_{\nu_2} \tag{1}$$

The choice of specific frequency at 1.4 GHz is simply to make it easier to compare our results with other literature. We assumed a standard $\alpha = -0.7$ if the power law index measurement is not available. This assumption is valid median value for radio galaxies as explained by [17].

To determine the LF (density of sources as a function of luminosity) at certain range of redshift bins, we utilized the $V_{\max}$ method [18]. Basic principle of this method is to calculate the maximum observable volume of each object of interest, constrained with certain selection criteria. One advantage of this method is because it is independent from LF's shape so that selection and sample biases can be minimized. The LF $\Phi(L, z)$ describes the number of object as a function co-moving volume and luminosity and is expressed as

$$\Phi(L, z) = \frac{1}{\Delta \log L} \sum_{i=1}^{N} \frac{1}{V_{\max,i}} \tag{2}$$

where $\Delta \log$ is luminosity bin width, $V_{\max}$ is the maximum observable co-moving volume of i-th source, and the summation is calculated over each source i in certain luminosity and redshift bin. To consider various effects and biases (e.g. non-uniform noise in the radio observation and luminosity limited sample) which results in incompleteness, the maximum observable volume $V_{\max}$ is calculated in general form equation which is

$$V_{\max,\mathrm{i}} = \sum_{z=z_{\min}}^{z_{\max}} [V(z+\Delta z) - V(z)] \tag{3}$$

where the summation begins at a certain redshift bin and adds together co-moving volume of spherical shells $V(z+\Delta z) - V(z)$ in small redshift intervals $\Delta z = 0.01$ until redshift bin end.

Following [19], the root-mean-square errors of LF in each luminosity and redshift bin is estimated by weighting the number of objects which contribute to the summation and expressed as

$$\sigma_\Phi = \frac{1}{\Delta \log L} \sqrt{\sum_{i=1}^{N} \frac{1}{V_{\max,\mathrm{i}}}} \tag{4}$$

However, if there are less than ten objects in a luminosity bin, we take the error as lower and upper 84% confidence interval. These intervals compatible with Gaussian $1\sigma$ errors so that $\sigma_\Phi = \Phi \times \sigma_N/N$. Here $\sigma_N$ is Poissonian asymmetrical error.

### *4.2. Radio luminosity function evolution*

Radio AGN population evolution can be parameterized by luminosity and density evolution of its local luminosity function as

$$\Phi(L, z) = (1+z)^{\alpha_D} \times \Phi_0\Big[\frac{L}{1+z^{\alpha_L}}\Big] \tag{5}$$

where $\alpha_L$ is luminosity evolution parameter and $\alpha_D$ is density evolution parameters, respectively, $L$ is luminosity, $\Phi(L, z)$ is the number density at redshift z, and $\Phi_0$ is the local luminosity function. Analytical equation for nearby radio AGN LF that we use is two power-laws model, taken from [20] as

$$\Phi_0(L) = \frac{\Phi^*}{(L^*/L)^{\alpha} + (L^*/L)^{\beta}} \tag{6}$$

By definition, the parameters are the scaling constant $\Phi^* = \frac{1}{0.4} 10^{-5.5}$ Mpc$^{-3}$ dex$^{-1}$ (normalized to the base of $d \log L$), the turning point $L^* = 10^{24.59}$ W Hz$^{-1}$, and the slopes of faint and bright end $\beta = -0.49$ and $\alpha = -1.27$, respectively. We take the radio AGN population luminosity function as determined by [20] and model its evolution by using Equation 6.

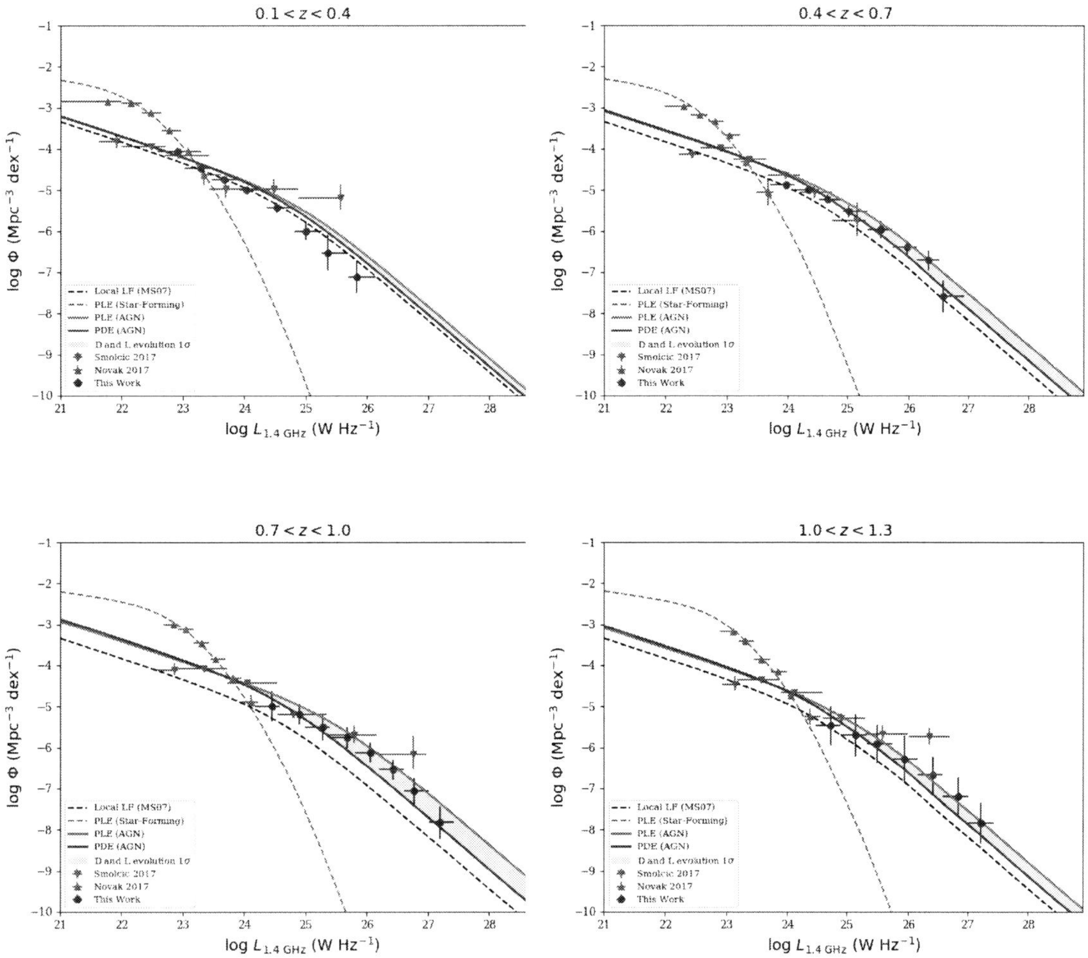


**Figure 1.** Best-fit for pure luminosity and pure density evolutions for each redshift bin. Our data is shown as blue dots with error bars while data from [1] and [17] are shown as magenta and green dots with error bars, respectively.

In Figure 1 we show the best-fit parameter of pure luminosity (PLE; $\alpha_D = 0$) and pure density (PDE; $\alpha_L = 0$) evolutions for each bins, which can be thought as cases of evolution in extreme regime. The calculated parameters are also tabulated in Table 2. The median pure density evolution and pure luminosity power-law indexes for all considered redshift bins are $\langle \alpha_L \rangle = 2.37 \pm 0.14$ and $\langle \alpha_D \rangle = 1.27 \pm 0.05$, respectively. Our result is consistent with [1] finding.

| Median redshift ($z$) | PDE ($\alpha_D$) | PLE ($\alpha_L$) |
|---|---|---|
| 0.32 | $1.15 \pm 0.22$ | $2.17 \pm 0.44$ |
| 0.59 | $1.39 \pm 0.05$ | $2.56 \pm 0.15$ |
| 0.88 | $1.73 \pm 0.04$ | $3.09 \pm 0.12$ |
| 1.10 | $0.94 \pm 0.03$ | $1.58 \pm 0.07$ |

**Table 2.** The value of best-fit LF evolution parameters for each redshift bin.

## 5. Summary and Conclusion

We derived the 1.4 GHz radio luminosity function of 6900 radio AGN at $0.1 \leq z \leq 1.3$ in this work. The radio data were taken from deep VLA observation of Stripe 82 field while the optical/near-infrared data were taken from Dark Energy Survey DR1. By using $1/V_{\rm max}$ method, we calculated the luminosity function, then put stringent constraints on the the evolution of the population via PLE or PDE model. At median z = {0.31, 0.59, 0.88, 1.10}, we found the power-law index of PDE is $\alpha_D = \{1.15, 1.39, 1.73, 0.94\}$ while for PLE is $\alpha_L = \{2.17, 2.56, 3.09, 1.58\}$. Our result is consistent with previous studies and gives better constraint to radio AGN luminosity and density evolution power-law indexes due to larger number of sources (6900) and wider covered sky fraction (92 deg$^2$). Furthermore, we found that high-luminosity radio AGN follow the double power-law model, similar to the the low-luminosity AGN, indicating that there isn't any difference of evolution between high- and low-luminosity AGN population.

## References

[1] Smolčič, V., Novak, M., Delvecchio, I. et al, A&A 602 (2017), A6
[2] Hodge, J. A., Becker, R. H., White, R. L., Richards, G. T., Zeimann, G. R. 2011, AJ, 142, 1
[3] Abbott, T., Abdalla, F., Allam, S. et al. 2018, arXiv e-prints, arXiv:1801.03181
[4] Abolfathi, B., et al. 2018, ApJS, 235, 42
[5] Davis, M., et al. 2003, SPIE Proc., 4834, 161
[6] Garilli, B., et al. 2014, A&A, 562, 23
[7] Le Févre, O., et al. 2013, A&A, 559, A14
[8] Liske, J., et al. 2015, MNRAS, 452, 2087
[9] Drinkwater, M. J., et al. 2010, MNRAS, 401, 1429
[10] Bradshaw, E. J., et al. 2013, MNRAS, 433, 194
[11] McLure, R. J., et al. 2013, MNRAS, 428, 1088
[12] Sadeh, I., Abdalla, F. B., Lahav, O. et al, PASP 128 (2016), 968: 104502
[13] Collister A. A., Lahav O., 2004, PASP, 116, 345
[14] Hoecker A. et al., 2007, ArXiv Physics e-prints
[15] Oyaizu H., Lima M., Cunha C. E., Lin H., Frieman J., 2008, ApJ, 689, 709
[16] Lima, M., Cunha, C. E., Oyaizu, H., Frieman, J., Lin, H., & Sheldon, E. S. 2008, MNRAS, 390, 118
[17] Novak, M., Smolčič, v., Delhaize, J. et al. 2017, A&A 602, A5
[18] Schmidt, M. 1968, ApJ, 151, 393
[19] Marshall, H. L. 1985, ApJ, 299, 109
[20] Mauch, T., & Sadler, E. M. 2007, MNRAS, 375, 931
[21] Newman, J. A., et al. 2013, ApJS, 208, 5

# Scientific Requirement of Radio Astronomy Observation in the Philippines

**R Masong**[1], **and R Pobre**[1,]

[1]De La Salle University, Manila; College of Science, Physics Department Optics/Photonics Technology, Instrumentation and Conservation Science (OPTICS) Research Unit Manila, National Capital Region, Philippines

Email: rizchel_masong@dlsu.edu.ph

**Abstract.** Planning to perform radio astronomy related observation necessarily requires a specific device known as radio telescope. These specially designed instruments observe the longest wavelengths of light, ranging from 1 millimeter to over 10 meters long. It offers various avenue of observation such as continuum observation, solar wind observation, up to spectral line observation. In archipelagic countries like Philippines, opportunities for radio astronomy observations can offer wide variety of location. That features the regions of high frequency and low frequency were identified. The researcher was able specify the characteristics of each regions as suitable site for radio observations including the types of observation to be made.

***Keywords:*** *Philippines, Radio astronomy, and Radio telescope*

## 1. Introduction

When we started to identify the selection of the most suitable site in which where we can conduct radio astronomy observation in the Philippines. One of its primary goals is to select the best location where a radio telescope can be constructed including the bandwidth that can be observed in the locality chosen. The construction of the radio telescope can be used to improve the radio frequency coverage of different networks within the Southeast Asia. Radio frequencies within the bandwidth of 73 – 74.6 MHz, 80.5 – 82.5 MHz, and 150.05 – 153 MHz were identified that is specified for both solar and continuum observation [1]. Considerations must be clear before performing land based astronomy observations. Bandwidth identification plays a significant role; this will specify the target radio observation and will make the determined frequency set the standard and desired outcome. Also note that archipelagic, and topographic characteristics of Philippines in which can provide a wide variety of observational outcome for both high and low frequency. For high frequency observation, the site should have low water vapor contents and lower dust density; for a low frequency observation should be distant from man-made radio noise signal [2]. This study offers a brief, precise, and cost time effective way for setting the scientific requirement needed for a mentioned bandwidth for radio astronomy observation. It also gives the practically of observations for each region. For validation, measuring the RFI dimensions needed to do in order to see the classify background of interference this will verify the analysis that is provided in this study. As assumption, the region will be described depending on the criterion specified.

## 2. Methodology

### *2.1 Criterion*

One of the key goals is to have a better field of view in the radio frequency spectrum within the various parts of the Philippine archipelago. Two general factors ponder in order to provide added information about the electromagnetic condition in the radio frequency spectrum. These are the geographical and anthropogenic factors [3].

### *2.2 Population density*

The radio astronomy spectrum is increasingly polluted by intentional and unintentional human generated Radio Frequency Interference (RFI). This RFI is primarily produced by urban activity in dense residential and commercial areas. Contrarily, on rural areas one need to take into account the absorption, reflection, and scattering of radio energy around objects such as trees and other plants, and their impact in the various directions near the receiving antenna of telecommunication link. It is important to quantify the effects of these RFIs on the radio astronomy spectrum [4].

### *2.3 Climate Type*

Climate type is one of important variable to radio propagation and must be considered. It will affect penetration of radio wave propagation where the water occupied in atmosphere will reflect or refract attenuate as well as the signal penetration in the atmosphere. We must consider a dry place in Philippines [5].

### *2.4 Altitude*

Radio astronomy uses specified radio spectrum to recognize different emissions from sources outside the earth and the others signal known as Radio Frequency Interference (RFI). It is an essential issue to address in order to characterize a type radio astronomy observation. Different altitude of observation for example will have an effect on RFI level or noise environment to radio observation.

### *2.5 Atmospheric Conditions*

The atmosphere is transparent or partially transparent within the radio wave range from about 15 m – 1 mm, or 200 m – 1 mm in terms of its wavelength. The density of molecules in the atmosphere could absorb, scatter, reflect, and penetrate radio signals that will lead to the reduction of signal's amplitude. This is through the analysis of amount of rainfall.

### *2.6 Topography*

In locating an observatory and avoid spectrum pollution most of radio observatories are often placed in valleys to further shield them from ground interference. So that altitude is a one factor in the site selection of radio astronomy observatory. Isolated land form can also be used for radio astronomy observatory.

### *2.7 Site Selection*

Luzon Island has an area of 109,964.9 $km^2$. Regions in Luzon were primarily selected in order to determine what suitable radio observation can be performed within the northern part of the Philippine archipelago over each region with features different feasible characteristics [6].

## 3. Result and Disscusion

Each parameters specified on Table 1 are important factors in conducting a radio astronomy observation on or before determining the frequency bandwidth. These are the altitude, population density, climate type, atmospheric conditions, and topography. Stated parameters are vital in order to associate the radio signal that can be detected in a particular area of interest. This will unlock the radio windows to be observed.

**Table 1.** Regions in Luzon and its Characteristics

| Region | Altitude (Highest point) | Population Density (persons per $km^2$) | Climate Type | Atmospheric Conditions (amount of rainfall per year in mm) | Topography |
|---|---|---|---|---|---|
| Ilocos Region | 1,154 m | 388 | I | 2000 - 2750 | At east Composed of mountain ranges, at west composed of seas |
| Cagayan Valley | 1,850 m | 116 | IV | 1700 - 3000 | At east seas and oceans, and at west composed of mountain ranges |
| Central Luzon | 1,252 m | 512 | I, III | 1800 - 3800 | Marked by an extensive central plain surrounded by three mountain ranges |
| CALABARZON | 2,328 m | 870 | I, II | 1550 - 3500 | Marked by scattered mountains, hills, plains, and mixtures of valleys |
| MIMAROPA | 2,133.6 m | 100 | II, IV | 1550 - 3500 | Mostly composed of islands and islets |
| Bicol | 2,462 m | 320 | II, III, IV | 1450 - 3750 | Marked by a peninsula and surrounded by seas |
| CAR | 2,922 m | 87 | I | 2500 - 3600 | Composed of mountain ranges |
| NCR | 127 m | 20,875 | III | 1800 - 3800 | Marked with coastal margin, plateau, and valley |

**Table 2.** Essential RFI Characteristics for each region

| Region | Essential Characteristics | |
|---|---|---|
| | Advantage | Disadvantage |
| Ilocos Region | Advisable for low and high frequency observation | Observations can only be made from November to April |
| Cagayan Valley | Advisable for low and high frequency observation | Inaccessibility of possible locations for observations |
| Central Luzon | Not advisable for radio observation | |
| CALABARZON | Advisable for low frequency observation | The region has the most number of total rainfall |
| MIMAROPA | Advisable for Low frequency observation | Inaccessibility of possible locations for observations |
| Bicol | Not advisable to conduct radio observation | |
| CAR | Advisable for low frequency observation | Inaccessibility of possible locations for observations |
| NCR | Not advisable to conduct radio observation | |

Advantages and disadvantages as specified in Table 2 describe the essential RFI characteristics for each regions. The primary classifications of the characteristics for advantages are specified as (1) low population density; (2) Climate type which is IV; (3) Altitude, and topography, which naturally isolates manmade radio noise; (4) atmospheric conditions that requires low amount of rainfall per year with the disadvantages. The classification of low frequency and high frequency observation. Regions suitable for low frequency observations are Ilocos, Cagayan, CALABARZON MIMAROPA, and CAR. On the other hand, for high frequency observation, Cagayan is suitable by having a low water vapor contents.

## 4. Conclusion

The parameters has been chosed and comparing it with other region. Majority of regions in Luzon were identified as a suitable location to conduct radio astronomy observation. The advantages and disadvantages showed the essential RFI characteristics of each region for suitable radio astronomy observation. This determines that Luzon is appropriate for low frequency and high frequency observation. This is also vital in determining the best place to construct radio telescope.

**Acknowledgments**

My deepest gratitude comes first to Dr. Romeric F. Pobre, my adviser for graduate school thesis who provided his upmost support for me to pursue my research interest and enabled me to finish this one. To Dr. Rogel Mari Sese and Dr. Reinabelle Reyes, they were the persons who inspired me to push astronomy research career and this paper together with other aspect of consultancy in both research and contextualization. My appreciation also extends to my laboratory colleagues, Francis, Ric Johh, Froilan, Paul, and the rest of the OPTICS research group who suggested some ideas and methods on how I should tackle each part of my research. To my friends, who provided enough criticism in improving my paper Sir Alden, Marvin, Wilson, John, Marlon. I would also like to thank my institution De La Salle University for funding my travel in this presentation including the support of DLSU-OPTICS, and DOST-SEI as well for my scholarship! Appreciation comes with financial and administrative support. I really love these institutions. Above ground, I am indebted to my family, whose value to me only grows with age. And finally, I acknowledge my partner, Juri, who stayed with me and provided enough sunshine when I'm alone staying in the dark – who serves coffee during my overnight research writing.

**References**

[1] European Science Foundation; Committee on Radio Astronomy Frequencies (n.d.), retrieved from https://www.britastro.org/radio/spectrum.html.
[2] Cheng, J., 2009. *New York, NY, USA*: Springer.
[3] Umar, R., 2014. *Doctoral dissertation*, University of Malaya.
[4] Philippine Statistics Association Population Density 2015, from https://psa.gov.ph/content/philippine-population-density-based-2015- census-population.
[5] Philippine Atmospheric Geophysical Astronomical Services Association, from https://pubfiles.pagasa.dos.gov.ph/climps/climate forum/climateoutlook.pdf
[6] NAMRIA Topographic Map, (2003) from DENR-NAMRIA

IOP Conf. Series: Journal of Physics: Conf. Series **1231** (2019) 012019 doi:10.1088/1742-6596/1231/1/012019

# The Measurement of Solar Differential Rotation from Proper Motion of Individual Sunspots

**K Permata**[1,*] **and D Herdiwijaya**[1]
[1]Astronomy Department, Bandung Institute of Technology, Ganesha 10 Bandung, Indonesia

Email: kamilapermata13@gmail.com

**Abstract** The differential rotation is the result of the interaction between rotation and convection and causes dynamo circulation that affect the cycle of solar activity. Tracer method using features in the photosphere such as sunspot is a simple method to measure the differential rotation. In this study, 98 individual sunspots on January 8 - 22, 2013 and August 25 - September 7, 2013 were used to measure the differential rotation of the Sun. The Sun's continuum images were obtained from HMI (Helioseismic Magnetic Imager) instrument at SDO (Solar Dynamic Observatory). Coordinate and area of sunspot were measured using ImageJ software and converted to Carrington coordinates. From the measurement, we derived the differential rotation equation, the relation of velocity and area of sunspot, and the relation of sunspot's velocity and Zurich classification. The differential rotation equation obtained in this study is $\omega$ (B) = (14.376 ± 0.04) + (0.6 ± 0.38) sin2B (°/day). For the relation between velocity and area of sunspot, we got the sidereal rotation rate of sunspots with area < 100 pixels$^2$ is about 0.8% higher than sunspots with area > 100 pixels$^2$.

***Keywords:*** *Solar Differential Rotation and Individual Sunspots*

## 1. Introduction

In 1630, Christoph Scheiner discovered that sunspots on the equator had shorter rotation periods and moved faster than sunspots at higher latitudes[1]. Scheiner has many observational data and he derived differential rotation equations in the function of latitude. However, the idea of differential rotation has not been received and appreciated at that time. In 1860, Richard Christopher Carrington observed the Sun and determined the rotation period of the Sun based on latitude. Sunspots around the equator have 25-day rotation period, while sunspots at 45° latitude have 28-day rotation period.

Differential rotation shows that the Sun is not a solid body that has different rotational rate of each latitude. Differential rotation is the result of the interaction between rotation, magnetic fields, and convection flow that causes dynamo circulation that affect the cycle of solar activity. The factors in the interior in the Sun that causes differential rotation are still a big question. Knowledge of the activity in Sun's interior is still limited. Therefore, rotation studies are important to get more information about the interior of the Sun.

There are several methods of measuring differential rotation, the tracer method and spectroscopic method. In the tracer method, photosphere features such as sunspots and magnetic fields are used to determine the differential rotation. In spectroscopic method, spectral measurements and the Doppler equation is used to obtain differential rotation values [2]. This method is very precise but very sensitive because it has a small wavelength order. A slight error in positioning will cause a large error in getting the velocity value. In this research, we used tracer method with sunspot as a tracer. Sunspots have advantages for easier to identify due to their high contrast intensity to the surroundings, and have long life time. Differential rotation equations are expressed in the following equation [2].

$$\omega\,(B) = a - b\,\sin 2B - c\,\sin 4B \tag{1}$$

For the calculation of the mean longitudinal drift based on the sunspot tracer, the third term is small, so the equation that used is $\omega\,(B) = a - b\,\sin 2B$ where $\omega\,(B)$ is the sidereal rotation rate (°/day), a and b are the constant, and B is the average latitude. For the calculation of latitudinal drift, the equation is

$$vm\,(B) = c\,\sin 2B \tag{2}$$

where vm is the latitudinal drift (°/day), c is the constant, and B is the average latitude.

The Carrington heliographic coordinates are coordinate systems to determine the position of features such as sunspot on the surface of the Sun. To find heliographic longitude and latitude, the equation that used is [3]

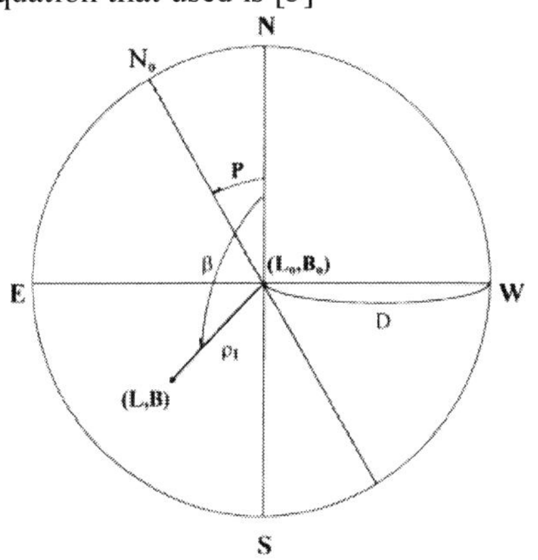


**Figure 1.** Sketch of the solar disk to Illustrate the heliographic coordinates [3]

$$\sin B = \sin Bo\,\cos\rho + \cos Bo\,\sin\rho\,\cos(P-\beta) \tag{3}$$
$$\sin(L-Lo) = \sin\rho\sin(P-\beta)\,\sec B \tag{4}$$
$$\sin(\rho + \rho_1) = \frac{\rho_1}{D} \tag{5}$$
$$\rho_1 = \sqrt{x^2 + y^2} \tag{6}$$

where (L, B) is the heliographic longitude and latitude, (Lo, Bo) is the heliographic coordinate at the centre of the disk, $\rho$ is the heliocentric angular distance, $\rho_1$ is the distance of the radial angle, $\beta$ is the angle of position measured from the north pole of the Sun (N) to eastward, D is the angular semi-diameter of the Sun, and P is the position angle of the Sun's north pole (N) relative to the direction of the celestial pole (No).

## 2. Objectives

The objectives of this research is measuring the proper motion of individual sunspots to determine the differential rotational of the Sun using continuum images data from the HMI instrument (Helioseismic and Magnetic Imager) onboard the SDO (Solar Dynamic Observatory) satellite. We determine the relation of proper motion's velocity of individual sunspot with area of the sunspot, and determine the relation of the proper motion's velocity of individual sunspot with the Zurich classification.

## 3. Method

The continuum image used in SDO is the image of the Sun on the continuum wavelength which has been corrected with the limb darkening. Image resolution is 4096x4096 pixels (resolution 0.5 arcsecond degree) with image size 4 – 5 MB. The data are downloaded on the https://sdo.gsfc.nasa.gov. We used ImageJ (https://imagej.nih.gov/ij/) to measure the area and coordinates of sunspot. ImageJ is a Java-based software that able to process and analyse images. ImageJ can read various image formats, measure distance, angle, area, pixel value. The HelioViewer software is used to find the values of Bo, Lo, P, and semi-diameter of the Sun. These parameters are needed to convert the coordinate in pixel (x, y) obtained from ImageJ to Carrington's heliographic coordinates in degrees.

The data in this research are sunspots on August 25 to September 7, 2013 and January 8 - 22, 2013. On those dates, the Sun's activity is high so there are many sunspots from various groups based on Zurich classification. From the data on August 25 to September 7, 2013, 27 individual sunspots from 7 groups of sunspots were observed and from the data on January 8 - 22, 2013, 71 individual sunspots of 10 sunspots groups were observed. For each date, 5 images are measured at 00.30 UT, 01.30 UT, 02.30 UT, 03.30 UT, and 04.30 UT. Sunspots that we used for the measurements are sunspots that have life time more than two days. To avoid the appearance of asymmetrical sunspot on the limb of the Sun or the Wilson effect, measurements were taken for the group of sunspots with positions less than 70° from the meridian. We checked the sunspot group's position from https://solarmonitor.org.

## 4. Result and Discussion

### *4. 1 Differential Rotation Equations*

To find the equation of differential rotation, we used linear regression between the sidereal rotation rate denoted by ω (B) and average latitude for each individual sunspots denoted by sin2(B). The values of a and b from the linear regression entered into the equation ω (B) = a + b sin2 (B). The Carrington velocity (Vc) is 14.184 °/day[3]. The value of the sidereal rotation rate is obtained from the following formula.

$$\omega\,(B) = \frac{(14.184°/\text{day} \times T) + (dL/dt)}{T} \tag{7}$$

with T is the duration or age of individual sunspot and longitudinal drift (dL/dt) is the longitude velocity of each individual sunspot.The differential rotation equation from 98 sunspots is

$$\omega\,(B) = (14.376 \pm 0.04) + (0.6 \pm 0.38)\sin 2B\ (°/\text{day}). \tag{8}$$

However, the constant b is positive. This is caused by small sample of individual sunspots. If the number of samples is multiplied, the constant b will have a negative value, according to the differential rotation phenomenon that the rotation rate will be slower in the higher latitude. In addition, the majority of sunspots measured in this study have small area. The existence of small sunspots with high velocity and short life periods causes the value of differential rotation equations to still be seen uniformly at various latitudes.

**Table 1.** The measurement of solar differential rotation that used sunspot as tracer

| Ref. | a (μrad/s) | b (μrad/s) | Period | Note |
|---|---|---|---|---|
| [4] | 2.905 | -0.598 | 1934-1944 | 136 sunspots with lifetime > 1 solar rotation |
| [5] | 2.939 ±0.001 | -0.5796 ±0.01 | 1921-1982 | White-light images of sunspot with area < 5 (mHem) |
| [5] | 2.917 ± 0.002 | -0.5284 ± 0.018 | 1921-1982 | White-light images of sunspot with 5 < area < 10 (mHem) |
| [5] | 2.885 ±0.0036 | -0.5325 ± 0.034 | 1921-1982 | White-light images of sunspot with area > 10 (mHem) |
| This study (2018) | 2.9118 ±0.008 | 0.12 ± 0.07 | 2013 | 98 individual sunspots |

In table 1, there are differential rotation measurements using different sunspots. The constant a obtained in these studies is near with the references (2.885 to 2.939) μrad/s. For the constant b, the value is far from the references (-0.436 to -0.732) μrad/s. The difference in the value of b in this study with references was caused by a small sample of sunspots, compared to other 96,283 sunspots [5], 276 sunspots [3], and 136 spots [4]. Actually, the number of sunspots used by the latter [4] was not very different from the number of sunspots used in this study. However, they used selected sunspots that large, almost symmetrical, and have lifetime more than one solar rotation. The majority of the sunspots measured in this study are less than 15° in latitude, so they are less representative for higher latitudes. For better differential rotation equation results, more sample sunspots in various latitudes are needed. It is also estimated that an error occurred because many individual sunspots have very small area. The movement of small sunspots are fluctuating, shrinking, enlarging, appearing, and disappearing in a short time. This causes the measured centre coordinates to be different and not constant.

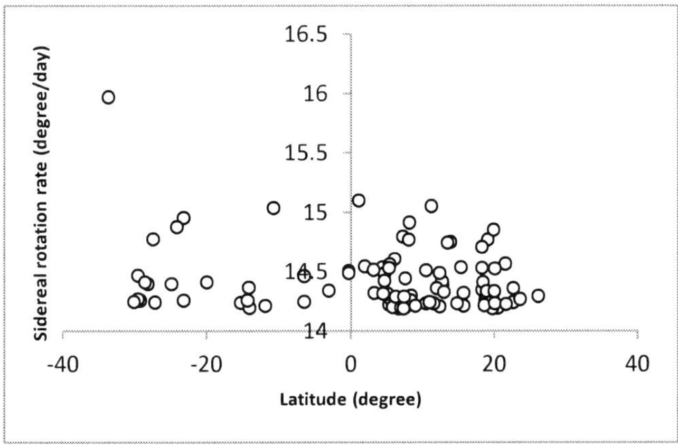


**Figure 2.** Sidereal rotation rate for 98 sunspots

Figure 2 shows the sidereal rotation rate for 98 individual sunspots. There is one sunspot at high latitude that has high velocity 15.38°/day. This high velocity may be due to the small size of the sunspot at 58.7 pixels$^2$. This sunspot showed a short duration and evolved from one small sunspot and enlarged. The shifting of the centre point of a sunspot from one small sunspot to the larger one cause a difference in the position, so the value of sidereal rotation rate is high.

Sunspots around the equator (<5 °) have average velocity 14.449°/day. Sunspots at moderate latitudes (10°-20°) have average velocity 14.392°/day, lower than the velocity at the equator. By removing a sunspot that has highest velocity than other sunspots, the velocity at high latitudes (> 25°) is 14.337°/day. Velocity at high latitudes is lower than velocity at the equator and lower latitude.

*4.2 Latitudinal Drift*

To find the latitudinal drift, we used linear regression between latitude and sin (2B). The latitudinal drift equation obtained is as follows.

$$vm\ (B) = (-0.72 \pm 0.32)\ \sin\ (2B)\ (°/\text{day}) \qquad (9)$$

The latitudinal drift equation obtained with 276 individual sunspots as follows [3].

$$vm\ (B) = (-0.05 \pm 0.03)\ \sin\ (2B)\ (°/\text{day}) \qquad (10)$$

The small area of sunspots in this study led to a greater latitudinal drift value because small sunspots have high velocity. There is also a difference in latitudinal drift values caused by the number of different sunspots. More sunspots lead to the smaller errors. Possibility of misidentifying small sunspots also causes a large error in the latitudinal drift equation obtained.

*4.3 Relation between Velocity and Area of Sunspots*

Figure 3 shows relation between longitudinal drift and latitudinal drift with area of sunspot. The biggest sunspot on this measurement has average area 1061.7 pixels$^2$ and the average area of smallest sunspot is 3.5 pixels$^2$. Using linear regression, the relation between longitudinal drift and area of sun spots is

$$\omega = -1.9 \times 10^{-4} A \pm 0.46\ (°/day) \tag{11}$$

with ω (B) is the longitudinal drift and A is the area of individual sunspot. The relation between latitudinal drift with area of sunspot obtained from linear regression is

$$vm = -\ 0.001\ A \pm 1.74\ (°/day) \tag{12}$$

with vm is latitudinal drift and A is the area of individual sunspot.

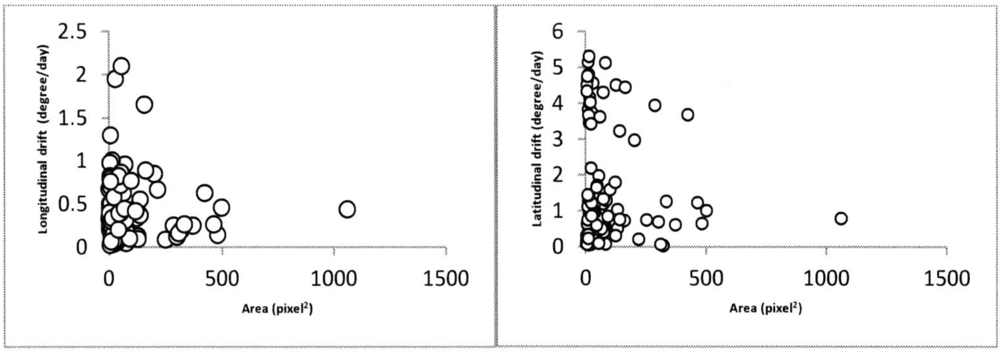


**Figure 3.** Relation between longitudinal drift and latitudinal drift with area of sunspots

These two equations show that smaller sunspots have a greater velocity. Small sunspots with area less than 100 pixels² has average sidereal rotation rate 14.451 °/day, while large sunspots with area more than 100 pixels² has average sidereal rotation rate 14.333 °/day. Sunspots with area less than 5 mHem rotate about 0.8% faster than large sunspots with area 15 mHem [2]. In this study, it was found that sunspots with area <5 mHem was 0.9% faster than sunspots with area >5 mHem. Sunspots with area <5 mHem rotate faster around 1.5% than sunspots with area > 15 mHem. This value is almost two times greater than the previous results [2]. However, there are only 7 sunspots with area > 15 mHem in this study, less representative of the velocity of sunspots which had a very large size.

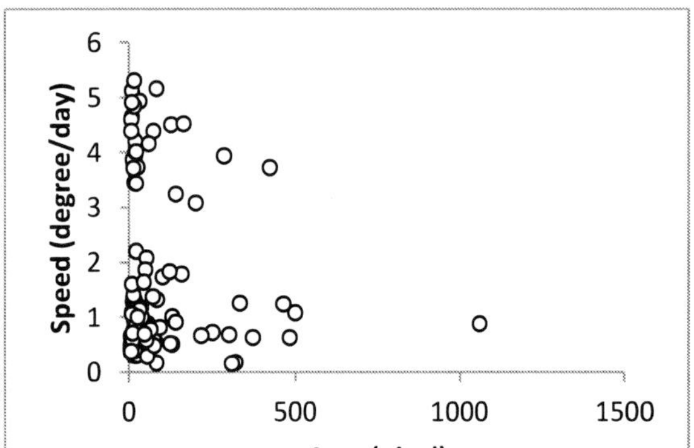


**Figure 4.** Relation between speed and area of sunspots.

For the value of speed, with the formula $v = \sqrt{\omega^2 + vm^2}$ (13)

We get the speed value as follows.

$v = -0.001A \pm 1.8$ (°/day)

Small sunspot area that has a weaker magnetic field tends to move faster than large spots that has greater magnetic field, which show similar results from the above relation. The difference in velocity is caused by the different depths of the magnetic flux. Large sunspots may also have greater shift in viscosity [2, 7].

*4.4 Relation between Velocity and Zurich Classification*

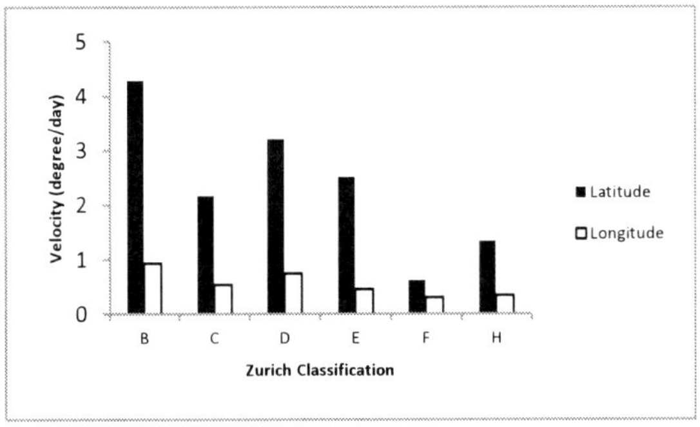


The velocity of longitude and latitude obtained is the result of the average velocity of individual sunspots in each class. Velocity for high classes (F and H) is smaller than the velocity of sunspots in the earlier class. Class H is a class in the sunspot group that is at the end of its life and will disappear on the west. The sunspot group also has an old age, so the movement is slow.

**Figure 5.** Relation between velocity and Zurich classification

Class F is the largest group compared to other classes. Class F also has a complex magnetic structure and large magnetic field strength, so the movement of sunspots is slower. Early classes like class B have high velocity because the area of the sunspots is small, and the magnetic field is weaker. The sunspot group in class B also has a young age so the velocity is high. The high latitude velocity is caused by the large number of small sunspots with high velocity measured in this study.

## 5. Conclusions

We have measured 98 individual sunspots on August 25 - September 7, 2013 and January 8 – 22, 2013 to find differential rotation equations, relation between velocity and area of sunspots, and relation between velocity and Zurich classification. For the differential rotation equation, we get $\omega$ (B) = (14.376 ± 0.04) + (0.6 ± 0.38) sin2B °/day. The constant a of (14.376 ± 0.04) has a value that near to the references. However, the constant b of (0.6 ± 0.38) has a different value with the references and shows the uniformity of velocity in all measured sunspots. In fact, the sunspots at higher latitudes should have lower velocity. This is caused by less number of sunspots measured. In addition, 93 sunspots are small (<15 mHem) and have high velocity and short life periods so at any latitude, the differential rotation equation shows almost the same value. In this study, the majority of the sunspots measured were at latitude less than 15° so there was less diversity of sunspots to show differential rotation. By using more diverse sunspot's position and greater number of sunspots, the value of the differential rotation equation obtained will be much better. Sunspots with area less than 100 pixels$^2$ have average sidereal rotation rate 14.451°/day, while large sunspots with the size of more than 100 pixels$^2$ have sidereal rotation rate 14.333°/day. Sidereal rotation rate of sunspots with area < 100 pixels$^2$ is about 0.8% higher than sunspots with area > 100 pixels$^2$. Small sunspots have a weaker magnetic field and move faster than large sunspots that have greater magnetic field strength and move more slowly. Sunspots in high classes (F and H) have low velocity, sunspots in class B and D have high velocity, sunspots in class C and E have medium velocity. Sunspots in class F have the lowest velocity because the magnetic field structure is the most complex and the size is the largest. Class B have the highest velocity because they are young, have the smallest size, and the magnetic field structure is still not complex. However, there is only 1 sunspot in class B, while class F has 47 sunspots. More number of sunspots are needed on this measurement to get a better relation between velocity of sunspot and Zurich classification.

## References

[1] Paterno L 2009 *Astrophysics and Space Science* **328** 269
[2] Howard R 1984 *Annual Review of Astronomy and Astrophysics* **22** 131
[3] Herdiwijaya D, Makita M, and Anwar B 1997 *Publ. Astronomical Society of Japan* **49** 235
[4] Newton H W and Nunn M L 1951 *Monthly of The Royal Astronomical Society* **111** 413
[5] Howard R, Gilman P A, and Gilman P I 1984. *Astrophysics Journal* **283** 373
[6] Bhatnagar A and Livingston W 2005 *Fundamentals of Solar Astronomy* (Singapura: World Scientific Publishing Co. Pte. Ltd.)
[7] Pulkkinen P and Tuominen I 1998 *Astronomy and Astrophysics* **332** 748

# Photospheric Vector Magnetic Field Parameters as A Predictor of Major Solar Flares

**M Z Nurzaman [1,*] and T Dani [1]**

[1]Space Science Centre, National Institute of Aeronautics and Space (LAPAN), Bandung, West Java, Indonesia 40173

Email: mzn5412@gmail.com

**Abstract.** Photospheric vector magnetic field data which have several Space-weather HMI Active Region Patches (SHARP) parameters are used to study active regions that produced major solar flares. SHARP parameter data obtained from the Helioseismic Magnetic Imager (HMI) instruments onboard Solar Dynamics Observatory (SDO) have a good spatial and temporal sampling. We consider three SHARP parameters with high F-scores, namely total unsigned vertical current, total photospheric magnetic free energy, and total unsigned current helicity as a useful predictor for major solar flares. In this paper, we present the data analysis procedure and sample results focusing only on major solar flares (M and X class flares). The preliminary results showed in some cases, one of SHARP parameters for the M-class flare almost has the same value as X-class flare in which ideally should be higher.

***Keyword:*** *Photospheric Vector Magnetic Field and Major Solar Flares*

## 1. Introduction

Solar flares are common and powerful phenomenon in the sun which can release huge energy (Kopp et al. 2015). But there is no single model or method yet that can predict correctly when flare will erupt. Many studies of prediction of solar flare have been done by physicist. These studies exploit all possible physical process and all relevant data using various method of simulations and data analysis. Yokoyama and Shibata (2001) studied flare through magnetohydrodynamics simulations. Others more concern on analyse observational data from photospheric level to coronal level. For instances, Leka & Barnes (2003) analyse photospheric magnetic field data using discriminant analysis, Falconer (2009) studied magnetogram dynamic of active region to make a prediction of flare and CME. Muhamad (2017) combine simulation and observational data to find parameters that can be used for a prediction of flare.

Many people believed that the enormous energy from solar flare is generated by magnetic field mostly in the active region. Thus, studying the nature of magnetic field on active region is critical to understand and ultimately predicting solar flares. The physical parameters that can be used to study the magnetic field is called non-potential parameters which is described as the deviation of a potential (current-free) field to a non-potential state, i.e. becomes more sheared and/or twisted, the energy will be stored in the stressed magnetic field and finally it will be released as flares, eruptive filaments and/or coronal mass ejections (CMEs) (Gary et al. 1987; Sakurai 1989; Schrijver et al. 2008; Sun et al. 2012). These parameters also well known as photospehric vector magnetic field parameters and widely used to study the solar flare. For examples, Leka & Barnes (2003) became the first to classify photospheric magnetic field properties of flaring versus flare-quiet active regions. Mason & Hoeksema (2010) followed their work but employed different method. Bobra & Couvidat (2015) applied non-linear algorithm classifier, called support vector machine, to determine the features that useful for discriminating between flaring and non-flaring active regions from vector magnetic field data.

These parameters can be obtained through observing magnetic activity in the sun surface using magnetogram. Since 2010, there is an instrument called Helioseismic Magnetic Imager (HMI) onboard Solar Dynamic Observatory which can detect magnetic activity every 12 minutes. This high cadence data is great dataset to make a good result to understand the nature of magnetic field before the flare empirically. From the previous research, we only know which parameters that can be used to predict solar flare but we still do not know yet its value. And in this work, we try to extract the value of non-potential parameters before the occurrence of the flare. These values are can be treated as the threshold energy of active region before its produce energy. If an active region passes these values, it became a warning that in the near time the flare will erupt.

## 2. Data and Methodology

There are two dataset that have been used in this research, i.e. peak time of flares and magnetic features of active region. For data flares, we used flares with peak magnitude above the M1.0 which correspond to $10^{-5}$ W $m^{-2}$ or only major flares based on data from *Geostationary Operational Environment Satellite* (GOES) X-ray flux. We do not include C-class flares because its impact on earth environment not to significant. Active region 's magnetic features can be described by non-potential parameters that represented by SHARPs (Space-weather HMI Active Region Patches, Bobra 2015) parameters. These parameters are the derivation product of HMI data which automatically identifies active region patches in the vector magnetic field data. From the work of Leka & Barnes (2003) and Bobra (2015), we know several non-potential parameters to identify the behavior of magnetic field on the active region, and we only choose three parameters with highest F-score (a statistical value to determine the best value) on predicting flare i.e. total photospheric magnetic free energy, total unsigned current helicity, and total unsigned vertical current. The SHARP data can be downloaded from Joint Science Operation Center (JSOC) that operated by Stanford University. Table 1 is the mathematical expression of these parameters. We use the data from May 2010 until November 2016 and test the result using data on 2017. We only select SHARP parameters at 3, 6, 9, 12, and 24 hours prior to the peak time of major flares. The diagram below shows the step of our work.

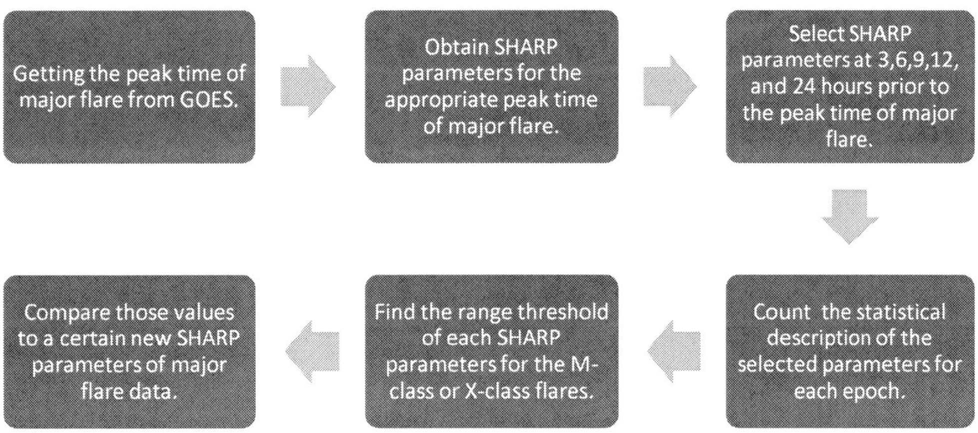


Figure 1. The detail steps to analyze the data.

Table 1. SHARP Active Region Parameter Formulae

| Keyword | Description | Formula |
|---|---|---|
| **TOTUSJH** | Total unsigned current helicity | $H_{ctotal} \propto \sum \lvert B_z \cdot J_z \rvert$ |
| **TOTPOT** | Total photospheric magnetic free energy density | $\rho_{tot} \propto \sum (B^{obs} - B^{Pot})^2 dA$ |
| **TOTUSJZ** | Total unsigned vertical current | $J_{ztotal} = \sum \lvert J_z \rvert \, dA$ |

## 3. Results

We divided the analysed data into two part, the data for M-class flare and X-class flare. So, we can get the threshold for each major flare. At figure 2 to figure 7, the single dot represented the mean data for each epoch (24, 12, 9, 6 and 3 hours) prior to the peak flare. As we can see, TOTPOT and TOTUSJZ show the similar pattern at all selected epoch. At the 24-hours before flare, the value has the lowest point around 9.4 x $10^{23}$ Ergs $cm^{-1}$ and 5.9 $10^{13}$ Amperes, respectively (figure 2 and figure 3). And reach the highest at 9 hours prior to the peak. The TOTUSJH parameter (figure 4) more fluctuate but still 24-hours before flare is the lowest value and 9-hours and 3-hours have the highest. This parameter span around 100 $G^2m^{-1}$ from the lowest to the highest with lowest value at 3415 $G^2m^{-1}$.

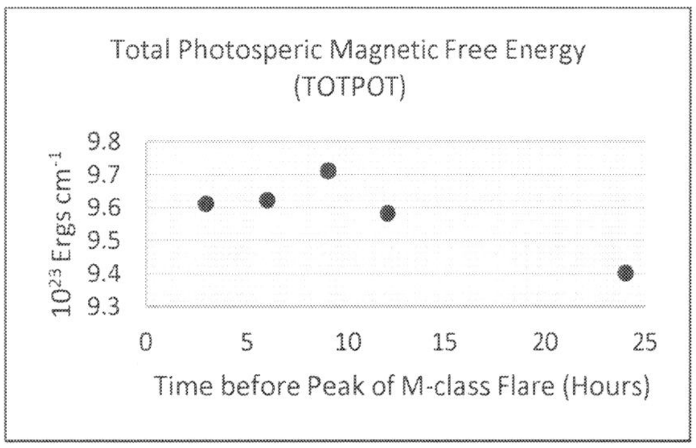


**Figure 2.** Total photospheric magnetic free energy for M-class flare until 24 hours before peak time.

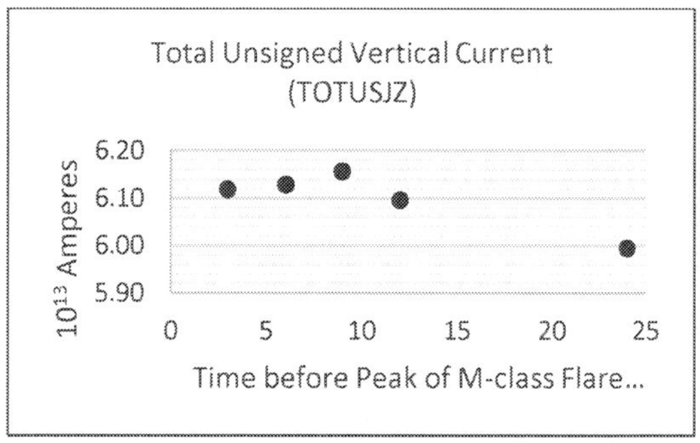


**Figure 3.** Total unsiged vertival current for M-class flase until 24 hours before peak time.

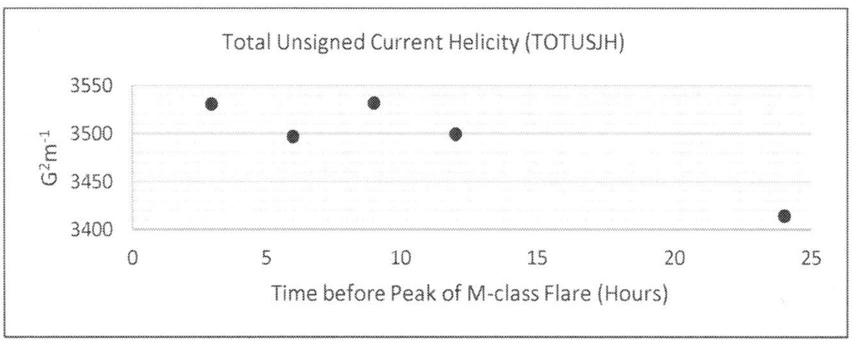


**Figure 4.** Total unsigned current helicity for M-class flare until 24 hours prior to peak time

As for non-potential parameters for X-class flare (figure 5 to figure 7), we see the same pattern for all those three physical parameters with the lowest value at 12-hours prior to peak time and the highest at 9-hours before peak time. TOTPOT and TOTUSJZ parameter (figure 5 and figure 6) have the exact pattern at all epoch and this pattern arouse suspicion. Maybe there is positive correlation between these parameters. The TOTPOT parameter has a value that ranged from 1.27 to 1.39 x $10^{24}$ Ergs $cm^{-1}$ and TOTUSJZ parameter has a value that ranged from 7.05 to 7.55 x $10^{13}$ Amperes. TOTUSJH parameter (figure 7) has different pattern compare to the two at 6-hours prior to peak time. The patterns closer to 3-hours while other closer to pattern at 9-hours before peak time.

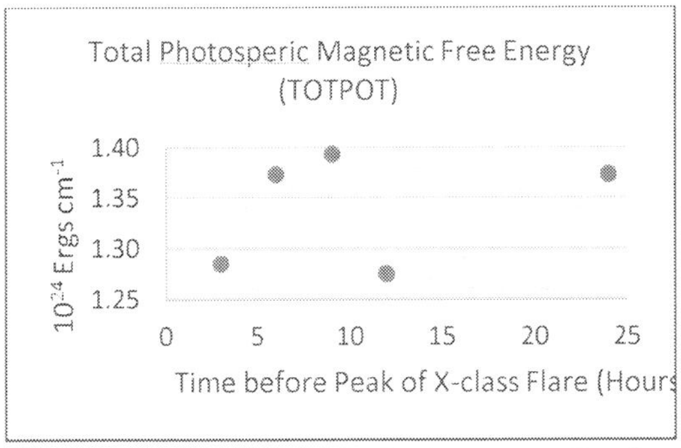


**Figure 5.** Total photospheric magnetic free energy for X-class flare until 24 hours before peak time.

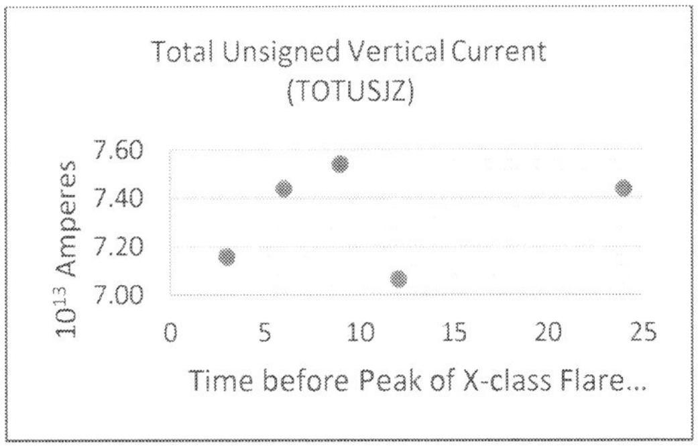


**Figure 6.** Total unsiged vertival current for X-class flase until 24 hours before peak time.

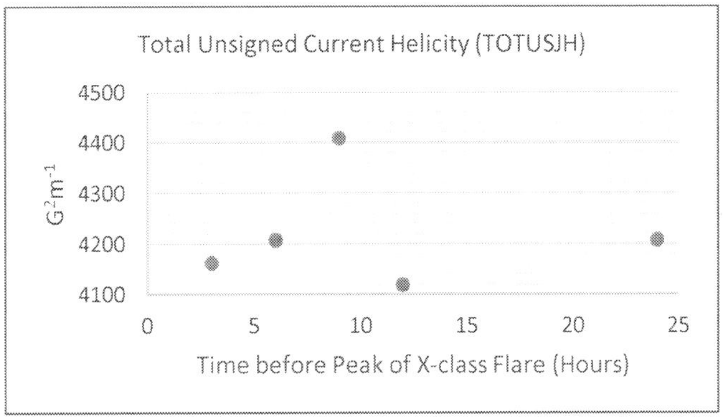


**Figure 7.** Total unsigned current helicity for X-class flare until 24 hours prior to peak time

The summary value of these parameters can be seen at table 2.

**Table 2.** The range value of non-potensial from 24-hours to 3-hours prior to peak time flare.

| Parameters | M-class flare | X-class flare |
|---|---|---|
| **TOTPOT** | 9.4 to 9.7 x $10^{23}$ Ergs cm-1 | 1.27 to 1.39 x $10^{24}$ Ergs $cm^{-1}$ |
| **TOTUSJZ** | 5.9 to 6 $10^{13}$ Amperes | 7.05 to 7.55 x $10^{13}$ Amperes |
| **TOTUSJH** | 3415 to 3530 $G^2m^{-1}$ | 4125 to 4410 $G^2m^{-1}$ |

Comparing these results to the recent flare that occurred at 2017 (figure 8), we can see that for M-class flare case, the recent flare value for all those parameters higher than our results at peak time. TOTUSJH and TOTUSJZ almost 1.5 times more than results. And for TOTPOT is slight high. For the case of X-class flare (figure 9), TOTPOT and TOTUSJH parameter show double value than our result. But TOTUSJZ has quite similar value as M-class flare case.

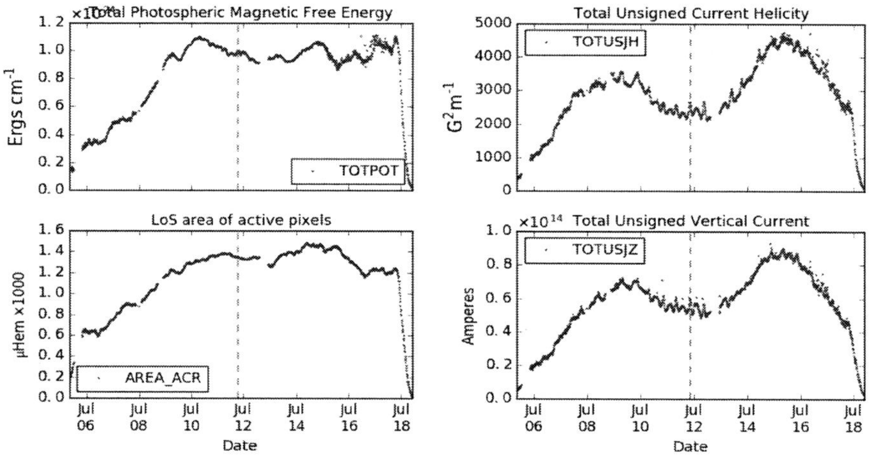


**Figure 8.** All three parameters for active region AR 12665 that appeared at mid-year of 2017. This active region produced M-class flare at July 15, 2017. AREA_ACR indicate the area of active region. Dash red line is the centre meridian of the sun which passed by this active region approximately at the end of July 11, 2017.

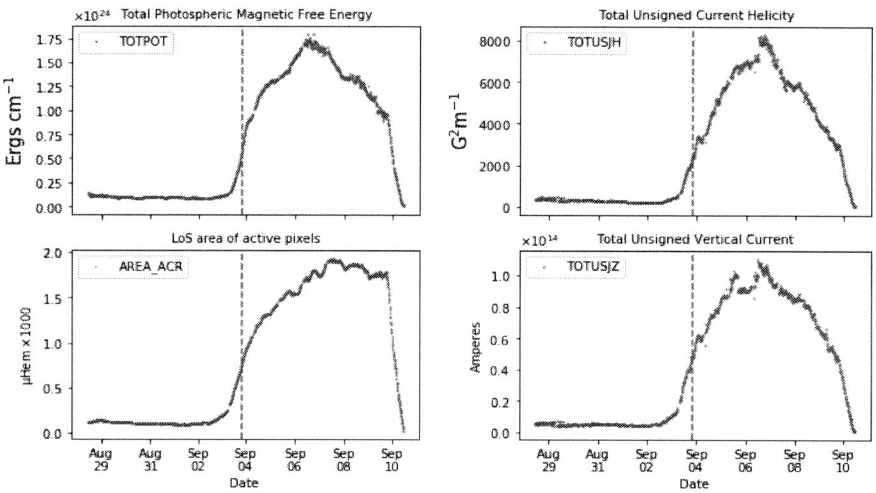


**Figure 9.** All three parameters for active region AR 12673 that appeared at September 2017. This active region produced X-class flare at September 08, 2017. AREA_ACR indicate the area of active region. Dash red line is the centre meridian of the sun which passed by this active region approximately at the end of September 04, 2017.

## 4. Conclusion

Solar flare is the common phenomenon in the sun surface which can release very huge energy but we cannot predict it yet, correctly. One of the effort to make a prediction of flare is through the analysis of magnetogram data which can be described by vector magnetic field parameters. Only the top three parameters which statistically have a higher chance for flare prediction have been used in this research namely total photospheric magnetic free energy, total unsigned current helicity, and total unsigned vertical current. The first two parameters have similar pattern at 24, 12, 9, 6, and 3 hours prior to M and X-class flare peak time. While the last one is quite different at 6-hours prior to the peak time. The ranged values of those parameters for X-class flare cases is higher than M-class flare. But if we compare it to the 'unseen' flare data which occurred in the last year, the total unsigned vertical current in quite same for X and M class flare.

**Acknowledgments**

Authors wishing to acknowledge the Space Weather Information and Forecast Services (SWIFtS) for the valuable discussion, and to Space Science Center LAPAN which provide the funding for this research.

**References**

[1] Bobra, M. G., and S. Couvidat. 2015. "Solar Flare Prediction Using Sdo /Hmi Vector Magnetic Field Data With a Machine-Learning Algorithm." The Astrophysical Journal 798 (2): 135.
[2] Bobra, M. G., Xudong Sun, J.T. Hoeksema, M. Turmon, Y. Liu, K. Hayashi, G. Barnes, and K.D. Leka. 2014. "The Helioseismic and Magnetic Imager (HMI) Vector Magnetic Field Pipeline: SHARPs-Space-Weather HMI Active Region Patches." Solar Physics 289 (9): 3549.
[3] Falconer, D. A., R. L. Moore, G. A. Gary, & M. Adams, 2009. The Main Sequence Of Explosive Solar Active Regions: Discovery and Interpretation. The Astrophysical Journal, 700(2), L166-- L169.
[4] Gary, G.A., Moore, R.L., Hagyard, M.J., Haisch, B M. 1987. Nonpotential features observed in the magnetic field of an active region. The Astrophysical Journal 314:782
[5] Muhamad, J., Kusano, K., Inoue, S., Shiota, D. 2017. Magnetohydrodynamic Simulations for Studying Solar Flare Trigger Mechanism. The Astrophysical Journal 842:86.
[6] Mumford, Stuart J, Steven Christe, David Pérez-Suárez, Jack Ireland, Albert Y Shih, Andrew R Inglis, Simon Liedtke, et al. 2015. "SunPy—Python for Solar Physics." Computational Science & Discovery 8 (1).
[7] Kop, G., Lawrence, G., Rottman, G. 2005. The Total Irradiance Monitor (TIM): Science Results. Solar Physics 230: 129-139.
[8] Leka, K. D., & G. Barnes, 2003. Photospheric Magnetic Field Properties of Flaring versus Flare- quiet Active Regions. I. Data, General Approach, and Sample Results. The Astrophysical Journal, 595(2), 1277–1295.
[9] Mason, J. P., & J. T. Hoeksema, 2010. Testing Automated Solar Flare Forecasting With 13 Years of Michelson Doppler Imager Magnetograms. The Astrophysical Journal, 723(1), 634–640.
[10] Pesnell, W. Dean, B. J. Thompson, and P. C. Chamberlin. 2012. "The Solar Dynamics Observatory (SDO)." Solar Physics 275 (1–2): 3–15.
[11] Sakurai, T. 1989. Computational modeling of magnetic fields in solar active regions. Space Science Reviews (ISSN 0038-6308), vol. 51, Oct. 1989, p. 11-48.
[12] Schou, J., P. H. Scherrer, R. I. Bush, R. Wachter, S. Couvidat, M. C. Rabello-Soares, R. S. Bogart, et al. 2012. Design and Ground Calibration of the Helioseismic and Magnetic Imager (HMI) Instrument on the Solar Dynamics Observatory (SDO). The Solar Dynamics Observatory. Vol. 9781461436.
[13] Schrijver, C. J., DeRosa, M. L., Metcalf, T., Barnes, G., Lites, B., Tarbell, T., McTiernan, J., Valori, G., Wiegelmann, T., Wheatland, M. S., Amari, T., Aulanier, G., Démoulin, P., Fuhrmann, M., Kusano, K., Régnier, S., Thalmann, J. K. 2008. Nonlinear Force-free Field Modeling of a Solar Active Region around the Time of a Major Flare and Coronal Mass Ejection. The Astrophysical Journal, Volume 675, Issue 2, pp. 1637-1644.
[14] Sun, Xudong, Hoeksema, J. Todd, Liu, Yang, Wiegelmann, Thomas, Hayashi, Keiji, Chen, Qingrong, Thalmann, Julia. 2012. Evolution of Magnetic Field and Energy in a Major Eruptive Active Region Based on SDO/HMI Observation. The Astrophysical Journal, Volume 748, Issue 2, article id. 77, 15 pp.
[15] Yokoyama, T., & K. Shibata, 2001. Magnetohydrodynamic Simulation of a Solar Flare with Chromospheric Evaporation Effect Based on the Magnetic Reconnection Model. The Astrophysical Journal, 549, 1160–1174.

# Atmospheric drag force on LAPAN-TUBSAT during geomagnetic storm 2015

**N Pratiwi[1*], M I Ikhsan[2]**

[1]Department of Atmospheric and Planetary Sciences, Institut Teknologi Sumatera, Lampung Selatan, Lampung, Indonesia
[2]Faculty of Science, Institut Teknologi Sumatera, Lampung Selatan, Lampung, Indonesia

E-mail: nindhita.pratiwi@aps.itera.ac.id

**Abstract.** The purposes of this research are (1) determine the relationship between LAPAN-TUBSAT orbital elements and geomagnetic activity; (2) estimate daily drag acceleration; (3) determine the relationship between drag acceleration and geomagnetic activity. We used STK (System Tool Kit), python, and CCMC (Community Coordinated Modeling Center) interactive model to process the data. To calculate the drag acceleration, it takes data of satellite's velocity that obtained from STK and atmospheric density. To generate atmospheric density, we used CCMC with MSISE-90 (Mass Spectrometer – Incoherent Scatter Extended 1990) and IRI (International Reference Ionosphere) atmospheric model. In addition, it takes the satellite's position (latitude, longitude, and height – taken from STK) as input. On March 2015, geomagnetic storm index (daily Dst) had a minimum value of -105 nT on March 18, 2015. According to hourly Dst data, the minimum value of Dst reached -223 nT at 23:00 UT on March 17, 2015. Changes of orbital elements were influenced by geomagnetic storms through the Dst index. Maximum changes of semi-major axis, inclination, argument of perigee, and longitude of ascending node occurred on March 17-21, 2015 (geomagnetic storm period). Minimum change of eccentricity occurred on March 15-16, 2015 when Dst reached a maximum value. Atmospheric model is influenced most by solar conditions both during periods of calm or storm. Atmospheric density generated by MSISE-90 and IRI, both of them has $10^{-6}$ order difference. MSISE-90 is $10^{-16}$ while IRI $10^{-22}$. For MSISE-90, maximum drag acceleration reached 451,4416 m/hari$^2$ occurred on March 17, 2015, along with a strong geomagnetic storm phenomenon. For IRI, the value of drag acceleration during peak geomagnetic storm (March 18, 2015) reached 0,000373 m/hari$^2$ (close to the minimum value in March 2015). Drag acceleration of LAPAN-TUBSAT is linear to the Dst index with a correlation coefficient of -0,51.

***Keyword*** *: satellite, orbital elements, drag atmosphere, geomagnetic storm*

## 1. Introduction

Drag force is a force which happens on the satellite's body in the opposite direction of satellite movement. It is the biggest non-gravitational perturbation on low earth orbit satellites. Satellites move through an atmosphere that has a total mass density of all components (proton, electron, and ion). Atmospheric ionization process increases extremely when a geomagnetic storm occurred. During extreme condition like an intensive geomagnetic storm, there were significant atmospheric density fluctuations that caused orbital perturbations. Perturbations of orbit caused changes of satellite's orbital elements [1]. First intensive geomagnetic storm of Solar Cycle 24 occurred on March 17, 2015. SOHO/LASCO C3 recorded partial halo Corona Mass Ejection (CME) that followed by C9.1/1F (S22W25) flare at 02:10 UT on March 15, 2015. The velocity of CME reached ~668 km/s. Interplanetary shockwave controlled by CME, arrived in the Earth at 04:45 UT on March 17, 2015. It caused a sudden geomagnetic storm, and then became a strong geomagnetic storm until Dst decreased to -73 nT at 10:00 UT. It caused by CME sheath crossing. After that, Dst of the storm increased to -44 nT because of Interplanetary Magnetic Field. On March 17, 2015, the storm became stronger than previous, at 23:00 UT Dst reached -223 nT. Geomagnetic storm power measured using Dst (Disturbance storm time) index. Dst is an hourly average value of horizontal magnetic field and expressed in nT [2]. Determining the relationship between atmospheric drag, orbital elements, and geomagnetic storm are purposes of this research. We used the Dst index and TLE (Two Line Elements) provided by *www.celestrak.com* and *https://omniweb.gsfc.nasa.gov*. Daily drag acceleration is calculated by using formula (1). LAPAN-TUBSAT was launched on January 10, 2007, from Satish Dhawan Space Center, India. Its orbit is sun-synchronous circular orbit with an inclination of 97.60°, eccentricity of 0.0014, period of 99.039 minutes and an altitude of 630 km. Kim *et al.* [3] found that the orbital path of KOMPSAT-1 was significantly perturbed during an extremely geomagnetic storm.

## 2. Method

Two Line Elements consist of semi-major axis (a), inclination ($i$), eccentricity ($e$), argument of perigee ($\omega$), and right ascension of the ascending node ($\Omega$). To determine the relationship between LAPAN-TUBSAT orbital elements and geomagnetic activity, we calculate the daily changes of orbital elements value. Next, we compare this value with daily Dst index (geomagnetic storm index). After that, we calculate satellite acceleration using this formula [4]:

$$a_{drag} = -\frac{1}{2}\rho\frac{C_D A}{m}v^2 \tag{1}$$

$a_{drag}$ is the satellite acceleration, $\rho$ is the local atmospheric total mass density, $C_d$ is the dimensionless drag coefficient, $A$ is the satellite's cross-sectional area perpendicular to the direction of the motion, $m$ is the satellite mass, and $V$ is the satellite's orbital velocity. We used CCMC (Community Coordinated Modeling Center) and STK (System Tool Kit) to calculate $\rho$ and $V$ value. Finally, we compared drag acceleration and daily Dst index.

## 3. Results and analysis

Figure 1(a) shows the daily Dst index in 2015 and figure 1(b) shows the daily Dst index on March $1^{st}$-$31^{th}$, 2015. The daily Dst index data are provided at *http://wdc.kugi.kyoto-u.ac.jp*. Based on hourly Dst index, severe geomagnetic disturbances occurred on March 17 with a minimum value of Dst is -223 nT. The first intensive geomagnetic storm of Solar Cycle 24 occurred on March 17, 2015. The source of this storm could be traced back to solar phenomena in the middle of March 2015 [2]. Figure 2 (a)-(c) are the changes of semi-major axis, inclination, and eccentricity (Δa, Δi, Δe). In the plot, we include the slope determined by least squares fitting and linear correlation coefficient (cc). There is a linear relationship between Δa and Dst, with a correlation coefficient of 0.0206. During the geomagnetic storm period, inclination value decreased significantly, it means that the orbital plane becomes so close to the equator plane. There is a linear relationship between Δi and Dst, with a correlation coefficient of 0.186. For eccentricity, the value was relatively small almost zero, which means the orbit is nearly circular. The change of eccentricity decreased on 15-16 but increased on 17-18 (the geomagnetic storm period). The relation between drag acceleration and Dst index during geomagnetic storm period is plotted in Figure 3. The maximum value of drag acceleration was 451,44 m/day$^2$ on March 17, 2015, when the severe geomagnetic storm happened. The drag acceleration increased well correlated with geomagnetic activity (Dst). There is a clear linear relationship between drag acceleration and Dst index, with a correlation coefficient of -0,51.

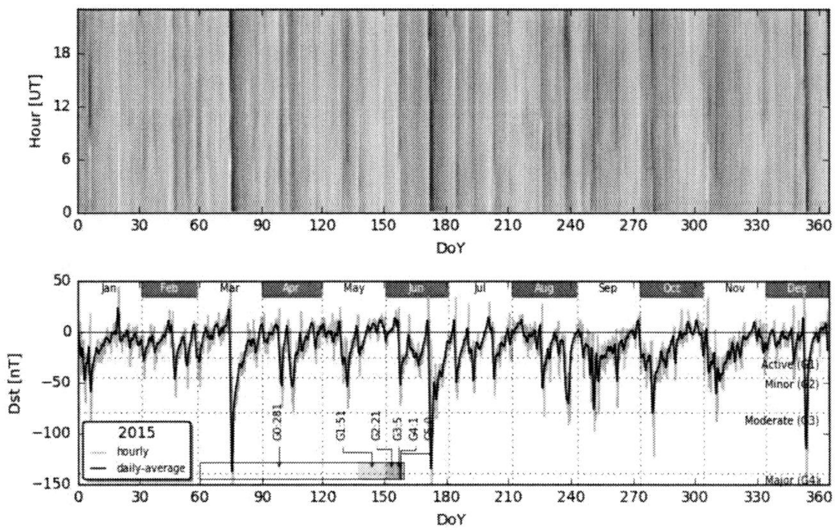


(a)

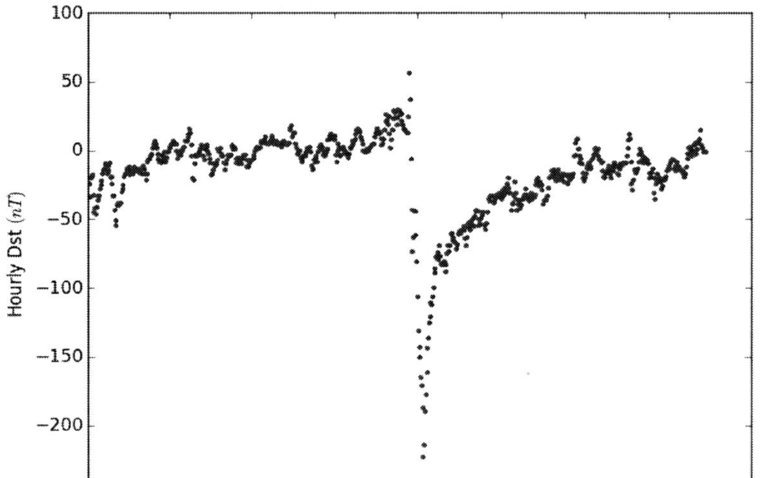


(b)

**Figure 1(a).** Daily Dst index (nT) in 2015. Source: *http://wdc.kugi.kyoto-u.ac.jp* **(b).** Daily Dst index (nT) on March 1st-31th, 2015

IOP Conf. Series: Journal of Physics: Conf. Series **1231** (2019) 012021 doi:10.1088/1742-6596/1231/1/012021

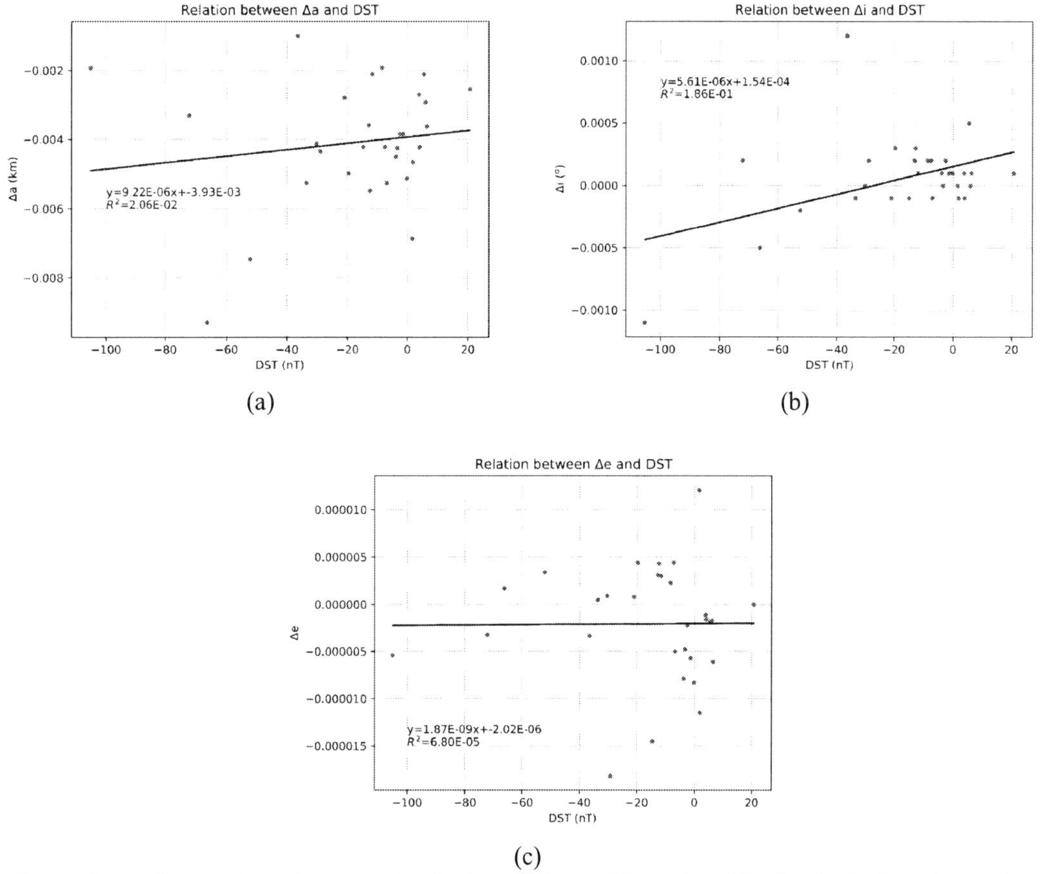


**Figure 2(a).** Comparison between $\Delta a$ (km) and Dst (nT) during March 2015. **(b).** Comparison between $\Delta i$ (°) and Dst (nT) during March 2015. **(c).** Comparison between $\Delta e$ and Dst (nT) during March 2015.

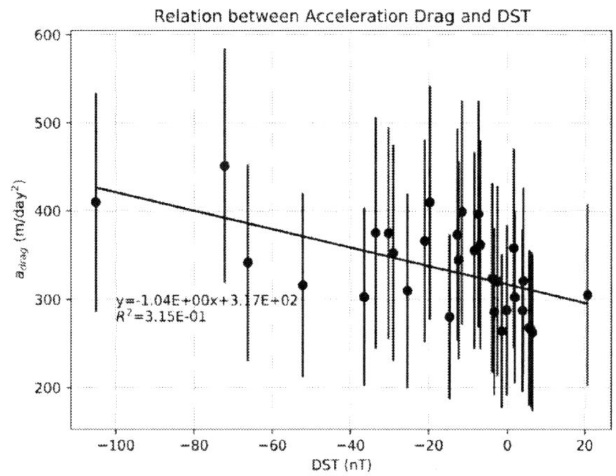


**Figure 3.** Relation between drag acceleration (m/day$^2$) and Dst index (nT) during March 2015.

## 4. Conclusion

We had determined the relationship between LAPAN-TUBSAT drag acceleration and geomagnetic storm in 2015. Our conclusion can be summarized as follows: (1) LAPAN-TUBSAT drag acceleration correlated strongly with the strength of geomagnetic disturbance. Drag acceleration of LAPAN-TUBSAT is linear to geomagnetic activity; (2) During severe geomagnetic storm period, the change of orbital elements were influenced most by geomagnetic storm through the Dst index. The change of orbital elements is linear to the Dst index.

## References

[1] Knowles S H, Picone J M, Thonnard S E, and Nicholas A C 2001 *Sol. Phys.* **204** 387
[2] Mavromichalaki H, Gerontidou M, Paouris E, Paschalis P, Lingri D, Laoutaris A, and Kanellakopoulos A 2015 The extended geomagnetic storm of March 17, 2015 *12th Hel.A.S Conference* on 28 Juny – 2 July, 2015 National and Kapodistrian University of Athens
[3] Kim K H, Moon Y J, Cho K S, Kim H D, and Park J Y 2006 *Earth Planets Space* **58** 25-28
[4] Montenbruck, O. and E. Gill 2001 *Satellite Orbits: Models, Methods, and Applications* (New York Springer Press)

# Prediction of maximum amplitude of solar cycle 25 using machine learning

**T Dani[1,*] and S Sulistiani[1]**

[1]Space Science Center, LAPAN, Indonesia

Email: tiar.dani@lapan.go.id

**Abstract**. Prediction of maximum amplitude of Solar Cycle 25 is obtained by using four different machine learning regression methods, i.e. Linear Regression (LR), Random Forest (RF), Radial Basis Function (RBF) and Support Vector Machine (SVM). Monthly mean sunspot number data during the 1856-June 2018 (solar cycles 10-24) from the World Data Center SILSO, Royal Observatory of Belgium, Brussels are used as machine learning inputs. According to LR, RF, RBF and SVM, the maximum of Solar Cycle 25 is predicted to occur in September 2023 (sunspot number of 159.4 ± 22.3), in December 2024 (sunspot number of 110.2 ± 12.8), in December 2024 (sunspot number of 95.5 ± 21.9) and in July 2024 (sunspot number of 93.7 ± 23.2), respectively. The prediction using LR method suggested that Solar Cycle 25 maximum will be slightly higher than the current cycle, while RBF and SVM suggested much lower cycles. RF prediction suggested a lower maximum with well-constructed double-peak. It was also found that the Solar Cycle 25 is predicted to begin in the late 2019 or early 2020 according to all four methods.

***Keywords**: solar cycle and machine learning*

## 1. Introduction

The prediction of solar activity in term of sunspot number is very important in the mitigation of space weather effects on the Earth's technology systems. The prediction of peak time and value of the solar cycle can give an overview of the solar activity in the next solar cycle compared to the current cycle. The solar activity in term of sunspot number during Solar Cycle 24 was considerably low. NOAA suggested that Solar Cycle 24 has reached its maximum in April 2014 with a mean sunspot number of 82 and is currently in the declining phase. This value is lower than Solar Cycle 23 maximum with sunspot number around 120.

Sunspot number is one of the parameters of solar activity describing the measure of the number of sunspots and groups of sunspots on the visible surface of the Sun. The larger the number, the bigger the probability of flare and CME occurrences on the Sun. These events can cause the geomagnetic and ionospheric storms on Earth and eventually can disrupt the ground-based and space-based technology systems. Previous studies have predicted the peak time and value of Solar Cycle 25 by means of various methods, e.g. [1-8]. This study is predicting the peak time and value of Solar Cycle 25 by using four different machine learning methods, i.e. Linear Regression (LR), Random Forest (RF),

Radial Basis Function (RBF) and Support Vector Machine (SVM). The comparison of this study and previously published works were also performed.

## 2. Methods

Our study is using the SILSO (Sunspot Index and Long-term Solar Observations) [9] data of monthly mean total sunspot number from 1855 (Solar Cycle 10) through May 2018 prepared by the Royal Observatory of Belgium, Brussels as inputs to machine learning program for prediction. We also made use of the Waikato Environment for Knowledge Analysis (WEKA) software to predict the peak time and value of the sunspot number. WEKA is developed by the University of Waikato, New Zealand, equipped by various machine learning algorithm written in Java programming language [10][11]. By using Time-Series Forecasting package available in WEKA, we utilized four different techniques, i.e. LR, RF, RBF, and SVM, to predict the monthly sunspot number for Solar Cycle 25.

## 3. Result and Discussion

WEKA provides time-series forecasting packages for regression schemes which include lagged variables and date-derived periodic variables. It is also able to do closed-loop forecasting. In this paper, we used four base learner machine learning algorithms for the monthly sunspot number forecast, i.e. Linear Regression, Radial Basis Function for Regression, Support Vector Machine for Regression and Random Forest. We used sunspot number historical data from 1855 until May 2018 as input.

### *3.1. Linear Regression (LR)*

Figure 1 shows the prediction of the monthly sunspot number for Solar Cycle 25 by using the linear regression technique. In WEKA, this technique is using the Akaike criterion for model selection. The root means squared error for this prediction is 22.3.

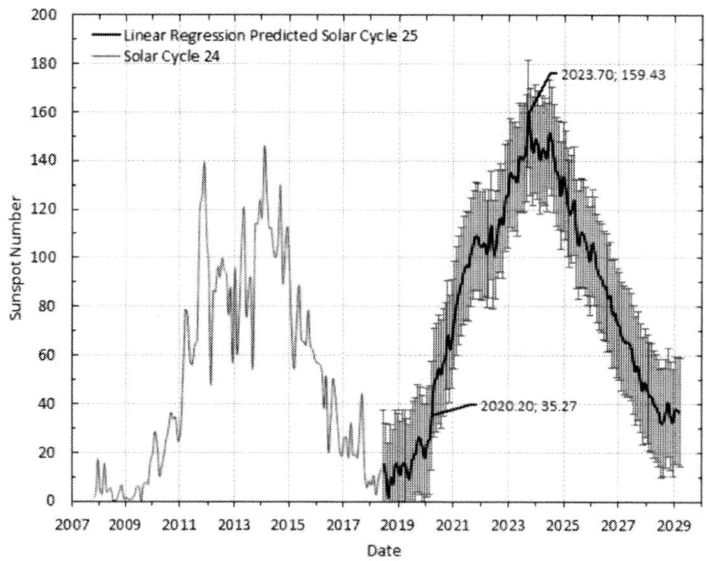


**Figure 1.** Monthly sunspot number during Solar Cycle 24 (red) and Solar Cycle 25 (black). The latter values are the prediction based on linear regression technique.

Based on this technique, Solar Cycle 25 will start in early 2020 and reach the maximum sunspot number of 159.4 ± 22.3 in September 2023. The value is higher than the previous cycle and this prediction was unable to reconstruct the common double-peak feature.

### *3.2. Random Forest (RF)*

In WEKA, this technique constructed a tree with random attributes K for each node which perform no pruning and backfitting option to allow estimation for class probabilities. Figure 2 shows the prediction of the monthly sunspot number for Solar Cycle 25 by random forest technique. The root means squared error for this prediction is 12.8.

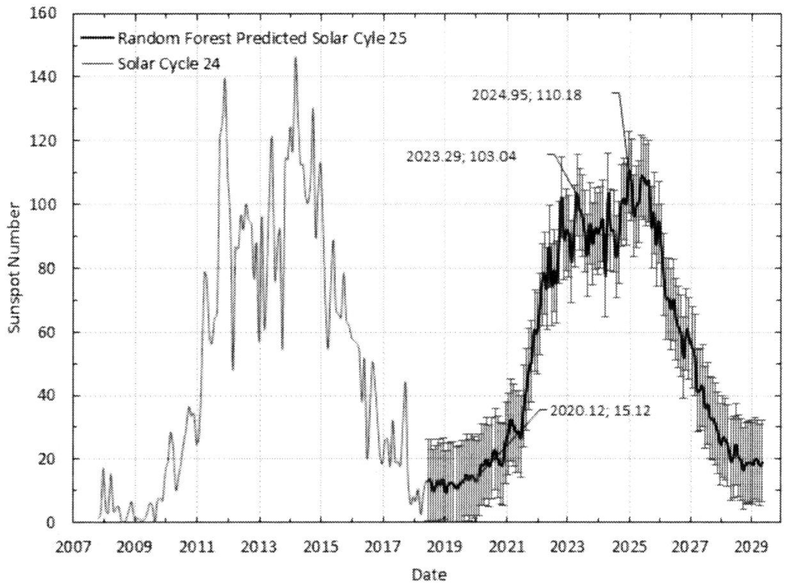


**Figure 2.** Monthly sunspot number during Solar Cycle 24 (red) and Solar Cycle 25 (black). The latter values are the prediction based on random forest technique.

Based on this technique, Solar Cycle 25 will start in early 2020 and reach the maximum value of 110.2 ± 12.8 in December 2024. This value is lower than that of Solar Cycle 24. The prediction of Solar Cycle 25 using random forest technique shows the double-peak feature commonly found in solar cycles. The first peak is predicted to reach a value of 103 ± 12.8 in April 2023.

### *3.3. Radial Basis Function (RBF)*

RBF in WEKA uses the k-means clustering algorithm to provides a logistic regression or linear regression learning model. Figure 3 shows the prediction of the monthly sunspot number for Solar Cycle 25 by radial basis function technique. The root means squared error for this prediction is 21.9.

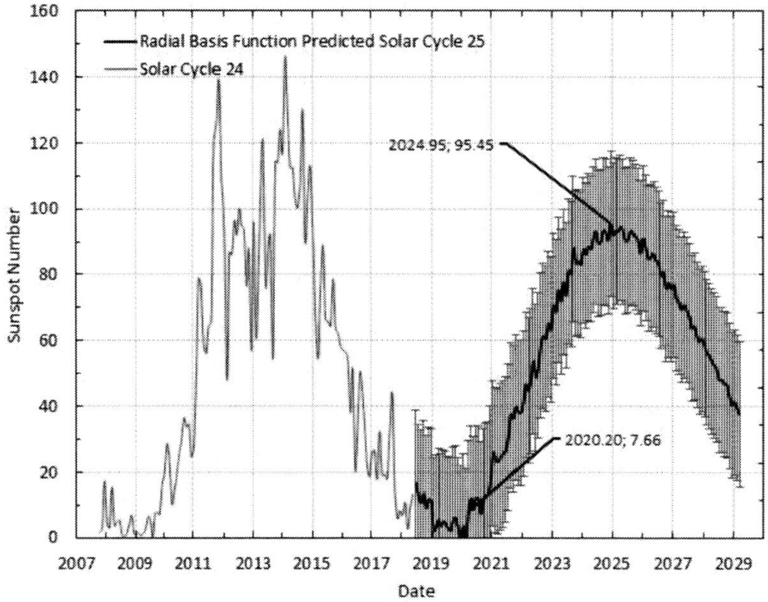


**Figure 3.** Monthly sunspot number during Solar Cycle 24 (red) and Solar Cycle 25 (black). The latter values are the prediction based on radial basis function technique.

Based on this technique, Solar Cycle 25 will start in early 2020 and reach the maximum value of 95.5 ± 21.9 in December 2024. This value is lower than that of Solar Cycle 24 and even lower than that predicted by random forest technique. The prediction of Solar Cycle 25 by radial basis function technique unable to construct double-peak feature commonly found in solar cycles.

### *3.4. Support Vector Machine (SVM)*

SVM in WEKA using SMOreg function which its parameters can be set using a various algorithm. Figure 4 shows the prediction of the monthly sunspot number for Solar Cycle 25 by support vector machine. The root means squared error for this prediction is 23.2. As can be seen from Figure 4, based on this technique Solar Cycle 25 is predicted to start in the early 2020 and to reach the maximum value of 93.7 ± 23.2 in July 2024. This value is the lowest among all machine learning technique used in this study. Similar to linear regression and radial basis function techniques, this one also unable to construct the double-peak feature on the prediction of Solar Cycle 25.

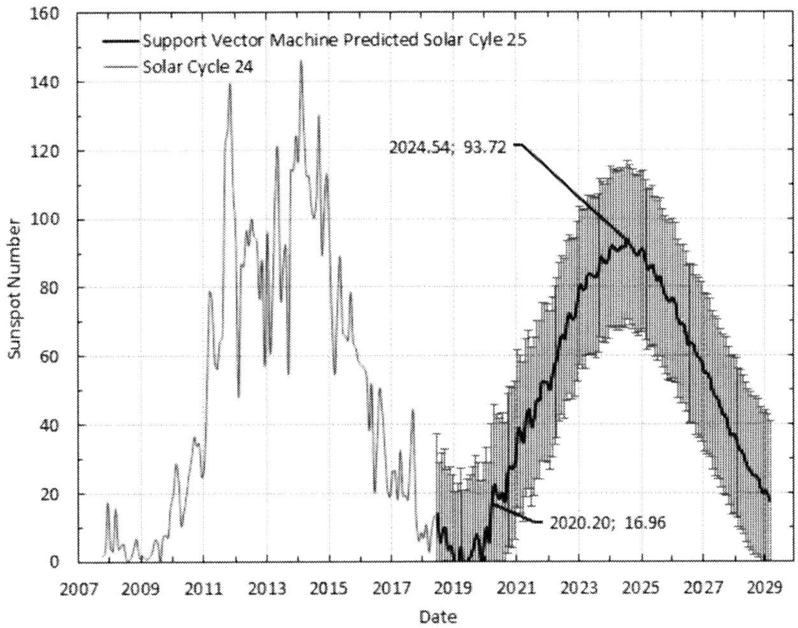


**Figure 4.** Monthly sunspot number during Solar Cycle 24 (red) and Solar Cycle 25 (black). The latter values are the prediction based on support vector machine technique.

### 3.5. *Comparison with other studies*

In this study, we tried to compare our results with the prediction of Solar Cycle 25 obtained from various methods by previous works, as shown in Table 1. Most of the studies suggested that Solar Cycle 25 will be similar to or lower than the current cycle, while our result using linear regression technique suggested otherwise. Furthermore, all studies considered here predicted the peak time of Solar Cycle 25 to range between the year of 2023-2024.

**Table 1.** Comparison of Solar Cycle 25 peak predictions from several works using different methods.

| References | MAX | | | Methods |
|---|---|---|---|---|
| | Month | Year | SSN | |
| This study | Dec | 2024 | 110.2 ± 12.8 | Random forest |
| This study | Sep | 2023 | 159.4 ± 22.3 | Linear Regression |
| This study | Dec | 2024 | 95.5 ± 21.9 | Radial Basis Function |
| This study | Jul | 2024 | 93.7 ± 23.2 | Support Vector Machine |
| [1] | Apr - Jun | 2023 | 112.3 ± 33.4 | Correlation Between Cycle Parameters |
| [2] | Apr | 2023 | 132.1 | Extrapolation of sunspot number spectral components |
| [3] | | 2022-2023 | 118.2 | Spotless |
| [4] | | | 42 ± 13 | Sunspot Group |
| [5] | Oct | 2023 | 109.1 | Classical statistical relations among feature parameters of solar cycle profiles |
| [6] | | | Cycle 25 will be similar in size to Cycle 24 | Surface Flux Transport |
| [7] | | | 116 (southern hemisphere-smoothed) | Solar polar microwave brightness temperature |
| [8] | | | 70 ± 65 | Ludendorff coronal flattening index of the Total Solar Eclipse (TSE) on March 9, 2016 |

## 4. Conclusion

Predictions of peak time and value of Solar Cycle 25 is done by using four machine learning technique, i.e. LR, RF, RBF and SVM. According to LR, RF, RBF, and SVM, the maximum of Solar Cycle 25 is predicted to occur in September 2023 (sunspot number of 159.4 ± 22.3), in December 2024 (sunspot number of 110.2 ± 12.8), in December 2024 (sunspot number of 95.5 ± 21.9) and in July 2024 (sunspot number of 93.7 ± 23.2), respectively. RF prediction is able to predict the double-peak feature of solar cycle. Comparison between this study and from other studies with different methods was also done. It was found that generally all predictions suggested that Solar Cycle 25 will be similar to or lower than the current cycle. Moreover, all works predicted Solar Cycle 25 will peak around the year of 2023-2024.

## 5. Acknowledgments

The sunspot records are courtesy of WDC-SILSO, Royal Observatory of Belgium, Brussels. We would like to thank the Indonesian Minister of Research, Technology and Higher Education (Kemenristekdikti RI) for funding this work through *Program Insentif Riset Sistem Inovasi Nasional* (InSINas) 2018.

## References

[1] Pishkalo M I 2008 *Kinematics and Physics of Celestial Bodies* **24** 242–247
[2] Rigozo N R, Echer M P S, Evangelista H, Nordemann D J R and Echer E 2011 *Journal of Atmospheric and Solar-Terrestrial Physics* **73** 1294-99
[3] Helal J H R, and Galal A A 2013 *Journal of Advanced Research* **4** 275–278
[4] Javaraiah J 2015 *New Astronomy* **34** 54-64
[5] Li K J, Feng W and Li F Y 2015 *Journal of Atmospheric and Solar-Terrestrial Physics* **135** 72–76

IOP Conf. Series: Journal of Physics: Conf. Series **1231** (2019) 012022 doi:10.1088/1742-6596/1231/1/012022

[6] Hathaway D H and Upton L A 2016 Journal of Geophysical Research:Space Physics **121** 11
[7] Gopalswamy N, Mäkelä P, Yashiro S and Akiyama S 2018 *Journal of Atmospheric and Solar-Terrestrial Physics*
[8] Dani T, Priyatikanto R and Rachman A 2016 *J. Phys.: Conf. Ser.* **771** 012005
[9] SILSO, World Data Center 1855-2018 *International Sunspot Number Monthly Bulletin and online catalogue* http://www.sidc.be/silso/
[10] Hall M, Frank E, Holmes G, Pfahringer B, Reutemann P and Witten I H 2009 *SIGKDD Explorations* **11** 1
[11] Frank E, Hall M A and Witten I H 2016 *The WEKA Workbench* (Morgan Kauffman) 4$^{th}$ Edition

# Zenithal sky glow measurement in Bandar Lampung as consideration in drafting the regulation of light pollution-free areas around the Lampung Astronomical Observatory (LAO)

**R Muztaba[1,*], B Saryantono[3], A N I Putri[1], and T D Pratiwi[2]**

[1] Departement of Atmospheric and Planetary Sciences & Lampung Astronomical Observatory, Institut Teknologi Sumatera, Lampung Selatan, Indonesia, 35365
[2] Physics Study Program, Geophysics Laboratory, Universitas Gadjah Mada, Yogyakarta, Indonesia, 55281
[3] Mathematics Education Study Program, STKIP PGRI Bandar Lampung, Bandar Lampung, Indonesia, 35214

Email: robiatul.muztaba@aps.itera.ac.id

**Abstract.** Urban development to big cities generally will be accompanied by excessive use of artificial light, such as street lamps, billboards, and building lighting systems. Ineffective and incorrect lighting installation design causes environmental degradation, i.e light pollution. Today, light pollution is one indicator of environmental degradation and energy waste behavior. Study on light pollution has progressed in many fields of science, extending from traditional fields of astronomy to atmospheric physics, environmental science, natural science, and social life. Measurement of sky brightness is also an indicator of the feasibility of an observatory development plan. The location of the observatory is located at coordinates latitude -05° 27' 71" and longitude 105° 09' 39" with a height of 1030 above mean sea level. The construction of an observatory requires a study of the sky's brightness conditions as a matter of consideration to obtain the best observation result. Therefore, to support the Lampung Provincial Government, Institut Teknologi Sumatera (ITERA) and Institut Teknologi Bandung (ITB) in carrying out the construction of observatories in TAHURA WAR, Gunung Betung, Lampung. We did the sky brightness measurements as far as 15 km from the location point of the observatory. We use SQM to measure the brightness of the sky towards the zenith in every crowded area in the city of Bandar Lampung. Then, from the measurement results, we make a map of light pollution. From the mapping results, there are four locations that are indicated to be contaminated by light pollution, namely Tanjung Senang, Teluk Betung, Kemiling, and Gedong Tataan with respective values of 15.8 mpas, 16.6 mpas, 16.8 mpas, and 17.00 mpas..

***Keywords:*** *Artificial light at night; Light Pollution; Environment; Sky Brightness; Sky Quality Meter; Urban City; Observatory*

## 1. Introduction

In the year 2016 Institut Teknologi Sumatera (ITERA), Institut Teknologi Bandung (ITB), and Provincial Government of Lampung formed a consortium to initiate construction of a new observatory. Based on topographic and astronomical site surveys, the best position to construct a new astronomical observatory has been obtained. The location of the observatory is located at coordinates latitude -05° 27' 71" and longitude 105° 09' 39" with a height of 1030 above mean sea level. In Indonesia, the Ministry of Public Works regulation no. 29/PRT/M/2006 concerning guidelines for technical requirements for building construction, not specifically mentioned about observatories. The observatory is a special building that requires consideration from the results of multi-disciplinary surveys, such as topography surveys, astronomical surveys, and geophysical surveys at construction sites. One of the key parameters to select a site to build an observatory is the night sky brightness because some astronomical research could not be performed with the required quality if the sky is not dark enough. Thus, we can find that astronomical observatories are always in areas far from sources of light pollution. Therefore, to support the Lampung Provincial Government, Institut Teknologi Sumatera (ITERA) and Institut Teknologi Bandung (ITB) in carrying out the construction of observatories in TAHURA WAR, Gunung Betung, Lampung. We did the sky brightness measurements as far as 10-30 km from the location point of the observatory. We use SQM to measure the brightness of the sky towards the zenith in every crowded area in the city of Bandar Lampung. Its result can be used as consideration in drafting the regulation of light pollution-free areas around the Lampung Astronomical Observatory (LAO).

In Section 2, the details of the survey are discussed, including the setup of the measurement devices (SQM) and collecting data. The resulting database forms the backbone for a study of the overall light pollution conditions in Bandar Lampung. Analyses and discussion of the results can be found in Section 3, focusing on the contribution of artificial lighting sources observed in an urban environment and location points that could potentially be disturbing of astronomical observations at the observatory site in the future.

## 2. Methodology

The Sky Quality Meter (SQM) has been widely used in studies related to light pollution [1,2,3,4]. We have used the SQM-LU model photometer for gathering data in the field. The SQM has installed in the housing it appears as at left in figure 1 (a USB based SQM-LU with only one cable). The SQM housing not only protects the device from the environment but also traps the heat generated by the operation of the device. The methodology was improved to a drive-through approach of integrating the SQM-LU with GPS via a laptop computer to find out the location point during the sky brightness measurement process. We combined with a photographic tripod for mapping a geographical in 10-30 km area from observatory site (see Fig. 1).

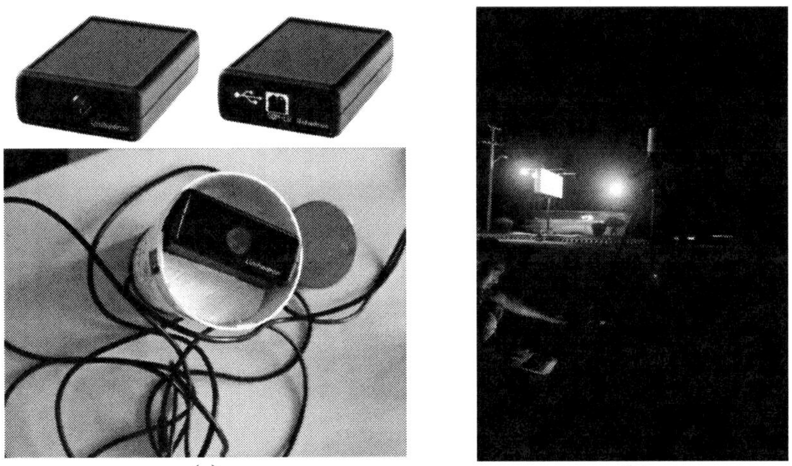

(a) (b)

**Figure 1.** (a) SQM installed in the housing (b) image when during the sky brightness measurement process.

The system is composed of a silicon photodiode as detector (ams-TAOS TSL237S). The spectral sensitivity of the photodiode, combined with the transmission of the HOYA CM-500 near-infrared cut-off filter, provides a final spectral response overlapping the Johnson B and V bands used in astronomical photometry. The SQM photometers have a quoted systematic uncertainty of 10 percent (0.1 mag arcsec−2) [5,6]. We moved to selected locations of a grid, taking the measurements one by one. As for the selected locations, namely Pahoman – A, Teluk Betung – B, Teluk Betung 2 – C, Gruntang – D, Kantor Gubernur – E, Lembah Hijau – F, Tahura – G, Kemiling – H, Bukit Kemiling Permai – I, Mesjid Agung Pesawaran – J, Gedong Tataan – K, Gading Rejo – L, Kedaton – N, Bukit Randu – O, Panjang – P, Panjang2 – Q, Gunung Betung – R, UBL – S, Rajabasa – T, and Alfurqon – U (see Fig. 2). All locations permitting access and reachable by motorized vehicles were surveyed. In this study, we have adopted from several measurement techniques to make a map of light pollution [7, 8, 9, 10, 11, 12, 13]. We measured at zenith angles and build a map of the all-sky brightness from the result the measurements selected locations. Many locations of target areas were covered on the same night to maximize the comparability of data obtained under similar conditions.

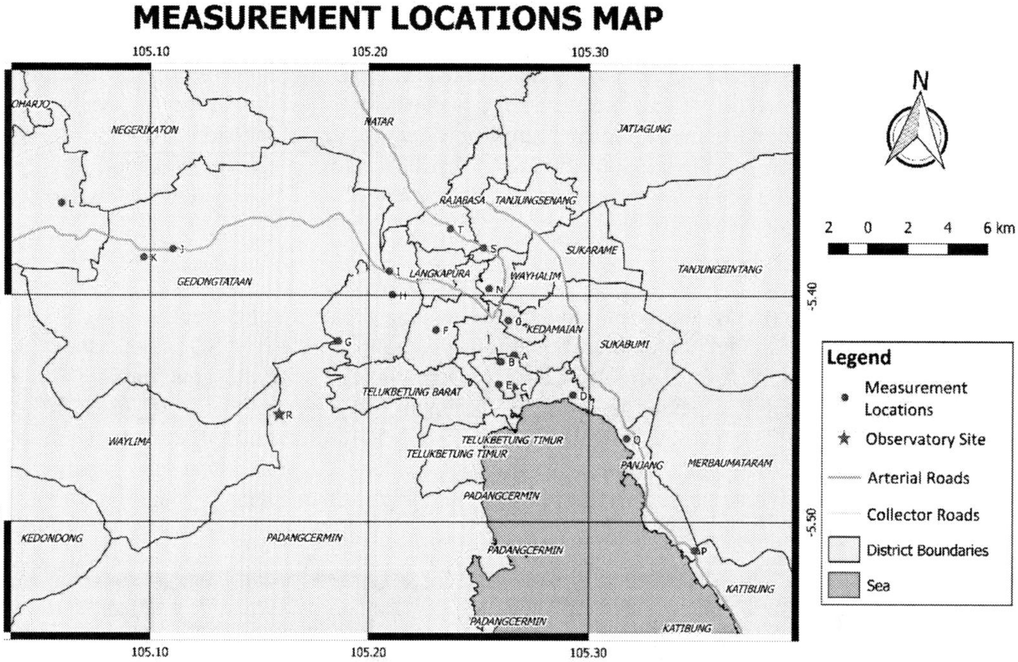


**Figure 2.** Map of measurement locations marked with red dots and observatory site marked with a red star shape

The expected values of light energy loss from Bandar Lampung can be estimated by using Walker's relation [14]:

$$\log R = -4{,}7 - 2{,}5\log D + \log F \tag{1}$$

where $R$ are the ratio between result of measurement by observation and natural background radiation value, $D$ is the distance (in km) between the observatory site and the location of source, and $F$ is the total luminance (in lumen) of the brightness sky at night. From Walker's relation can be obtained the total luminance as follows:

$$F = 10^{(4{,}7+2{,}5\log D+\log R)} \tag{2}$$

$E$ is the electric power loss (in Watts) can be calculated as follows:

$$E = \frac{F}{L} \tag{3}$$

where L is the average lighting efficiency (in lumen/Watt). From relations (2) and (3) we can put E in the following form:

$$E = \frac{10^{(4,7+2,5\log D+\log R)}}{L} \tag{4}$$

If we know the average lighting efficiency ( $L$) and assumed the average time for lighting at night is 10 hours per night during the year, then the average electric energy loss per year ($T_y$) can be calculated as follows:

$$T_y = E(W) \times 10\text{hour} \times 365\,\text{day} \quad (\text{Wh}) \tag{5}$$

Or

$$T_y = \frac{C\left[10^{(4,7+2,5\log D+\log R)}\right]}{L} \quad (\text{kWh}) \tag{6}$$

where C = 3.65.

## 3. Results and Discussions

Lampung Province is the region with the highest population density on the Sumatra Island reaching 234 people / km$^2$, for comparison of other provinces such as North Sumatra reaching 191 people / km$^2$, and West Sumatra reaching 124 people / km$^2$ (adopted in The BPS-Statistic Indonesia, 2013). Urbanization is driving the rapid growth of cities. With the increasing number of population living in urban areas, the demand for new residential areas will also increase too. In addition, the need for lighting will also gradually increase too. The fact is that much outdoor lighting used at night is inefficient. Many sources of light contribute to sky glow such as street light, advertisements and signage, and Several stars like Wezen, Adhara, and Sirius become hard to see because of the effects of the bright sky due to light pollution (see Figure 3). This light and the electricity used to create it is being wasted by spilling it into the sky, rather than focusing it on to the actual objects and areas that people want to be illuminated. The process of developing an area cannot be avoided. However, to minimize the occurrence of light pollution can be prevented through local regulations that control the type of lights and the technical installation in their respective places.

(a) (b)

**Figure 3.** (a) Several stars like Wezen, Adhara, and Sirius become hard to see because of the effects of the bright sky due to light pollution (b) Many sources of light contribute to sky glow such as street light, advertisements, and signage.

We did the sky brightness measurements as far as 15 km from the location point of the observatory to be used as a light pollution free area. We use SQM to measure the brightness of the sky towards the zenith in every crowded area in the city of Bandar Lampung. Then, from the measurement results, we make a map of light pollution (see Figure 4). The result from this research there is four locations that are indicated to be contaminated by light pollution, namely Tanjung Senang, Teluk Betung, Kemiling, and Gedong Tataan with respective values of 15.8 mpas, 16.6 mpas, 16.8 mpas, and 17.00 mpas.

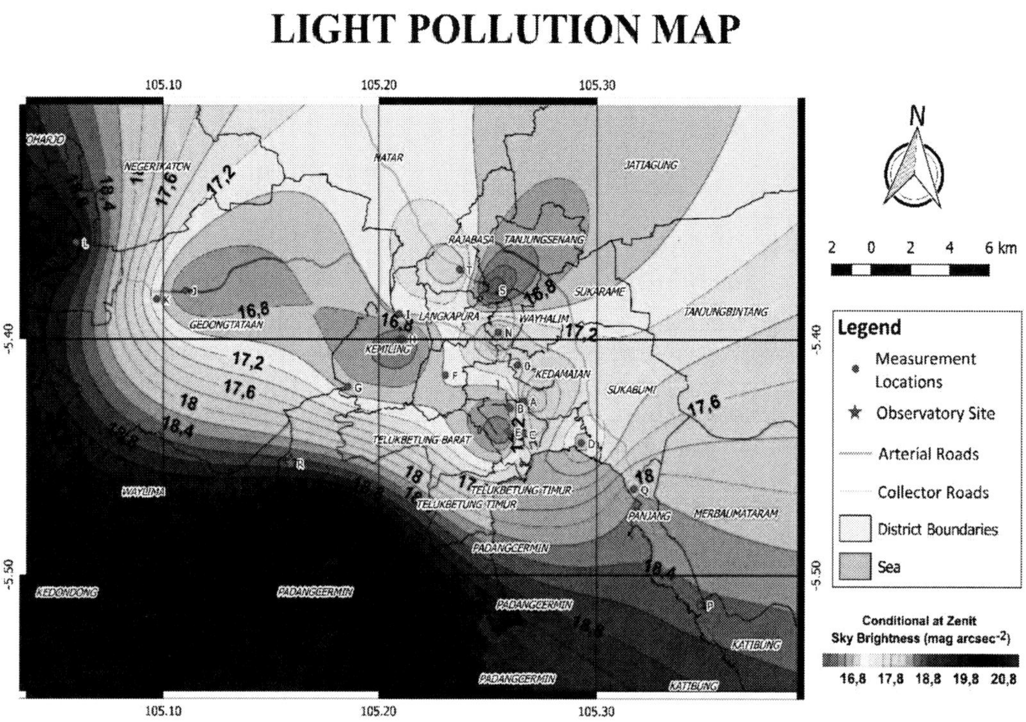


**Figure 4.** Map of measurement locations marked with red dots, and observatory site marked with a red star shape

In this study, we also determined the amount of loss cost due to light pollution. Based on the effective installation of lights to minimize glare, which is less than 70°. The energy used effectively is assumed to be 30%, while the energy loss is 70%. Hereafter, according to the regulation of Minister of Energy and Mineral Resources Regulation No. 28 of 2016 concerning electricity tariffs for the purposes of public street lighting at low voltage (P-3/TR) of Rp 1352, - / kWh. Therefore, it can be determined the amount of the loss of energy and cost due to the poor installation and lighting system (see Table 1). Overall, the average loss cost for lighting at night in Bandar Lampung is 0.793 Billion IDR per year

**Table 1.** Electric energy loss from Bandar Lampung City per hour and year.

| Date | Locations Point Measurement | Average MPSAS (in mpas) | Ratio (R) | Distance (D in km) | Electric power loss (E in kWh) | The average electric energy loss per year (Ty in kWh) | Ineffective 70% Energy loss per year (Ty in kWh) | Loss of Cost (in Billion IDR) |
|---|---|---|---|---|---|---|---|---|
| 3/16/2018 | G. Rejo - L | 19.17 | 0.92 | 15.24 | 379.56 | 1385386.62 | 969770.60 | 1.311 |
| 4/7/2018 | Pahoman - A | 18.11 | 0.87 | 12.25 | 207.93 | 758924.78 | 531247.30 | 0.718 |
| 4/7/2018 | Alfurqon - U | 16.97 | 0.81 | 11.44 | 163.98 | 598528.10 | 418969.70 | 0.566 |
| 4/7/2018 | T. Betung - B | 16.29 | 0.78 | 11.56 | 161.60 | 589839.75 | 412887.80 | 0.558 |
| 4/7/2018 | T. Betung2 - C | 17.59 | 0.84 | 11.92 | 188.61 | 688413.45 | 481889.40 | 0.651 |
| 4/15/2018 | Rajabasa - T | 17.63 | 0.85 | 12.69 | 220.86 | 806132.02 | 564292.40 | 0.763 |
| 4/15/2018 | UBL - S | 15.86 | 0.76 | 13.22 | 220.39 | 804409.18 | 563086.40 | 0.761 |
| 4/15/2018 | Kedaton - N | 17.99 | 0.86 | 12.32 | 209.40 | 764300.11 | 535010.10 | 0.723 |
| 4/28/2018 | G. Tataan - K | 17.05 | 0.82 | 10.51 | 133.34 | 486678.59 | 340675 | 0.461 |
| 4/28/2018 | Pesawaran - J | 16.56 | 0.79 | 9.89 | 111.31 | 406293.88 | 284405.70 | 0.385 |
| 4/28/2018 | BKP - I | 17.06 | 0.82 | 9.01 | 909.35 | 331911.99 | 232338.40 | 0.314 |
| 4/28/2018 | Kemiling - H | 16.29 | 0.78 | 8.21 | 688.06 | 251140.32 | 175798.20 | 0.238 |
| 4/29/2018 | Tahura - G | 16.87 | 0.81 | 4.64 | 170.76 | 62328.20 | 43629.74 | 0.059 |
| 4/29/2018 | L. Hijau - F | 17.04 | 0.82 | 8.96 | 894.49 | 326488.44 | 228541.90 | 0.309 |
| 4/29/2018 | K. Gub. - E | 16.20 | 0.78 | 11.23 | 149.42 | 545374.18 | 381761.90 | 0.516 |
| 4/29/2018 | Gruntang - D | 16.82 | 0.81 | 14.88 | 314.01 | 1146121.63 | 802285.10 | 1.085 |
| 5/5/2018 | B. Randu - O | 17.30 | 0.83 | 12.48 | 207.90 | 758843.93 | 531190.80 | 0.718 |
| 5/6/2018 | Panjang - P | 18.44 | 0.88 | 21.98 | 912.24 | 3329666.96 | 2330767 | 3.151 |
| 5/6/2018 | Panjang2 - Q | 18.19 | 0.87 | 17.61 | 517.04 | 1887193.16 | 1321035 | 1.786 |
| 9/9/2017 | Betung - R | 20.85 | 1 | 0 | - | - | - | - |

## 4. Conclusions and Recommendations

In Bandar Lampung region their many sources of light contribute to sky glow such us street light, spotlight, advertisements and signage. There are four locations that are indicated to be contaminated by light pollution, namely Tanjung Senang, Teluk Betung, Kemiling, and Gedong Tataan with respective values of 15.8 mpas, 16.6 mpas, 16.8 mpas, and 17.00 mpas. The average loss cost for lighting at night in Bandar Lampung is 0.793 Billion IDR per year. So, the following recommendations should be done to overcome the light pollution problem in the future especially the observatory site in the Mount Betung. Set restrictions and laws that require correct light systems by the government of Lampung Province and replace all bad lighting systems with new systems that consume less energy and give adequate lighting as needed of each place.

## Acknowledgments

Authors would like to thank the following for their moral and/or technical support: To the Ministry of Research, Technology and Higher Education because this research has been financial support through Simlibtabmas Program with number 007/SP2H/LT/DRPM/2018. To the Lampung Astronomy Community and local residents who have given permission to take measurements at the research location

**References**

[1] Falchi, F., Cinzano, P., Elvidge, C.D., Keith, D.M. and Haim, A., 2011. *Journal of environmental management*, ***92*(10)**, pp.2714-2722.
[2] Kyba, C.C., Tong, K.P., Bennie, J., Birriel, I., Birriel, J.J., Cool, A., Danielsen, A., Davies, T.W., Peter, N., Edwards, W. and Ehlert, R., 2015. *Scientific reports*, ***5***, p.8409.
[3] Pun, C.S.J. and So, C.W., 2012. *Environmental monitoring and assessment*, ***184*(4)**, pp.2537-2557.
[4] Puschnig, J., Posch, T. and Uttenthaler, S., 2014. *Journal of Quantitative Spectroscopy and Radiative Transfer*, ***139***, pp.64-75.
[5] Cinzano, P. 2013a, http://unihedron.com/projects/darksky/sqmreport_v1p4.pdf accessed 3 June 2017
[6] Cinzano, P. 2013b, http://unihedron.com/projects/sqm-l/sqmreport2.pdf accessed 3 June 2017
[7] Biggs, J.D., Fouché, T., Bilki, F. and Zadnik, M.G., 2012. *Monthly Notices of the Royal Astronomical Society*, ***421*(2)**, pp.1450-1464.
[8] Falchi, F., Cinzano, P., Duriscoe, D., Kyba, C.C., Elvidge, C.D., Baugh, K., Portnov, B.A., Rybnikova, N.A. and Furgoni, R., 2016. *Science advances*, ***2*(6)**, p.e1600377.
[9] Cinzano, P., Falchi, F., & Elvidge, C. D., 2001, Mon. Not. R. Astron. Soc., **328**, 689-707
[10] Zamorano, J., Nievas, M., Sánchez de Miguel, A., Tapia, C., García, C., Pascual, S., Ocaña, F. and Gallego, J., 2015. *IAU General Assembly*, *22*.
[11] Puschnig, J., Schwope, A., Posch, T. and Schwarz, R., 2014. *Journal of Quantitative Spectroscopy and Radiative Transfer*, ***139***, pp.76-81.
[12] Ges, X., Bará, S., García-Gil, M., Zamorano, J., Ribas, S.J. and Masana, E., 2018. *Journal of Quantitative Spectroscopy and Radiative Transfer*, ***210***, pp.91-100.
[13] Jechow, A., Kolláth, Z., Ribas, S.J., Spoelstra, H., Hölker, F. and Kyba, C.C., 2017. *Scientific Reports*, ***7*(1)**, p.6741.
[14] Osman, A.I.I., Isobe, S., Nawar, S. & Morcos, A.B. 2001, *IAU Symposium*, Vol.196.

IOP Conf. Series: Journal of Physics: Conf. Series **1231** (2019) 012024 doi:10.1088/1742-6596/1231/1/012024

# Internet of Things (IoT) For Galactic Cosmic Ray Application Over Astronomical Observatory in Near Future: A Review Study

**W. S. Putro[1,2,*], R. Muztaba[1], N. Pratiwi [1], A. N. I. Putri[1], and W. Birastri[1,2]**

[1]Department of Atmospheric and Planetary Sciences, Institut Teknologi Sumatera (ITERA), 35365, Lampung, Sumatera, Indonesia

[2]Center of Meteorology, Climatology and Geophysics, Institut Teknologi Sumatera (ITERA), 35365, Jati Agung, Lampung, Indonesia

E-mail: wahyu.putro@aps.itera.ac.id

**Abstract**. Galactic Cosmic Ray (GCR) it's very harmful for human life. The Ultra high Neutrino radiation GCR energy from outside galaxy were intimidated Earth and Environmental. The measurement GCR energy using neuron monitor is successful to capture Ultra-high Neutron and Gamma ray, respectively. During Fourth Industrial Revolution (4IR), the current neutron monitor was expanded into Internet of Things (IoT) system to obtain Ultra-high Neutron and Gamma ray in near future. Thus, in this study aimed to develop conceptual study of Neutron Monitor with IoT system over Astronomical Observatory. The result shows Neutron Monitor with IoT system need to deploy in Equator and Antarctic region due to Gamma ray distribution. Here, Neutron Monitor data with IoT system were applied to analyse Natural Hazard potential with predictor from GCR flux parameter. By using auto send status from IoT system, Natural Hazard potential will be updated per-hour into stakeholder (e.g. hospital, Astronautics, and meteorology service). Thus, a disaster victim from Natural Hazard potential caused by Ultra-high Neutron and Gamma ray will be minimized using IoT system during evacuation process in near future.

***Keywords**: Galactic Cosmic Rays Application (GCR), Internet of Things (IoT), and Natural Hazard.*

## 1. Introduction

Space radiation is composed by heavier charged particle, neutron, and electron up to $10^{20}$ eV. During several years the space radiation are increased especially Galactic Cosmic Ray (GCR) flux due to inversion earth magnetic field [1]. A massive GCR flux it's very harmful for human life e.g. communication technology, human health, and space exploration (satellite radiation dose) [2]. Many researchers have been studied to observe GCR flux using neutron monitor over Astronomical observatory [3]. The observation system from neutron monitor has been recorded and shared using File Transfer Protocol (FTP) technology [4]. However, the FTP technology has disadvantages based on service and transfer speed of the data. Thus, this study aimed to design Internet of Thing (IoT) for GCR application based a review study to increased service and transfer speed of the data observation. During Fourth Industrial Revolution (4IR), the Internet of Thing (IoT) have a function to increased service and transfer speed of the data observation using cloud system [5]. The cloud system was applied in many applications including prediction and forecasting natural hazard disaster [6]. The preliminary study about GCR distribution particle energy using GEANT4 application was sucessful studied [7]. Here, the GCR flux assessment was calculated over four stations during Solar Particle Event (SPE) from ground-based station on 14 July 2000 and 28 October 2003, respectively. However, in this study the average error of GCR simulation less than 3%. Thus, the soft sensor modeling using Artificial Neural Network (ANN) for GCR application was proposed to estimate average error for simulation GEANT4 over previous study [8]. The estimation result using ANN was improved GCR average error $\pm$ 3% to $\pm$ 1% on November 2001 to December 2012 over two ground-based stations (Forth Smith and Inuvik). However, the training data to estimate GCR result using ANN need processing time due to observation data. Thus, the cloud system (including storage and processing data) were proposed to improve processing time over training data using ANN [9]. The integrated system based on Internet of Things (IoT) was suggested to integrated communication system and processing data in cloud computing [10]. Based on previous study, the integrated system cloud computing and IoT it's powerful to develop forecast, prediction, and nowcasting in many application especially for GCR flux parameter over astronomical observatory in near future.

## 2. Methodology

### *2.1. Cloud Computing*

In order to explain the understanding cloud computing system, we compare with two adopted system e.g. Cluster and Grid Computing. The assessment Cloud, Cluster, and Grid computing based on characteristics is showed in Table 1 [11]. Here, the characteristics between Cloud, Cluster, and Grid system are described in seven parameters. The seven parameters are assessed based technical aspect and service to develop system.

**Table 1.** Assessment characteristics of clusters, Grids, and Cloud systems [11]

| Characteristics | Systems | | |
|---|---|---|---|
| | Clusters | Grids | Clouds |
| Population | Commodity computers | High-end computers (servers, clusters) | Commodity computers and high-end servers and network attached storage |
| Size/scalability | 100s | 1000s | 100s to 1000s |
| Node Operating System (OS) | One of the standard Oss (Linux, Windows) | Any standard OS (dominated by Unix) | A hypervisor (VM) on which multiple Oss run |
| Ownership | Single | Multiple | Single |
| Interconnection network speed Security/privacy | Dedicated, high-end with low latency and high bandwidth Traditional login/password-based. Medium level of privacy – depends on user privileges. | Mostly Internet with high latency and low bandwidth Public or private key pair-based authentication and mapping a user to an account. Limited support for privacy. | Dedicated, high-end with low latency and high bandwidth Each user/application is provided with a virtual machine. High security/privacy is guaranteed. Support for setting per-file access control list (ACL). |
| Discovery | Membership services | Centralized indexing and centralized info services | Membership services |
| Service negotiation | Limited | Yes, SLA based | Yes, SLA based |

IOP Conf. Series: Journal of Physics: Conf. Series **1231** (2019) 012024 doi:10.1088/1742-6596/1231/1/012024

As can be seen in Table 1, the cloud computing platform have similar characteristics from both cluster and grid. The advantages of cloud computing are integrated with web service interface with visualization and supported application service. Thus, the mechanism service of cloud computing is provided without reference to the infrastructure on which these are hosted [11]. In current architecture ground-based station of GCR observation, the File Transfer Protocol (FTP) was used as a portal to get the data. However, the FTP system has disadvantages due to access mode, cost, and security compared by cloud system.

### *2.2. Ground Base GCR Observation*

In order to improve the quality and processing data over ground base GCR observation, we explain the current observation system using Neutron monitor. Here, Neutron monitors are ground-based recorders of cosmic rays. Cosmic rays are highly energetic charged particles (predominantly protons, together with heavier nuclei of atoms, but also electrons and gamma rays) originating in the galaxy and traveling through interstellar space. The nuclear active components of the secondaries are detected by a neutron monitor (see Figure 1).

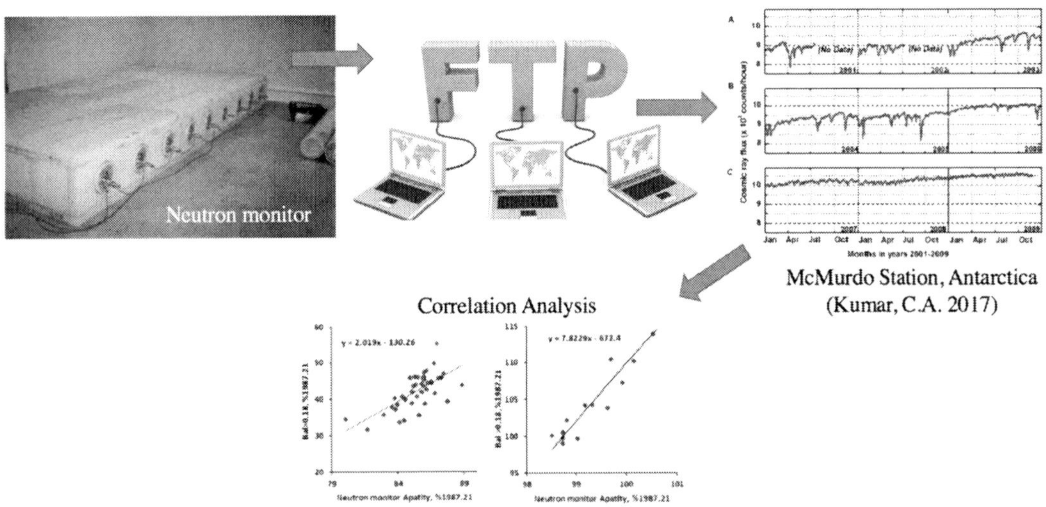


**Figure 1.** The current architecture observation system using Neutron Monitor over ground-based station in correlation analysis case study

As can be seen in Figure 1, the current system to observe GCR flux parameter were developed in several years ago. The highest cost issues to observe GCR flux parameter using Neutron monitor including monitoring system is very expensive. Moreover, the maintenance cost on monitoring system to guaranteed the life time over sensor, electronics, and communication system (server and storage). Thus, to reduce development cost issues and maintenance cost, the Internet of Things (IoT) was proposed in future system. The IoT system with cloud computing will be reduce maintenance server and storage cost in near future.

### *2.3. Internet of Things (IoT) system*

During Fourth Industrial Revolution (4IR), the Internet of Thing (IoT) have capability to reduce maintenance cost and processing time over observation data. In this study, IoT will be improve current architecture observation system using Neutron Monitor over ground-based station. IoT and cloud system it's powerful to process observation data also broadcast the GCR observation result (intensity) over ground-based station after calculation process. Figure 2 shows flowchart of IoT mechanism and cloud computing over calculation process also validation process.

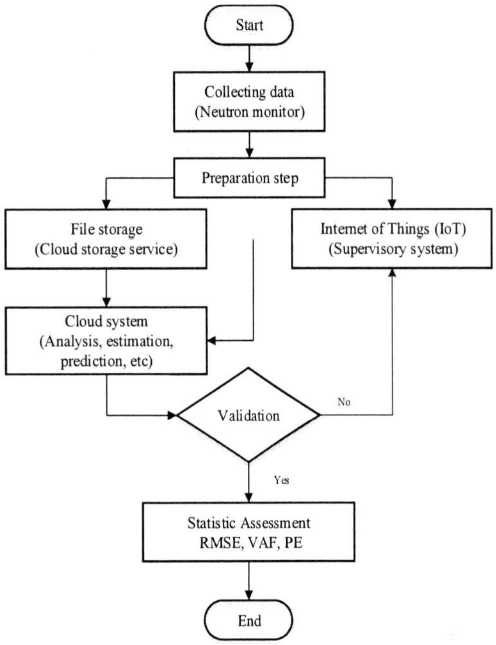


**Figure 2.** Flowchart IoT system for GCR observation over Astronomical Observatory

As can be seen in Figure 2, the future concept of observation GCR flux was designed using Neutron monitor over ground based station. Here, the preparation step collect GCR flux parameter using neutron monitor and recorded to cloud storage service while the IoT system will be manage whole of the system. In order to obtain the result, IoT proposed to cloud system to calculate GCR flux (get data observation) from cloud storage. The GCR flux data was selected to train and testing data, respectively. After the analysis/estimation/prediction of GCR flux obtained, the Root Mean Square Error (RMSE) and Variance Accounted For (VAF) are calculated to obtain performance model over statistic assesment to obtain model accuracy. Before the statistic assesment result obtained (RMSE and VAF), the analysis/estimation/prediction result must be validated by using data observation (observation). If the result status is invalid, all the data will be analyze to obtain good value. The advantages this application is reduced time process and budget limitation.

## 3. Result and Discussion

In order to obtain the fastest processing GCR observation over ground base station with low cost architecture, we analyse the whole of the system into part system. The current architecture GCR monitoring system using Neutron monitor has combined with FTP and transfer data (storage service). However, the processing observation data (e.g. estimation, prediction, forecast, etc.) were processed in the outside of observation system due to traffic transfer data on server. Thus, the processing GCR data need additional cost, storage system, and time to calculate observation parameter. During Fourth Industrial Revolution (4IR), the big data observation can be handled by cloud system to calculate observation parameter without additional cost, storage system, and time (see figure 3).

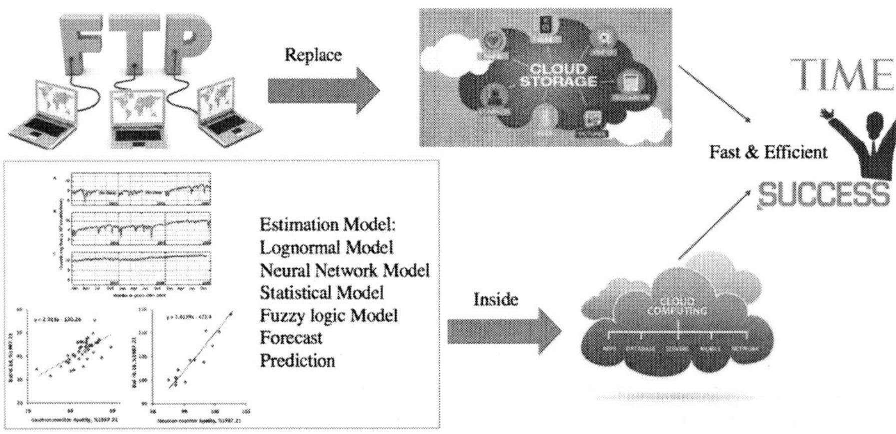


**Figure 3.** The future architecture observation system using Neutron Monitor over ground base station

IOP Conf. Series: Journal of Physics: Conf. Series **1231** (2019) 012024 doi:10.1088/1742-6596/1231/1/012024

As can be seen in Figure 3, the future architecture observation system using Neutron Monitor over ground base station are handled by cloud system. Here, FTP and processing observation data (e.g. estimation, prediction, forecast, etc.) are located inside cloud storage and cloud computing, respectively. The cloud computing was reduced maintenance and service cost due to handle by one cloud system. Moreover, the cloud system is very secure and fast to transfer data also processing observation data.

In order to obtain the low-cost system and processing time over observation data, the IoT system were proposed to manage and running the program called supervisory system. In this study, the supervisory system (including cloud and IoT system) are used to process observation data without lack of data and wasting the time. The new architecture of GCR observation using supervisory system were implemented over instrumentation building in Lampung Astronomical Observatory (LAO) formerly OAIL ITERA in near future. Figure 4 shows the new concept architecture supervisory system to monitor GCR flux parameter using Neutron monitor over Lampung Astronomical Observatory (LAO), Indonesia.

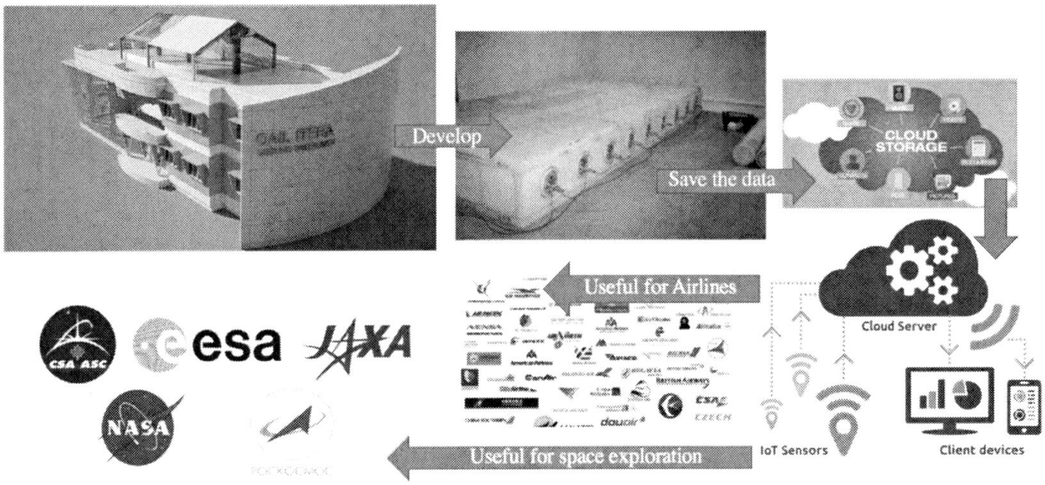


**Figure 4.** The new concept architecture supervisory system to monitor GCR flux parameter using Neutron monitor over Lampung Astronomical Observatory (LAO), Indonesia

As can be seen in Figure 4, the future architecture observation system using Neutron Monitor has been developed over Lampung Astronomical Observatory (LAO), Indonesia in near future. The advantages of supervisory system are auto update (calculation status) based on estimation or prediction model from GCR flux parameter. The IoT system were updated each of the calculation GCR flux parameter to airlines, national space service, and astronomical observatory network also stakeholder who use GCR flux parameter due to natural hazard protection in communication technology. Finally, the integrated cloud and IoT system was successful to design ground-based station over astronomical observatory in near future.

## 4. Conclusion

The review study in Internet of Things (IoT) for Galactic Cosmic Ray Application Over Astronomical Observatory successful studied. The replacement FTP system and processing mechanism to cloud system is reducing maintenance cost and service over Astronomical Observatory especially in the LAO, Indonesia. The cloud system and IoT has integrated called supervisory system to manage and processing the GCR observation data. Furthermore, the auto update system (calculation GCR flux) from IoT status are useful for airlines, national space service, and astronomical observatory network also stakeholder. Based on review study, the supervisory system using IoT and cloud computing have advantages compared by current system of GCR observation. Thus, based on the result we successful study new concept architecture supervisory system to monitor GCR flux parameter using Neutron monitor over astronomical observatory.

**Acknowledgments**
We would like to thank the Institut Teknologi Sumatera (ITERA) for research funding. SEAAN local organizing committee and MKG Research Assistant and the other parts were supporting this research activity.

**References**
[1] Tinsley B A 1996 *Journal of Geophysical Research: Atmospheres* **101** 29714.
[2] Hellweg C E and Baumstark-Khan C 2007 *Naturwissenschaften* **94** 526.
[3] Mishev A L, Kocharov L G, and Usoskin I G 2014 *Journal of Geophysical Research: Space Physics* **119** 679.
[4] Steigies C, Thomann M, Rother O, Wimmer-Schweingruber R, and Heber B 2007 In Proc. 30th ICRC.
[5] Hashem I A T, Yaqoob I, Anuar N B, Mokhtar S, Gani A, and Khan S U 2015 *Information Systems* **47** 115.
[6] Goswami S, Chakraborty S, Ghosh S, Chakrabarti A, and Chakraborty B 2018 *Ain Shams Engineering Journal* **9** 378.
[7] Sihver L, Matthiä D, Koi T, and Mancusi D 2008 *New Journal of Physics* **10** 105019.
[8] Suparta W and Putro W S 2014 *Journal of Physics: Conference Series* **495** 012038.
[9] Suparta W and Putro W S 2014 *Journal of Physics: Conference Series* **539** 012021.
[10] Wang C, Bi Z, and Da Xu L 2014 *IEEE Transactions on Industrial Informatics* **10** 1434.
[11] Buyya R, Yeo C S, Venugopal S, Broberg J, and Brandic I 2009 *Future Generation computer systems* **25** 616.

IOP Conf. Series: Journal of Physics: Conf. Series **1231** (2019) 012025 doi:10.1088/1742-6596/1231/1/012025

# Possibility of astronomical phenomena to be used to support tourism industry

**C Kunjaya**[1, 2, *], **Melany**[1], **A A Sukmaraga**[1] and **T Arsono**[3]

[1] Universitas Ma Chung, Villa Puncak Tidar N-01, Malang, Indonesia
[2] Institut Teknologi Bandung, Jl Ganesa 10, Bandung, Indonesia
[3] Amongputro Skyworld, Taman Mini Indonesia Indah, Jakarta, Indonesia

Email: chatief.kunjaya@machung.ac.id

**Abstract.** Several astronomical events in the past have shown their potential to attract tourists. This motivates some astronomers and tourism activists to create tourism related events when an interesting astronomical phenomenon occur. The most attractive recent astronomical phenomenon for tourists was the solar eclipse as was proven during the total solar eclipse of March 9, 2016. Similarly, other phenomena such as the lunar eclipse, meteor shower, Mars Opposition have some potential to be exploited as events to attract tourists. Belitong Geopark had organized a sky observation event, for example super blue blood moon event on January 31, 2018 and Mars opposition event on July 21, 2018. It successfully drew attention far and wide, but it needed more effort and creativity to make these events more attractive and marketable. Other cultural heritage sites which contain inherent astronomical tie-in knowledge, such as Borobudur Temple, are also strong prospective to be developed for astro-tourism destinations. In the future, astro-tourism permanent facilities may be developed, for example, sky observation facilities for amateurs in Dark Sky National Park, near Mount Timau National Observatory in Kupang, East Nusa Tenggara Province.

***Keywords:*** *Astro-tourism and Astronomical Phenomena*

## 1. Introduction

Astro-tourism is a tourism activity with astronomical phenomena or facilities as the main attraction. Astronomy related tourism activities have existed around the world for a long time. They are mostly sporadic; depend strongly on some very attractive sky phenomena, involving not so many professional astronomers, and almost without any ecosystem support. In recent years, the potential of astronomical phenomena to be used as a tourism attraction has begun to be explored more systematically. A few international conferences or workshops on this topic had been organized in some countries. The latest was held in the *Osservatorio Polifunzionale del Chianti*, Firenze, Italy, July 11-13, 2018. It discussed the existing astronomy related tourism activities and their development in the future.

Developing astro-tourism in Indonesia is in line with one of Indonesian government priority programs, namely, sustainable tourism. The concept of Astro-tourism has a big economic potential; it can contribute to support the existing tourism industry, provided a good ecosystem support exists. It also meets with the International Astronomical Union Strategic Plan 2010 – 2020 – "Astronomy for the Developing World" – where the appeal of astronomy is not only for astronomy world itself but also to support development in other fields and for the benefit of the community at large. To make attractive, effective and successful events based on celestial phenomena, for the benefit of the larger populous community, it is necessary to develop astro tourism ecosystem. This paper is written as a pioneering work to develop astro tourism ecosystem in Indonesia.

## 2. Potential of astro-tourism in Indonesia

In Indonesia and South East Asian countries in general, the ecosystem supporting astro-tourism has yet to be developed. There are sporadic activities related to sky phenomena which attract tourists, for example Total Solar Eclipse. However, this is a rare phenomenon, occurring only once in several decades over the same region. There are also infrastructure facilities providing astronomy related attraction, such as a planetarium and Skyworld in the capital city of Indonesia, to name a few, but these are independently run, and not collectively a part of an astro-tourism ecosystem. Tourism agencies in developed countries could better exploit the phenomena to create marketable events entertaining tourists, but Indonesian agencies are still far behind. Some hotels and tourist destinations do enjoy the spill over benefits from phenomena like the solar eclipse, in the form of increased room occupancy with corresponding higher number of tourists. There are many other celestial objects and phenomena which would be interesting enough to garner a wide appeal if special events connecting the common people and the sky are well planned, designed and marketed.

In the past, professional astronomers exposed and explained, through various observatory activities, in order to guide others to enjoy astronomical phenomena; resulting in intermittent educational exposure to the interested public and the local tourism industry inadvertently benefitting from the sky phenomena through such a guidance from these astronomers. However, these activities are not sustainable within the tourism industry, due of its voluntary nature and the astronomers themselves did not get comparable benefits for their expertise shared.

With the ominous lack of ecosystem connecting the astronomers and the tourism industry in Indonesia, information regarding specific sky phenomenon would usually be made available not long before the actual event, through newspapers, television, online news etc., except during Total Solar Eclipse. Consequently, this usually leave the tourism industry with very little time to prepare an attractive event which could satisfy tourists well, both from the appropriate accommodation availability to the public relatable understanding of the actual scientific concept of the phenomenon. Even when a sky event could only to be observed within Indonesia, foreign tour and travel agencies could immediately reap more benefits by exploiting the event as part of their tourism business because they have access to a more established system and information channel, much to the detriment of Indonesian tour and travel agencies which could not benefit from their own local event.

**Table 1.** International Visitors Arrivals to Indonesia 2015-2017 [1]

| | 2015 | 2016 | 2017 |
|---|---|---|---|
| January | 785,973 | 814,303 | 1,107,968 |
| February | 843,928 | 888,309 | 1,023,388 |
| March | 841,071 | 915,019 | 1,059,777 |
| April | 801,873 | 901,095 | 1,171,386 |
| May | 852,388 | 915,206 | 1,148,588 |
| June | 872,385 | 857,651 | 1,144,001 |
| July | 877,584 | 1,032,741 | 1,370,591 |
| August | 911,704 | 1,031,986 | 1,393,243 |
| September | 920,128 | 1,006,653 | 1,250,231 |
| October | 877,798 | 1,040,651 | 1,161,565 |
| November | 835,408 | 1,002,333 | 1,062,030 |
| December | 986,519 | 1,113,328 | 1,147,031 |

Notwithstanding, tourist destinations usually need events to attract additional volume of tourists – events like annual festivals, sports competitions, cultural and traditional ceremonies, and artistic performances must be staged either routinely or specifically scheduled. The Indonesian government has continuously placed the tourism industry as one of the priority sectors to be developed, because of its very rich natural resources, enhanced by its diverse cultural and historical heritage, which support the growth of the domestic tourism industry.

Ten tourist destinations have been chosen as the top priority to be developed, namely: Borobudur, Toba Lake, Bromo-Tengger-Semeru mountain complex, Komodo Island, Seribu Archipelago, Belitung Island, Mandalika, Wakatobi, Morotai, and Tanjung Lesung [2]. The development of these tourism areas is targeted to boost the country's income from the tourism sector to US$20 billion in the next five years, from the current range of US$10 billion. Table 1 shows that foreign tourists is in an increasing trend in the recent years.

**Figure 1.** Number of annual foreign tourist arrival in Indonesia [1]

## 3. Previous events

Over the last 4 decades, several great astronomical phenomena had occurred and quietly passed without any event aiming at attracting tourists, be they local and foreign. For example, in June 1983, one of the longest total solar eclipse, occurred over Java Island. At that time, no tourism events were staged, despite earlier socialization through activities by astronomers. The government even prohibited people to view the eclipse, warning about the risk of blindness due to extended exposure of the solar eclipse when observed with the naked eyes. Some astronomers who planned to observe this in public areas were even ordered to leave by the local police. In recent years, with more people being educated and the flow of information easily spread, more astronomy related events were staged, taking full advantage of opportunities available through astronomical phenomena.

### *3.1. Super Blue Blood Moon*

On January 31, 2018, a celestial phenomenon nicknamed Super Blue Blood Moon, occurred. It is quite a rare combination of three phenomena, namely moon near perigee, second full moon in a month and lunar eclipse. Such a combination is very rare and it is widely publicised in media. With such wide media exposure, this phenomenon made people more curious. Many institutions and astronomy clubs, in several cities organized observational events, and opened them to the public for a wider appreciation of this rare spectacle. One of the many events, stretched across the southern hemisphere, was held in Nyiur Melambai beach, Belitung island. It was held in cooperation with schools, travel bureau, local government and Belitong Geopark Management Agency. One of the motivations to hold such an event is to attract tourists and promote Belitung island as a new Geo-park and tourist destination. The success of the event motivated the tourism agency union in Belitung to hold another astronomical event in the same year - Mars Opposition event on July 21, 2018.

### *3.2. Mars Opposition and lunar eclipse*

Near the peak of Mars Opposition of July 27, 2018, the longest total lunar eclipse in a century occurred. Not only was the coincidence of the total lunar eclipse and Mars opposition happening simultaneously, but other bright planets, Venus, Jupiter, Saturn were also visible. Those with access to a telescope had a chance to observe many interesting objects in the galaxy in the same night. An observation event in Malang, East Java was held with the cooperation of Universitas Ma Chung, a local hotel with an open rooftop deck, a local newspaper as a media partner which featured a fortnight of background information about the phenomenon, and other participating universities hosting international students during their summer break.

### *3.3. Total solar eclipse*

Indonesia is the only big country traversed by the Total Solar Eclipse on March 9, 2016. Many foreign tourists and local visitors, including professional astronomers, flooded the designated cities for best unblocked observation views of the spectacular event. The Indonesian Ministry of Tourism promoted the event abroad to attract foreign tourists and inform local hotels and travel agents to prepare several months in advance. Many hotels in the cities traversed by the eclipse were even booked one year ahead of the event. Foreign travel agents arranged special eclipse tour to Indonesia.

**Table 2.** Hotel occupancy rate in few cities traversed by total solar eclipse [1]

| | March 2015 | March 2016 | March 2017 |
|---|---|---|---|
| Palu | 58.6% | 72.3% | 58.1% |
| Palangkaraya | 53.3% | 55.8% | 65.5% |
| Tanjung Pandan | 33.3% | 46.7% | 41.2% |
| Palembang | 55.4% | 61.6% | 53.2% |

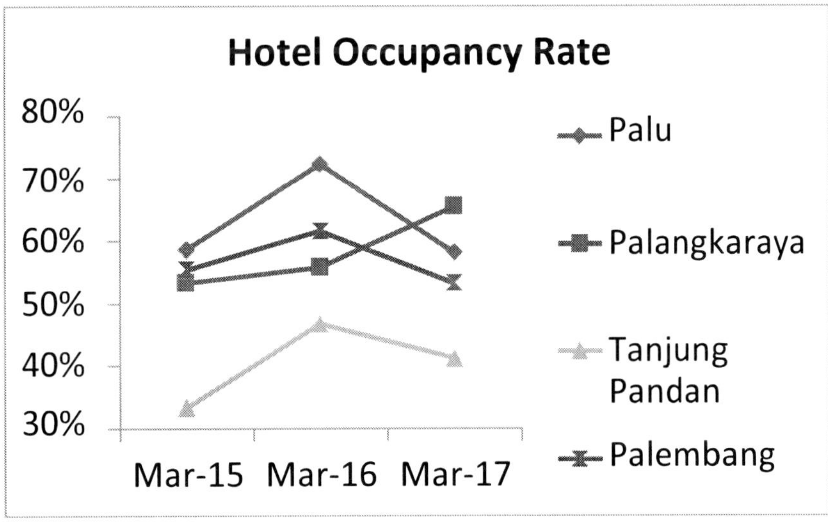


**Figure 2.** Hotel occupancy rate in few big cities traversed by the total solar eclipse, March 9, 2016 [1]

Astronomers, both amateurs and professionals, did public outreach informative programs and education activities prior to the event, while some of them freely provided assistance for the public to observe the eclipse in a correct way during the eclipse in order for them to enjoy the rare and spectacular phenomenon and to avoid possible blindness. Aside from the international media coverage of this event, a report on the eclipse event in the city of Palu, west Sulawesi and Malang, East Java had been published by Rachmadian et al. (2016) [3].

In every city traversed by the eclipse in totality, the crowd of local and foreign tourists could easily be found, much more than any earlier congregation of visitors previously gathered. Table 2 indicates that the solar eclipse provided a boost in hotel room occupancy. It is a year-to-year monthly data collected over the same period. Unfortunately, we don't have the daily data, which could be expected to show the drastic jump of hotel occupancy number around the day of the solar eclipse.

## 4. Cultural related astro-tourism

Traditionally, Indonesians of yore, had some form of ancient astronomical knowledge which were integrated into their lives, and this was used mainly to determine the seasons for agricultural and particularly harvest timing. Then fishermen and sailors needed celestial knowledge for navigation. Tracing the track of astronomical knowledge found in some ancient temples, especially Buddhist temples, it was believed that the architects of the Borobudur temple, a 9$^{th}$ century Mahayana Buddhist temple in Magelang Regency of Central Java, had a very astute knowledge of astronomy. A massive 15,000 sq m stone structure, its East-West and North-South gates were accurately aligned. The biggest stupa sitting atop at the highest level had a role as a gnomon, marking a new season when its shadow would fall onto one of the stupas on the second highest floor. Without such an in-depth knowledge of astronomy, it would be unthinkable, how the alignment of such a giant structure, laid brick by brick, could be so acutely precise, considering no modern civil engineering knowledge and tools existed in the 9$^{th}$ century Java Island.

These temples have already been tourist attractions till the present day, but only the cultural aspects are highlighted. The astronomical aspects of these ancient structures could be one of the astro-tourism objects. However, further research is needed to delve into the astronomical content of the archeo-cultural objects. From the extensive compilation of knowledge as a result of research, the tourism industry could then create well designed communicative materials to be used as attractive presentations for tourists.

## 5. Dark Sky Tourism

Dark Sky tourism refers to the utilizing of the night sky as the main attraction in local tourism. It is a new term in Indonesia, and has yet to be done systematically. It is comparable to Astro-tourism. To date, there is no permanent site nor facility for Dark Sky tourism in Indonesia. The idea of Dark Sky tourism surfaced when a group of Indonesian astronomers chose Timau Mountain, Kupang Regency, Timor Island, East Nusa Tenggara province, in south east Indonesia, for the future site of the Indonesian big telescope [4].

The large area surrounding the observatory site is mainly a savanna expanse, with some isolated trees, making it beautiful like a botanical garden. While the population density of the area is expectedly low, the area is under developed economically, with minimal infrastructural support including no electricity grid from the state electricity company (PLN) – this is an absolute advantage as there is almost no light pollution there. However, in near future, when the national observatory unit is established, electricity will be an inevitable necessity and it would be certain that the electricity grid will enter the region soon enough.

Establishing an observatory should provide positive impact for the local society, otherwise problems will arise in the future. The observatory needs to cooperate with the surrounding landowners and villagers to protect the sky from light pollution. Without any benefit for them, it will be difficult to persuade these locals to cooperate. The lesson learned from the Bosscha observatory taught a very important lesson – the importance of people participation. Bosscha observatory now suffers from increasingly bad light pollution slowly degrading the observation quality. The effort to communicate and educate those surrounding the observatory to reduce light pollution has not help much, and participation rate is low, because the local population there could not feel the tangible benefit of participating in the observatory sky protection program.

Learning from this particular experience, it is important to anticipate the potential 'worsening' sky condition surrounding Timau Mountain. It is expected that social development in that area should not be stopped, but the progressive intensity of light pollution in that area could be carefully monitored and inhibited (if necessary) so that the life of an expensively invested observatory could be sustained longer. Community development program since the preparation of the observatory can appropriately motivate people participation in the observatory protection program. Everyone concerned must feel the benefit of protecting the sky from light pollution. But how could the locals enjoy such benefits, if it is ever tangible? Would it be plausible that their dark sky be convertible to be one of their sources of income? Therefore cooperation between observatory management, Astro-tourism activists and surrounding community is critical and be made beneficial for all, in their effort to protect the dark sky for astronomical research and at the same time accelerating the local economy.

One of the ideas for Astro-tourism activity involving the local community is to create home-stay program in local residents' houses. The biggest challenge to execute this idea is to change the local lifestyle and habits from the local trappings into a more tourist-friendly environment, while keeping its traditional nuance. There must be a house upgrading program to modify their current houses including its interior and standard of sanitation, which needs considerable amount of investment. They also need hospitality training so that their homes can be comfortable for tourists.

One other necessary change is the design of lamps to be used in the city for it to be dark sky friendly – every lamp must have an upper cover so that each source emanating light must be directed downward, to minimize light pollution. The building material and exterior paint colour of the houses must also be restricted to, once again, reduce light pollution to a minimum.

To facilitate tourists enjoying the natural sky after dark, an observation plaza could be built near the observatory. The plaza could be a concrete flat space where tourists and interested parties could install their own small telescope and camera on a tripod easily and participate in their own observation activity, at their own pace. The plaza should also be equipped with electricity sources and Wi-Fi hotspots. Not so far from it, there should be a rest area or shelter for those wanting to lounge when the sky is cloudy or when it rains. To make it more interesting, cafés and a barbeque pit could be built to facilitate a star party event. Also, a telescope rental shop could be provided for the Astro-tourists who do not bring their own telescope or camera. It could sell telescope/camera accessories and to also provide information services related to astronomy and the incumbent telescopes.

**Figure 3.** Observation plaza for amateur astronomers to do independent observation, not so far from the observatory.

## 6. Conclusion

Indonesia has the infinite potential and absolute necessity to develop Astro-tourism. Previous celestial events which drew public attention have merely indicated the initial potential to attract tourists. Astro-tourism activities could be incorporated in the existing tourism destinations with careful analysis of climate and seasonal changes. One of the potential sites to be developed to become a permanent Astro-tourism site is Timau Mountain in Timor Island because of its dry climate, long dry seasons and an underdeveloped society. Other existing tourist destinations could also get additional benefits from astronomical phenomena when events are staged to attract tourists. Currently, there is negligible amount of research done in the field of Astro-tourism in Indonesia, while it has become necessary to develop Astro-tourism ecosystem in Indonesia. Moving forward, more research in this field is recommended for the benefit of the tourism industry.

## Acknowledgement

The authors wish to acknowledge the support from Universitas Ma Chung, Amongputro Skyworld, Belitong Geopark, Institut Teknologi Sumatera and LAPAN in this work.

## References

[1] http://www.kemenpar.go.id/asp/ringkasan.asp?c=110, last opened October 22, 2018
[2] Cholik MA 2017 *European J. of Research and Reflection in Management Sci* **5** 49
[3] Rachmadian A.P., Kunjaya C., Wahono W., Audito 2016 Journal of Physics Conf Series
[4] Hidayat et al. 2014 Experimental Astronomy 37 85

# Constraints on the total mass and mass hierarchy of neutrinos from cosmological observations with bayesian analysis

**N P A P Emas[1] and H Wulandari[1]**
[1]Astronomy Research Division, Faculty of Mathematics and Natural Sciences, Institut Teknologi Bandung, Jalan Ganesha 10, Bandung, 40132

E-mail: auditasisva@gmail.com

**Abstract.** In the standard model neutrinos consist of three species (flavors) and are massless. However, solar neutrino problem that was solved by neutrino oscillations requires that neutrinos have non-zero masses. Another support for non-zero neutrino masses comes from kinematic experiments. Oscillation experiments give rise to a further problem regarding the order of neutrino masses, known as the neutrino mass hierarchy problem, i.e. whether the hierarchy is normal or inverted. On cosmological scales, massive neutrinos affect Cosmic Microwave Background (CMB) and Large Scale Structure (LSS) formation. The impact of massive neutrinos is imprinted in the power spectrum of CMB, and in the LSS matter power spectrum as well. Since neutrinos affect events in the universe, it is possible to constrain neutrino properties based on cosmological observations. This work is aiming at constraining neutrino total mass from cosmological observations in the two hierarchy scenarios, i.e. normal and inverted hierarchies, and also at determining which neutrino mass hierarchy is favoured by cosmological observations. To constrain neutrino total mass we employed CosmoMC and performed Bayesian analysis with MCMC algorithm, while neutrino mass hierarchy was determined based on Bayes factors, in which likelihoods of the two hierarchy scenarios were compared. In addition, neutrino individual masses were estimated by combining results of oscillation experiments. Depending on the dataset used, different total mass constraints were obtained. The tightest constraint comes from a combination of CMB + LSS + BAO + Supernova data, i.e. $\sum m_\nu <$ 0.183 eV (normal hierarchy) and $\sum m_\nu <$ 0.188 eV (inverted hierarchy) with 95% C.L. The Bayes factors analysis prefers normal hierarchy, although with weak evidence.

***Keywords:*** *Mass hierarchy of neutrinos, Cosmological observation, and Bayesian analysis*

## 1. Introduction

Neutrinos are fermions that come in three flavours (electron, muon and tau neutrinos) and are massless in the standard model. However, oscillation experiments (solar and atmospheric) show that neutrinos change flavours from one to another during propagation, meaning that they have non-zero masses. The experiments give squared mass difference instead of absolute masses, which lead to another problem known as neutrino mass hierarchy problem. The neutrino mass hierarchy is called normal (NH) if $m_1 << m_2 < m_3$ and inverted (IH) if $m_3 << m_1 < m_2$. The total mass of neutrinos depends on the hierarchy. [1] for example, found a lower limit of 0.06 eV if the mass hierarchy is normal, and 0.1 eV if it is inverted. Beside oscillation experiments, kinematic experiments give us evidence for non-zero neutrino masses and set upper bounds on the individual mass of each neutrino, i.e. $m_{\nu_e} < 0.2$ eV, $m_{\nu_\mu} < 190$ keV and $m_{\nu_\tau} < 18.2$ MeV [2]. Massive neutrinos play important roles in the history of the universe. Neutrinos have significant impacts on the two most important observables: Cosmic Microwave Background (CMB) and large scale structures (LSS). Prior to last scattering, there were two important epochs influenced by the total mass of neutrinos: matter-radiation equality and recombination. If the total mass of neutrinos is greater than ~0.5 eV, they would contribute to the matter component, otherwise to the radiation component. The total mass of neutrinos affects the heights of the peaks in the power spectrum of CMB, especially the heights of the acoustic peaks (200 < l < 1000), and the integrated Sachs-Wolfe (ISW, 20 < l < 200) region. It also shifts the position of the acoustic peaks. If neutrinos contribute to the matter density, the height of the acoustic peaks in the power spectrum of CMB would decrease with the increasing total mass of neutrinos. On the contrary, if neutrinos contribute to the energy density of radiation, the height of the acoustic peaks would increase with the increasing total mass of neutrinos. The ISW region is affected only if neutrinos contribute to the matter density. The height of the power spectrum in this scale would decrease with the increasing neutrino total mass. As aforementioned, neutrino total mass has also impact on the position of the acoustic peaks. The higher the neutrino total mass, the more to the left (to the smaller multipoles) would the peaks be shifted. This effect can be seen either neutrinos contribute to radiation or matter density. Massive neutrinos also play important roles in the formation of large scale structures. On small scales, massive neutrinos would suppress structure growth. This effect shows up on the small scales of the matter power spectrum.

Since massive neutrinos influence the history of the universe, constraining the neutrino total mass can be done by analysing cosmological data. In this work, we used data from CMB, LSS, Baryon Acoustic Oscillation (BAO) and Supernova Ia observations. By combining these data, we aimed at better constraints on other cosmological parameters, so that we in turn get tighter constraint on the total mass of neutrinos. Here we have focused on constraining the total mass of neutrinos from cosmological data with Bayesian statistics by employing CosmoMC [3]. We discussed two cases of the neutrino mass hierarchy, i.e. normal hierarchy and inverted hierarchy. We also calculated likelihood probabilities for both cases and compared their Bayes factors, in order to choose which neutrino mass hierarchy is preferred by the observational data. Thereafter, we calculated neutrino individual masses by following the method described in [1]. In this project, we used the standard cosmological model (ΛCDM model).

This paper is organized as follows: we explain data and method used in this project in section 2, in section 3 we show and discuss the results, and we conclude our work in section 4.

## 2. Data and Method

We used Bayesian statistics to constrain cosmological parameters, especially the total mass of neutrinos. Mathematically, Bayes theorem is based on conditional probabilities:

$$P(\theta|d) = \frac{P(d|\theta)P(\theta)}{P(d)} \quad (1)$$

where $\theta$ is the parameter we search for, $d$ is data, $P(d|\theta)$ is likelihood probability, $P(\theta)$ is prior probability, $P(d)$ is evidence and $P(\theta|d)$ is posterior probability.

Bayesian inference can be carried out by using Markov Chain Monte Carlo (MCMC) algorithm. We used "efficient sampling of fast and slow" MCMC algorithm [4] that converges faster than other algorithms. This algorithm is included in CosmoMC, a widely used code to explore cosmological parameters. For each dataset, we run CosmoMC for two scenarios of neutrino mass hierarchy, i.e. normal and inverted. The priors we chose were based on ΛCDM model, as are summarized in Table 1. The first six parameters come from Planck Collaboration 2015 for ΛCDM model and the total mass of neutrinos is from Particle Data Group (PGD) results in 2011-2016.

**Table 1.** Priors for Bayesian analysis with CosmoMC

| Parameter | Physics | Prior |
|---|---|---|
| $\Omega_c \boldsymbol{h^2}$ | Density parameter of cold dark matter | 0.001 – 0.239 |
| $\Omega_b \boldsymbol{h^2}$ | Density parameter of baryon | 0.005 -0.035 |
| $100\Theta_s$ | Ratio between diameter of angular distance to sound horizon | 0.5 – 5 |
| $\tau$ | Thomson's optical depth at reionization | 0.01 – 0.5 |
| $n_s$ | Scalar spectral index | 0.8 -1.2 |
| log $A_s$ | Amplitude of primordial power spectrum | 2 – 4 |
| $\sum \boldsymbol{m_\nu}$ *(eV)* | Total mass of neutrinos | 0 – 1.5 |

After calculating the posterior probability of each mass hierarchy for each dataset, we stepped further with comparing the two neutrino mass hierarchies:

$$\frac{P(\theta|d)_1}{P(\theta|d)_2} = \frac{P(d|\theta)_1 P(\theta)_1}{P(d|\theta)_2 P(\theta)_2} = B_{1,2}\frac{P(\theta)_1}{P(\theta)_2} \quad (2)$$

Here $B_{1,2}$ is a Bayes factor,

$$B_{1,2} = \frac{P(d|\theta)_1}{P(d|\theta)_2} \rightarrow \ln B_{1,2} = \ln P(d|\theta)_1 - \ln P(d|\theta)_2 \quad (3)$$

To decide which hierarchy is favoured by the observational data, we used Jeffrey's scale as summarized in Table 2.

**Table 2.** Jeffrey's scale [5] for interpretation of Bayes factors.

| $\ln B_{1,2}$ | Odds | Strength of evidence |
|---|---|---|
| <1.0 | < 3 : 1 | inconclusive |
| 1.0 – 2.5 | (3 – 12) : 1 | weak |
| 2.5 – 5.0 | (12 – 150 ) : 1 | moderate |
| > 5.0 | > 150 : 1 | strong |

The datasets used in this project are summarized in Table 3. CMB [6] and LSS (WiggleZ [7], SDSS LRG DR 4 [8]) data come from processes that are directly influenced by the total mass of neutrinos. We included additional data that are less directly connected to the total mass of neutrinos, but would tighten constraints on other cosmological parameters, which in turn have impact on the neutrino total mass. For example, we used CMB lensing data [6] to constrain $A_s$ and $\sigma_8$, that are related more directly to the total mass of neutrinos. We included SDSS BAO DR12 Final consensus data [9] to constrain matter density, since neutrinos contribute to matter density if they are nonrelativistic. Pantheon Supernova data [10] were included to constrain $H_0$, that would affect distance scales.

Having decided on the favoured mass hierarchy from the analysis of Bayes factors, we proceeded to the determination of individual mass of neutrinos by using hierarchy parameters defined in [1] as

$$\Delta = \frac{m_3 - m_1}{m_3 + m_1} \quad (4)$$

combined with oscillation experiment results from Nu-Fit 2018:

$$\Delta m_{21}^2 = 7.40^{+0.21}_{-0.20} \times 10^{-5}\mathrm{eV}^2$$
$$\Delta m_{31}^2 = 2.494^{+0.033}_{-0.031} \times 10^{-3}\mathrm{eV}^2 \quad (5)$$

## 3. Results and Analysis

We summarize our results in Table 3 for each mass hierarchy and dataset. The triangle plots of the derived parameters are shown in Figure 1, for the case of normal hierarchy. Correlations of other cosmological parameters with the total mass of neutrinos are shown here only for $\sigma_8, \Omega_m$, and $H_0$. The figure shows that the higher the total mass of neutrinos, the higher is the matter density, but the lower is the Hubble constant, because higher matter density would slow down the expansion of the universe. In the small scales, the correlation between $\sigma_8$ (the fluctuation on a scale of 8Mpc/$h$) and the total mass of neutrino is negative, because neutrinos would obstruct structure growth in small scales due to free-streaming of neutrinos. We found that adding lensing to CMB data worsen the constraint on the neutrino total mass, which is not expected and we are still working on the explanation for this. Adding BAO SDSS BAO DR12 Final consensus data to CMB + lensing data improves the constraint significantly, better than adding LSS data. Even the combination of CMB + lensing + LSS + BAO does not differ from CMB + lensing + BAO. Finally, adding supernova data improves the constraint only slightly. The constraints on the total mass of neutrinos we obtained from cosmological observations are therefore stringent than those obtained from kinematic experiments, such as in [2]. This is expected, since a kinematic experiment measures the sum of the rest mass and the kinetic energy of neutrino, whereas from cosmology we get the rest mass, albeit the total rest mass.

Our Bayes factor calculations (see Table 4) show that we are not yet able to distinguish between normal and inverted mass hierarchies with strong evidence. It is expected to be solved by the next missions of LSS survey, such as EUCLID, that will provide us with higher precision data. Although we do not have strong evidence to decide which neutrino mass hierarchy is favoured by the observational data, we still have moderate evidence that favours normal hierarchy from the Bayes factors of CMB + lensing and CMB + lensing + LSS datasets (see Table 4). We therefore chose normal hierarchy and proceeded to the determination of the individual masses of neutrinos. The results are summarized in Table 5. As expected, the most stringent constraint on the individual mass comes from the combination of all data. The upper limit of each neutrino mass eigenstate is given for 95% C.L.

**Table 3**. Result of total mass of neutrinos of each mass hierarchy from six dataset

| **Data set** | **Hierarchy** | $\sum m_\nu$ **(eV) (68% CL)** | $\sum m_\nu$ **(eV) (95% CL)** |
|---|---|---|---|
| CMB TT, TE, EE + LowTEB (CMB) | normal | < 0.258 | < 0.593 |
| | inverted | < 0.160 | < 0.404 |
| CMB TT, TE, EE + LowTEB + lensing (CMB + lensing) | normal | < 0.335 | < 0.591 |
| | inverted | < 0.318 | < 0.540 |
| CMB TT, TE, EE + LowTEB + lensing+ WiggleZ + SDSS LRG DR4 (CMB + lensing + LSS) | normal | < 0.253 | < 0.434 |
| | inverted | < 0.267 | < 0.504 |
| CMB TT, TE, EE + LowTEB + lensing + SDSS BAO DR12 Final consensus (CMB + lensing + BAO) | normal | < 0.112 | < 0.205 |
| | inverted | < 0.109 | < 0.202 |
| CMB TT, TE, EE + | normal | < 0.104 | < 0.195 |

| | | | |
|---|---|---|---|
| LowTEB + lensing+ WiggleZ + SDSS LRG DR4 + SDSS BAO DR12 Final consensus (CMB + lensing + LSS + BAO) | inverted | < 0.110 | < 0.203 |
| CMB TT, TE, EE + LowTEB + lensing+ WiggleZ + SDSS LRG DR4 + SDSS BAO DR12 Final consensus +Pantheon (CMB + lensing + LSS + BAO +SN) | normal<br>inverted | < 0.098<br>< 0.105 | < 0.183<br>< 0.188 |

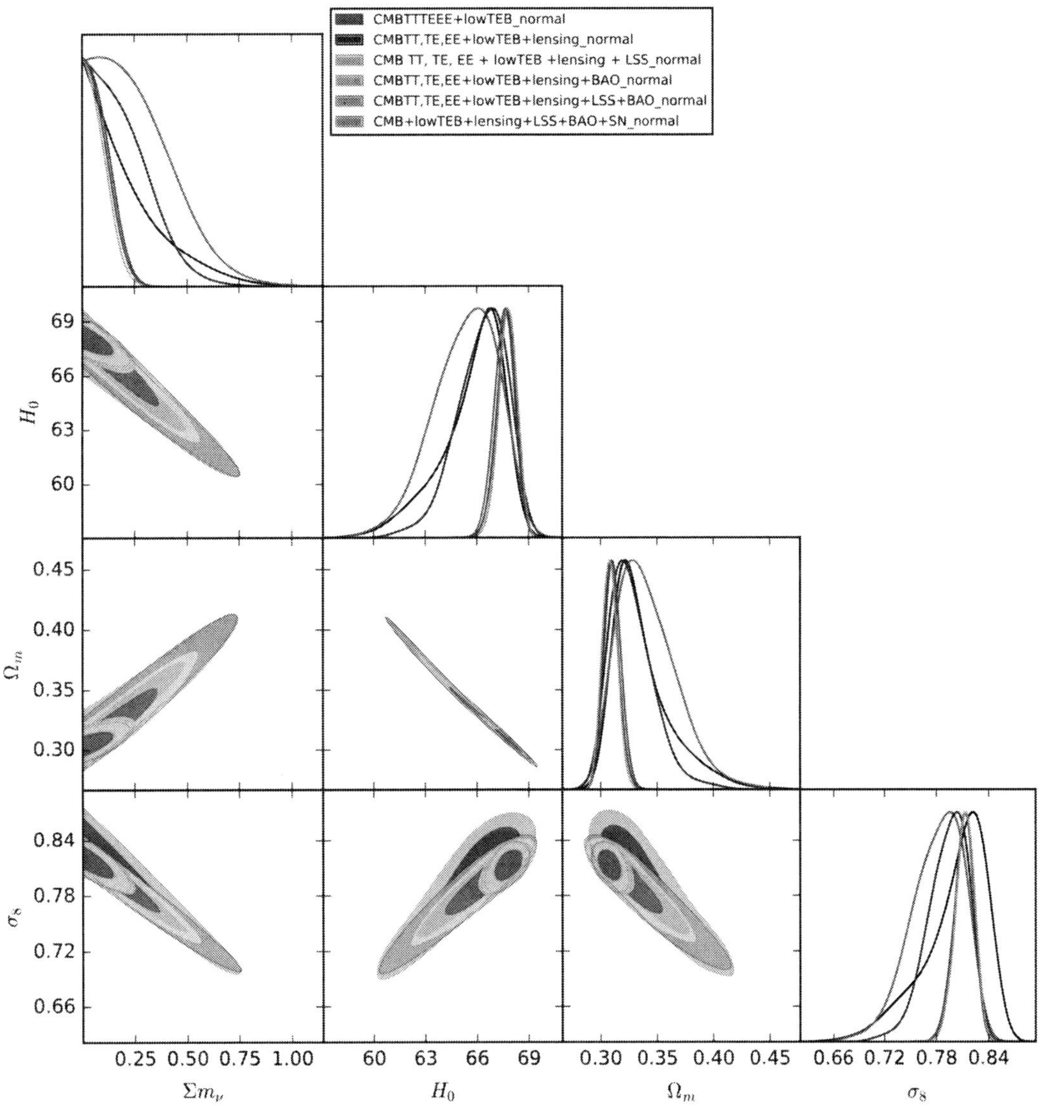


**Figure 1.** Triangle plots that show correlations between some derived parameters with total mass of neutrino for the case of normal hierarchy.

IOP Conf. Series: Journal of Physics: Conf. Series **1231** (2019) 012026 doi:10.1088/1742-6596/1231/1/012026

**Table 4.** Bayes factors calculated for different dataset

| Data Set | Hierarchy | Best-fit (log(likelihood)) | ln $B_{NH,IH}$ | ln $B_{IH,NH}$ |
|---|---|---|---|---|
| CMB | normal<br>inverted | -6472.011<br>-6472.467 | 1.050<br>(weak) | -1.050<br>(inconclusive) |
| CMB +lensing | normal<br>inverted | -6477.220<br>-6478.883 | 3.829<br>(moderate) | -3.829<br>(inconclusive) |
| CMB +lensing + LSS | normal<br>inverted | -6717.594<br>-6718.786 | 2.745<br>(moderate) | -2.745<br>(inconclusive) |
| CMB +lensing + BAO | normal<br>inverted | -6480.983<br>-6481.599 | 1.418<br>(weak) | -1.418<br>(inconclusive) |
| CMB +lensing + LSS +BAO | normal<br>inverted | -6721.289<br>-6721.237 | -0.120<br>(inconclusive) | 0.120<br>(inconclusive) |
| CMB +lensing + LSS +BAO+SN | normal<br>inverted | -7239.403<br>-7239.281 | -0.281<br>(inconclusive) | 0.281<br>(inconclusive) |

**Table 5**. Individual masses of neutrinos derived from different datasets

| Data set | $< m_1$(eV) | $< m_2$(eV) | $< m_3$(eV) |
|---|---|---|---|
| CMB | $0.19552 \pm 0.00437$ | $0.19571 \pm 0.00437$ | $0.20180 \pm 0.00426$ |
| CMB+lensing | $0.19484 \pm 0.00437$ | $0.19503 \pm 0.00437$ | $0.20114 \pm 0.00426$ |
| CMB+lensing+LSS | $0.14174 \pm 0.00434$ | $0.14200 \pm 0.00433$ | $0.15028 \pm 0.00426$ |
| CMB+ lensing+BAO | $0.06229 \pm 0.00464$ | $0.06288 \pm 0.00463$ | $0.07984 \pm 0.00364$ |
| CMB+lensing+ LSS + BAO | $0.05867 \pm 0.00468$ | $0.05929 \pm 0.00468$ | $0.07704 \pm 0.00359$ |
| CMB+ lensing +LSS +BAO + SN | $0.05428 \pm 0.00475$ | $0.05496 \pm 0.00474$ | $0.07376 \pm 0.00352$ |

## 4. Conclusion

Cosmology gives us constraints on neutrino total mass. In our project, we found that the most stringent constraint is obtained if we include SDSS BAO DR12 Final consensus. Cosmological data cannot yet distinguish between the two neutrino mass hierarchies with strong evidence, but we found moderate evidence that normal hierarchy is favoured from the Bayes factor calculated for the combination of CMB + lensing + LSS. A global analysis that combine different observations of neutrinos, i.e. cosmology, oscillations and kinematics would do the task better. In the future, observations such as EUCLID (cosmology) and KATRIN and MARE (kinematics experiments) will provide us with higher precession data, so that we can better improve the constraints on neutrino mass and hierarchy.

**Acknowledgments**

We thank to Institut Teknologi Bandung (ITB) for the P3MI 2018 grant.

## References

[1] Xu L and Huang Q 2018 *Sci. China-Phys. Mech. Astron.* **61** 03952
[2] Otten E W and Weinheimer C 2009 *Rept. Prog. Phys.* **71** 086201
[3] Lewis A and Briddle S 2002 *Phys. Rev.* D **86** 103511
[4] Lewis A 2013 *Phys. Rev.* D **87** 103529
[5] Trotta R 2008 *Contemp.Phys.* **49** 71-104
[6] Ade P A R et al 2016 *A&A* **594** A13
[7] Parkinson D et al 2012 *Phys. Rev.* D **86** 103518
[8] Tegmark M et al 2006 *Phys.Rev.* D **74** 123507
[9] Alam S et al 2017 *MNRAS* **470** 2617–2652
[10] Scolnic D M et al 2018 *ApJ* **859** 101

# Mass determination of open clusters using kinematics data

**M I Ikhsan[1,2,*], M I Arifyanto[2], Aprilia[2], M I Hakim[2], and E Soegiartini[2]**

[1] Faculty of Science, Institut Teknologi Sumatera, Lampung, Indonesia
[2] Department of Astronomy, Institut Teknologi Bandung, Bandung, Indonesia

Email: isnaenda.ikhsan@staff.itera.ac.id

**Abstract.** We estimate the mass of three open clusters, NGC 2168, NGC 188, NGC 6819, using kinematics data of open cluster stars member. Radial velocity data determined from WIYN survey and proper motion determined from Gaia DR2. Using virial theorem and projected-mass estimator method we estimate the mass of star open clusters. Both methods give a reasonable result for all clusters mass. The accuracy of estimated mass affected by some parameters such as membership probability of stars sample, the distance, age, and angular radius of open clusters. We estimate the mass of clusters within a central radius of the cluster. The mass of open clusters using virial theorem is 4056 $M_\odot$ (d < 7 pc) for NGC 2168, 2776 $M_\odot$ (d < 8 pc) for NGC 188, and 4244 $M_\odot$ (d < 1.6 pc) for NGC 6819. Using projected mass estimator we get 5833 $M_\odot$ (d < 7 pc) for NGC 2168, 3326 $M_\odot$ (d < 8pc) for NGC 188, and 3518 $M_\odot$ (d < 1.6pc) for NGC 6819.

***Keywords:*** *kinematics data and open clusters*

## 1. Introduction

There are a large number of open star clusters in our galaxy, especially in our galactic disk. Several study have been done to determine the mass of some star open clusters. Leonard & Merritt by using projected mass estimator method determine the mass of open clusters M35. They estimate the dynamical mass of M35 within the central radius 3.75 pc lies between 1600 and 3200 $M_\odot$ [1]. Geller et al. by using velocity dispersion of radial velocity determine the mass of open cluster NGC 188. They use radial velocity data from WIYN survey to estimate the mass of NGC 188 using virial mass theorem and estimate a mass of 2300 ± 460 $M_\odot$ [2]. Yang et al. calculate a tidal mass of NGC 6819 using modified version of the point-mass Galactic model and averaging it with photometric mass. They obtain the total mass of ~2700 $M_\odot$ for NGC 6819 [3].

GAIA DR2 produces high precision for kinematics data for large number of stars, especially for proper motion and parallax [4]. Combined proper motion data from GAIA DR2 and radial velocity from WIYN survey we could get kinematics data with high precision for three open cluster we choose; NGC 2168 (M35) [2], NGC 188 [5], and NGC 6819 [6]. This paper is organized as follows. Section 2 provides short description of the parameter of open clusters and the membership reduction. In Section 3, we explain the methods that we use to determine the mass using kinematics data. Section 4 shows our result and analysis. Section 5 present conclusion.

## 2. Clusters parameters and membership reduction

Table 1 shows the parameters of NGC 2168, NGC 188, and NGC 6819 that determined from MWSC (Milky Way Star Clusters) catalogue [7]. We choose these three open clusters because it had a huge number of data of radial velocity from WIYN and proper motion from GAIA DR2.

**Table 1.** Parameters of NGC 2168, NGC 188, and NGC 6819

| **Parameter** | **NGC 2168** | **NGC 188** | **NGC 6819** |
|---|---|---|---|
| RAJ2000 (h:m:s) | 06 09 12.6 | 00 47 24.0 | 19 41 16.8 |
| EJ2000 (d:m:s) | +24 21 36 | +85 15 18 | +40 11 42 |
| Distance (pc) | 830 | 2000 | 2360 |
| Age (log t) | 8.255 | 9.650 | 9.210 |
| r (deg) | 0.98 | 0.57 | 0.2 |

**Table 2.** Number of stars of NGC 2168, NGC 188, and NGC 6819 that meet required condition.

| | **NGC 2168** | **NGC 188** | **NGC 6819** |
|---|---|---|---|
| **Number of Stars** | 236 | 268 | 321 |

The membership reduction is done by using membership probability that provided by WIYN catalogue. The membership probability is calculated based on radial velocity data. We choose the stars with membership probability > 90% to become the member of open cluster. Beside the membership probability we also consider the distance of the stars using GAIA Distance catalogue [8]. We choose stars with distance within the distance from the centre of cluster ± 100 pc. Table 2 shows the number of stars for each cluster that meet both condition.

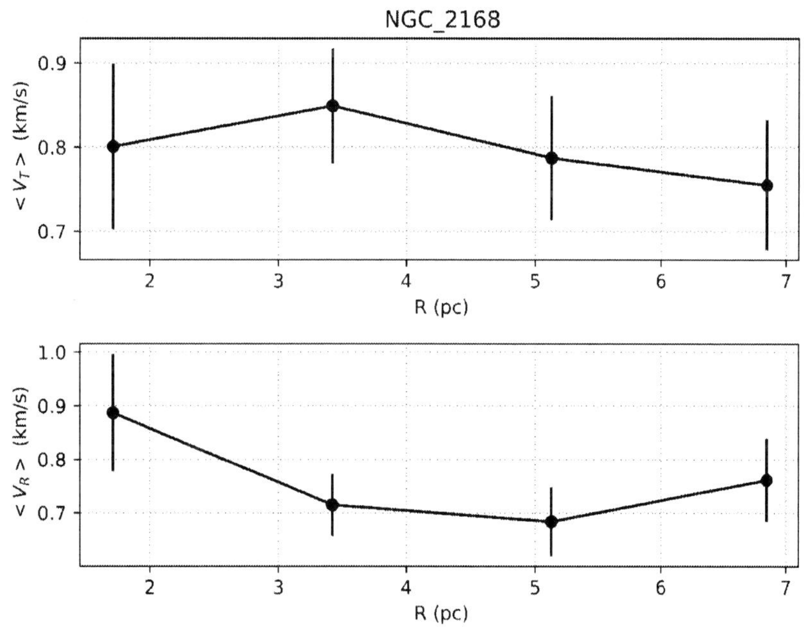


**Figure 1.** Velocity dispersion in tangential direction (top) and radial direction (bottom) for NGC 2168. R shows the distance from the centre of the clusters.

## 3. Method

We determine virial mass and projected mass of the three clusters. To determine the virial mass we need radial velocity dispersion that determined from radial velocity data. We use virial mass formula from equation (4) from [9] with assumption that all the bodies have equal masses. Projected mass determined using both radial and tangential velocity dispersion. We use projected mass estimator formula from equation (19) from [1].

To calculate the velocity dispersion we divide the cluster into several region by making rings with different radius from the centre. For each cluster we divide the cluster into four rings with the outer rings is centre radius of the cluster obtained from MWSC catalogue [7]. The reason we choose the outer rings is because most of the star are inside this radius, so we want to make sure that in every ring have at least 2 stars in it. After we sure that each of the ring have at least 2 stars, then we could calculate the velocity dispersion in that region for the tangential and radial direction.

## 4. Results and analysis

Figure 1 shows the radial and tangential velocity dispersion for NGC 2168. The value of velocity dispersion for both direction are less than ~1 km/s which is a normal value for open clusters. To calculate the error of the velocity dispersion, we use standard error formula, equation (1),

$$SE = \frac{\sigma}{\sqrt{2n}}, \tag{1}$$

Where $\sigma$ is standard deviation for all stars in each ring and $n$ is the number of stars.

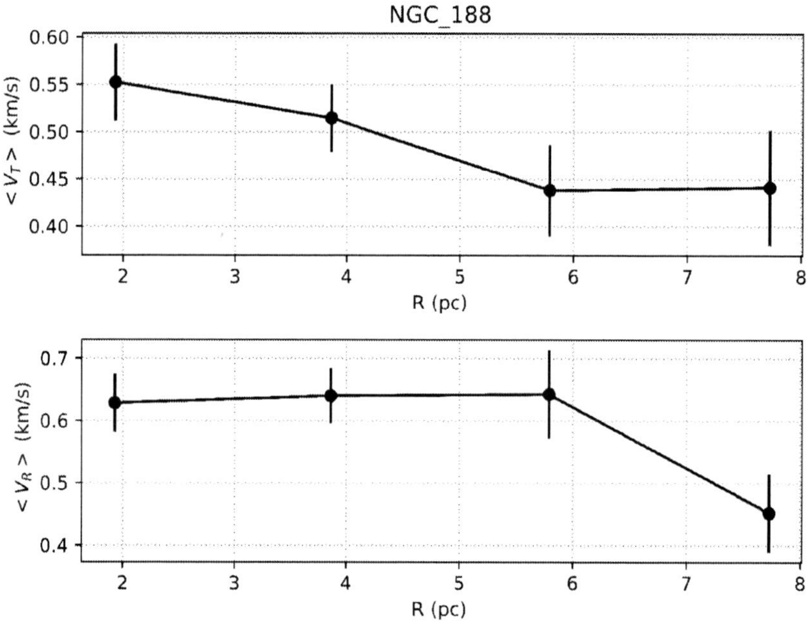


**Figure 2.** Velocity dispersion in tangential direction (top) and radial direction (bottom) for NGC 188. R shows the distance from the centre of the clusters.

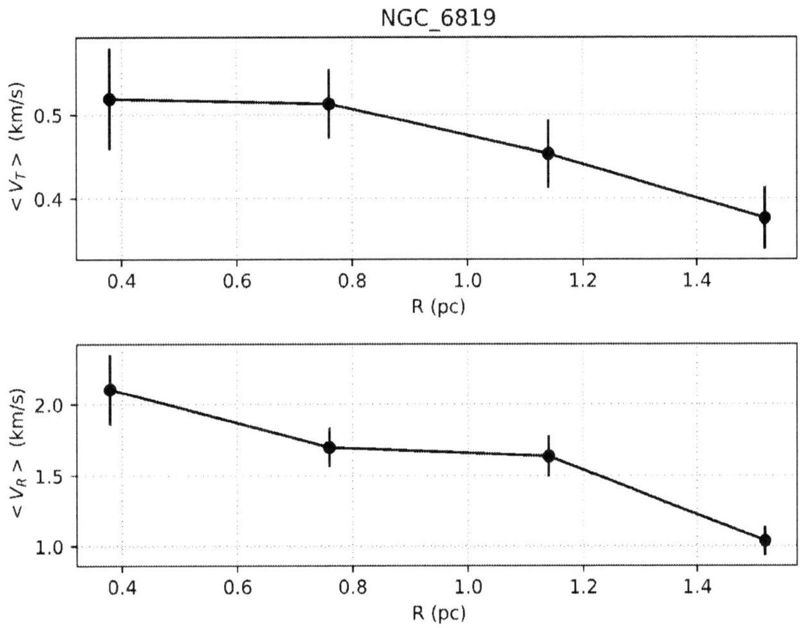


**Figure 3.** Velocity dispersion in tangential direction (top) and radial direction (bottom) for NGC 6819. R shows the distance from the centre of the clusters.

For both directions in Figure 1, the velocity dispersion have a similar value which is between 0.7~0.9 km/s. The similarity indicates that the velocity distribution on radial and tangential direction maybe isotropic. The trend of velocity dispersion for both direction is also similar, have high value close to the cluster core and decrease with the radius. The decreasing with the radius shows that the region near centre is more crowded than the outer region. This also show that the region with high population of stars or higher mass will make the velocity dispersion bigger than the less populated region. Using this velocity dispersion we obtain the virial mass of ~4056 $M_\odot$ and projected mass of ~5833 $M_\odot$ within the radius R < 7pc for NGC 2168. This mass is in the range of the mass that Leonard and Merritt [1] determined.

Figure 2 shows the velocity dispersion of NGC 188 on radial and tangential direction. The NGC 188 have a slightly bigger centre radius than NGC 2168 and the value of velocity dispersion is also smaller than NGC 2168. These might have a relation with the clusters age. The NGC 188 is older cluster than NGC 2168, so NGC 188 had already expanding more than NGC 2168. The expansion of the clusters make the radius bigger and decreasing the number density of stars, causing the velocity dispersion also smaller than more dense and younger cluster, NGC 2168. Velocity dispersion in radial and tangential direction also have slightly different value that indicate that the velocity distribution is not quite isotropic. This is maybe caused by the evolution of cluster that make an asymmetric expansion. Using the velocity dispersion we obtain the virial mass of ~2776 $M_\odot$ and projected mass of ~3326 $M_\odot$ within the radius R < 8pc for NGC 188. This mass is within the range of the mass that Bonatto et al. [10] determined. The velocity dispersion of NGC 6819 is shown in Figure 3. The value of tangential velocity dispersion is similar with the previous clusters, but the radial velocity dispersion is larger than the previous clusters. This large velocity dispersion could be caused by the error of radial velocity determination because the distance of NGC 6819 is the farthest among the three, so higher precision instrument is needed to determine same data with high accuracy. This also could be caused that the velocity distribution of NGC 6819 is not asymmetric that make the velocity at our radial direction is bigger than the tangential component. Using the velocity dispersion we obtain the virial mass of ~4244 $M_\odot$ and projected mass of ~3518 $M_\odot$ within the radius R < 8pc for NGC 6819. The mass of NGC 6819 that we obtain is still in the range of the mass that determined by Yang et al. (2013) [3].

## 5. Conclusion

In this paper, we have determined the virial mass and projected mass of NGC 2168, NGC 188, and NGC 6819. The mass is derived from velocity dispersion in radial and tangential direction. The velocity dispersion is similar for all cluster except for NGC 6819 which have a larger value of radial velocity dispersion. We obtain the virial mass of the three clusters ranging from ~2700 to ~4300 $M_\odot$ and the projected mass ranging from ~3300 to ~5900 $M_\odot$. Theoretically this mass range is normal for normal open clusters which have number of star from 100 to $10^4$ stars. We need to do further study to be sure that the masses we obtain is accurate, especially about the membership determination and the accuracy of kinematics data.

## Acknowledgments

This research are founded by 'Program Penelitian, Pengabdian kepada Masyarakat dan Inovasi ITB' (P3MI-ITB) research, Astronomy Research Group ITB.

## References

[1] Leonard P J T and Merritt D 1989 *The Astrophysical Journal* **339** 195
[2] Geller M A, Mathieu R D, Harris H C, and McClure R D 2008 *The Astrophysical Journal* **135** 2264
[3] Yang S C, Sarajedini A, Deliyannis C P, Sarrazine A R, Kim S C, and Kyeong J 2013 *The Astrophysical Journal* **762** 3
[4] Gaia Colaboration 2018 *Astronomy & Astrophysics* **616** 1
[5] Geller M A, Mathieu R D, Braden E K, Meibom S, Platais I, and Dolan C J 2010 *The Astrophysical Journal* **139** 1383
[6] Hole K T, Geller M A, Mathieu R D, Platais I, Meibom S, and Latham D W 2009 *The Astrophysical Journal* **138** 159
[7] Kharchenko N V, Piskunov A E, Schilbach E, Roser S, and Scholz, R D 2013 *Astronomy & Astrophysics* **558** 53
[8] Bailer-Jones C A L, Rybizki J, Fouesneau M, Mantelet G and Andrae R 2018 *The Astrophysical Journal* **158** 58
[9] Heisler J, Tremaine S, and Bahcall J N 1985 *The Astrophysical Journal* **298** 8
[10] Bonatto C, Bica E, and Santor Jr J F C 2005 *Astronomy & Astrophysics* **433** 917

# The Formation of Binary Star Cluster in Our Galaxy from Fractal Stellar Distribution

**R Darma[1,*], M I Arifyanto[1], and M B N Kouwenhoven[2]**

[1] Department of Astronomy, Faculty of Mathematic and Natural Sciences, Institut Teknologi Bandung, Jalan Ganeca no. 10, Bandung 40132, Indonesia
[2] Department of Mathematical Sciences, Xi'an Jiaotong-Liverpool University, 111 Ren'ai Rd, Suzhou Industrial Park, Suzhou 215123, P.R. China

Email: darmarendy@gmail.com

**Abstract.** Observations showed that there are star clusters in paired condition, called as binary star cluster. Some statistical studies attempted to identify the existence of binary and multiple clusters in our Galaxy and Magellanic Clouds. MIStiX survey also found that most of embedded younger star clusters (a few Myr) have clumpy structures, indicating star distribution in young clusters cannot be simply described as spherical shape. This structure can be the key-factor of the formation of primordial binary cluster in our Galaxy. In this work, we simulate the formation of binary star cluster from fractal distribution in isolated condition and under Galactic tidal field for 50 Myr. This aims to investigate the most probable condition to let the highly formation of binary star cluster in our Galaxy and the role of Galactic tidal field in that formation. We find that the clumpier structure of star cluster, the number of binary/multiple cluster will increase, and the same condition also occurs for the warmer condition or higher $\alpha_{vir}$ (the kinetic to potential energy ratio). But this number will decrease at larger time due to the merger processes between the star clusters. By these arguments, we conclude that the most probable condition of star cluster to let the highly formation of binary star cluster is the one with highly sub structured and warm condition. For such condition of star cluster, the fraction of binary/multiple in our Galaxy is $0.87 \pm 0.06$ at time 20 Myr and decreases to be $0.63 \pm 0.09$ at time 50 Myr, with reduction $0.24 \pm 0.08$. As comparison, this reduction is only $0.07 \pm 0.05$ for the same condition in isolated. The higher deflation shows that the Galactic tidal field has important role in the formation of binary star cluster, as well as the initial mass of star cluster. And we suggest the discovery of multiple clusters in this work is interesting to learn more deeply in the future work.

***Keywords:*** *Binary Star Cluster and Fractal Stellar*

## 1. Introduction

The most of stars are born in star cluster [1]. Some observations showed that there are star clusters (open clusters) in paired condition [2, 3]. Some of them are gravitationally interacting, known as binary star cluster. Besides that, many statistical studies attempted to identify binary and multiple star clusters in our Galaxy and Magellanic Clouds, see [2, 4, 5, 6, 7, 8]. They found that the number of binary/multiple clusters in our Galaxy is smaller than Large Magellanic cloud due to the difference of star formation rate both in the galaxies. Binary star clusters are systematically younger than single clusters, i.e., $\leq 25$ Myr [2]. Besides that, several previous studies such as [9, 10, 11] showed most of younger star clusters (a few Myr) have clumpy structures. Furthermore, MIStiX survey supports these arguments after observed embedded star clusters in our Galaxy, see [12]. The clumpy structure of star cluster at early age can trigger the formation of binary star cluster, especially primordial binary cluster that formed through simultaneous mechanism [2]. A few numerical studies such as [13] and [14]

IOP Conf. Series: Journal of Physics: Conf. Series **1231** (2019) 012028 doi:10.1088/1742-6596/1231/1/012028

attempted to model the clumpy structure of star cluster with fractal distribution to investigate the formation of binary star cluster. But their simulations were made in isolated condition without the effect of Galactic tidal field. In this work, we simulate the formation of binary star cluster not only in isolated condition, but also under Galactic tidal field. It aims to investigate the most probable condition of star cluster which let the highly formation of binary star cluster in the Milky Way. Furthermore, we also would like to investigate the role of Galactic tidal field in the formation of binary star cluster. In the section 2, we introduce our fractal model of star cluster and Milky Way potential. Our simulations and statistical method are presented in the section 3. Finally, we summarize our results in the section 4.

## 2. Model of Star Cluster

We follow [15] to model our star cluster from fractal distribution. To model the fractal cluster, we have to define a cube with length of $N_{div}$ and put a parent particle at the center of the cube (first generation parent particle). Then divide the cube into ${N_{div}}^3$ subcubes and put a child particle at the center of each subcubes. Child particle with probability ${N_{div}}^{D-3}$ is taken to be the second generation parent particle (survived particle), while others are deleted. That probability depends on fractal dimension ($D = 0 - 3$) which shows the clumpier structure of star cluster, the smaller value of $D$ (see Figure 1). A little noise is added to each position of survived particle, to avoid an obviously gridded structure. This process is repeated recursively until there is a sufficiently large survived particle. We then impose a spherical envelope of radius 2 pc within the cube (as radius of star cluster) to reduce the survived particles that located out of the radius. The remaining survived particles are then culled randomly until the required number is left. That's why, there should be more survived particles than the required number of stars before imposing the radius.

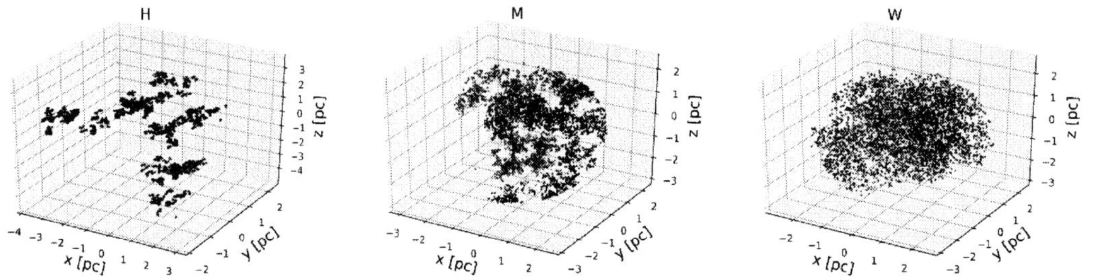


**Figure 1.** Model of star clusters from fractal distribution with $D$ = 1.6, 2.2, and 2.9 for $\alpha_{vir}$ =

**Table 1.** Physical parameters set to model star cluster.

| Parameters | Values | References |
|---|---|---|
| Number of stars in cluster ($N$) | 5000 | |
| Mass of cluster ($M_c$) | 3000 M⊙ | |
| Radius of cluster ($R_c$) | 2 pc | [14, 15, 16] |
| Kroupa [17] IMF | 0.1 – 50 M⊙ | [13, 14, 15] |

We set the star velocity dispersion is coherent, as suggested by [15]. Random velocity is given for the first generation parent particle. Then the second generation parent particles get their velocities inherited from the first one, plus a little noise. This process is repeated recursively and the noise is large at the higher levels, and then becomes smaller, in order to scale the velocities so that $\alpha_{vir}$ has prescribed value, see [14]. Here $\alpha_{vir}$ is the kinetic to potential energy ratio of star cluster and this parameter describes how large the star velocity dispersion in star cluster. We set some physical parameters to model our star cluster in Table 1. We follow [14] for the combination of $D$ and $\alpha_{vir}$ to describe various condition of structure and velocity dispersion of star cluster (see Table 2). For the simulations under Galactic tidal field, we use Milky Way axisymmetric potential as presented in [18].

**Table 2.** The combination of $D$ and $\alpha_{vir}$ to describe various conditions of structure and velocity dispersion of star cluster, adopted from [14]. Highly substructured ($D$ = 1.6), moderate substructure ($D$ = 2.2), and smooth structure ($D$ = 2.9) are denoted by letter H, M, and S. Cool ($\alpha_{vir}$ = 0.3), virialised ($\alpha_{vir}$ = 0.5), and warm ($\alpha_{vir}$ = 0.7) regions are denoted by letter C, V, and W.

| $D$ | | $\alpha_{vir}$ | |
|---|---|---|---|
| | 0.3 | 0.5 | 0.7 |
| 1.6 | HC | HV | HW |
| 2.2 | MC | MV | MW |
| 2.9 | SC | SV | SW |

## 3. Simulation and Identification of Binary Star Cluster

### 3.1. *N-Body Simulations*

We use AMUSE Framework to make our N-body simulations [19]. We assume that all gases are expelled from the cluster environment. Furthermore, stellar evolution and binary fraction are also not considered in the simulations. In every case, we run 30 simulations for each initial condition as set in Table 2 for 50 Myr with time step 0.1 Myr. For isolated condition, we use BHTree code [20] to calculate the gravitational interactions between stars in the cluster. For the condition under Galactic tidal field, we use Bridge code [21] to couple the BHTree code with the calculation of gravitational interactions between stars and Milky Way. There is no different distribution of stars' position, velocity, and mass in the cluster for simulations in isolated and under Galactic tidal field. We put the position of star cluster at $(x, y, z) = (8.5, 0.0, 0.02)$ kpc in the Milky Way, adopted from [22]. Then the velocity $(u, v, w) = (11.1, 12.4 + v_{lsr}, 7.25)$ km/s, adopted from [23]. Here $v_{lsr}$ = 220 km/s which is corresponding to our Milky Way model. By the physical parameters from Table 1 and the position of star cluster in the Milky Way, we calculate the initial tidal radius of star cluster at time 0 Myr is 8.5 pc. It is 4.25 times radius of the cluster.

### 3.2. *Identification of Binary Star Cluster*

We introduce weighted K-Means algorithm to identify binary star cluster at time 20 Myr and 50 Myr. K data points are chosen from dataset as initial centroids using K-Means++ algorithm. Here we use the position of stars as dataset. By calculating weighted euclidean distance between data points and every centroid, all those are closest to a centroid will create a cluster (clustering process). We use mass of stars as weight in this calculation. Now we have K clusters that need new centroids by calculating mean of all members' position in a cluster (moving centroids process). Repeat clustering and moving centroids processes until new centroids are no longer similar with previous ones.

K-Means++ algorithm is used in order to choose K initial centroids. In this algorithm, firstly we choose one data point from dataset randomly as an initial centroid. Then calculate squared weighted euclidean distance between each data point and the initial centroid, denoted by $d_i^2$ where $i$ is each data point. A data point will be an initial centroid when its probability is proportional to $\frac{d_i^2}{\sum_{i=1}^{n} d_i^2}$. Repeat this process until K initial centroids have been chosen. In this work, we set K = 1, 2, …, 10 as guessed number of centroids and then implement Bayesian Information Criterion (BIC) and angle-based method in order to optimize the best K centroids for each dataset. By using K-Means algorithm, we calculate the fraction of *single*, *binary/multiple*, and *merger*. The term of *single* refers to the star cluster that formed as a single star cluster. *Binary/multiple* refers to the star cluster that formed as binary/multiple cluster, both of the one that orbiting each other or separated in different directions. When the formed binary/multiple cluster then merges into a single star cluster, it is categorized as *merger*. We calculate these fractions at time 20 Myr and 50 Myr to investigate the fate of dynamical evolution of star cluster at different times.

## 4. Results and Discussion

We present the fractions of single, binary/multiple, and merger at 20 Myr and 50 Myr for isolated condition in Figure 2. At time 20 Myr, we can see that the clumpier structure of star clusters, the fractions of binary/multiple increase. The highest fraction of binary/multiple occurs in the H conditions ($D$ = 1.6), where the star clusters are highly substructured. Besides that, the velocity dispersion of stars in a cluster has also important role to form binary star cluster. From the same figure, we can see the W conditions ($\alpha_{vir}$ = 0.7) have the highest fractions of binary/multiple. The higher value of $\alpha_{vir}$, the higher star velocity dispersion in a cluster. That's why the warm condition tends to form agglomerative hierarchical clustering. We can conclude that the most probable condition to let the highly formation of binary star cluster is the one with highly substructured and warm condition (HW condition).

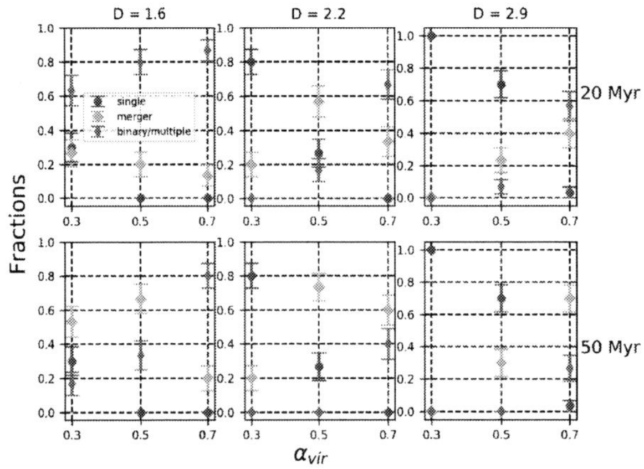


**Figure 2.** The fractions of single (green), binary/multiple (red), and merger (orange) for isolated condition at 20 Myr (top panel) and 50 Myr (bottom panel).

Our result for isolated condition at time 20 Myr is in a good agreement with [14], although it seems that our more massive star cluster models evolve faster than [14] which makes our fractions of binary/multiple are higher and the fractions of single are smaller. This shows that the initial mass of star cluster has important role in the formation of both binary and multiple cluster. If we compare the fractions at time 20 Myr to 50 Myr, we get there are decreasing fractions of binary/multiple for all conditions in Figure 2 due to merger processes between star clusters. It means that the older age of star cluster, the fractions of binary/multiple decrease and followed by the increasing fractions of merger for the same lifetime.

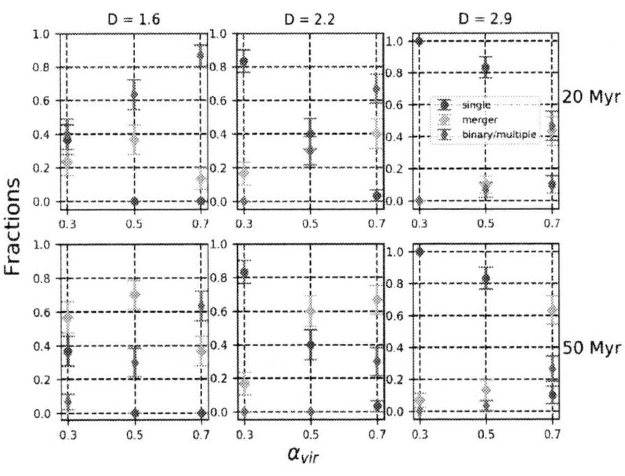


**Figure 3.** The fractions of single, binary/multiple, and merger for condition under Galactic tidal field at 20 Myr (top panel) and 50 Myr (bottom panel). The colors as explained in Figure 2.

We get the trends of fractions in condition under Galactic tidal field are similar with isolated. But the fractions of binary/multiple are a bit smaller and the fractions of single are a bit higher than isolated at time 20 Myr. For HW condition under Galactic tidal field, we get the fraction of binary/multiple is 0.87 ± 0.06 at time 20 Myr and then decreases to be 0.63 ± 0.09, with deflation 0.24 ± 0.08. While the HW isolated condition, the deflation is only 0.07 ± 0.05. The higher deflation we get in the condition under Galactic tidal field shows that our Galaxy has important role on the dynamical process of the formation of binary star cluster. The existence of multiple clusters that identified in our simulations are interesting to investigate more deeply how this systems can be formed. Besides that, the fractions of binary/multiple are still high enough at time 50 Myr and suggesting that binary and multiple clusters could have lifetime more than 50 Myr. It is interesting to learn in the future work of how do these binary and multiple clusters end their life after 50 Myr with considering the stellar evolution.

**Acknowledgements**
R.D. would like to thank H.R.T. Wulandari, H.L. Malasan, and M. Raharto from Astronomy Department of Institut Teknologi Bandung for helpful conversations and advices. R.D. and M.I.A. acknowledge support from Program Penelitian Pengabdian kepada Masyarakat dan Inovasi ITB (P3MI-ITB). M.B.N.K. acknowledges support from the National Natural Science Foundation of China (grant 11573004). This research was supported by the Research Development Fund (grant RDF-16-01-16) of Xi'an Jiaotong-Liverpool University (XJTLU).

**References**
[1] Lada C J, and Lada E A 2003 *Annu. Rev. Astron. Astrophys.* **41** 57
[2] de la Fuente Marcos R, and de la Fuente Marcos C 2009 *Astron. Astrophys.* **500** L13
[3] de la Fuente Marcos R, and de la Fuente Marcos C 2010 *Astron. Astrophys.* **500** L13
[4] Bhatia R K, and Hadzidimitriou D 1988 *Mon. Not. R. Astron. Soc.* **230** 215
[5] Subramaniam, A, Gorti U, Sagar R, and Bhatt H C 1995 *Astron. Astrophys.* **302** 86
[6] Pietrzynski G, and Udalksi A 2000 *Acta. Astron.* **50** 355
[7] Pietrzynski G, Udalski A, Kubiak M, Szymanski M, Wozniak P, and Zebrun K 1999 *Acta. Astron.* **49** 521
[8] Dieball A, Muller H, and Grebel E K 2002 *Astron. Astrophys.* **391** 547
[9] Elmegreen B G, Efremov Y, Pudritz R E, and Zinnecker H 2000 *in Protostars and Planets IV, ed. V. Mannings, A. P. Boss, & S. S. Russell (University of Arizona Press)* 151
[10] Williams J P, Blitz L, and McKee C F 2000 *in Protostars and Planets IV, ed. V. Mannings, A. P. Boss, & S. S. Russell (Tucson, AZ: Univ. Arizona Press)* 97
[11] de la Fuente Marcos R, and de la Fuente Marcos C 2006 *Astron. Astrophys.* 452 163
[12] Kuhn M A, Feigelson E D, Getman K V, Baddeley A J, Broos P S, Sills A, Bate M R, Povich M S, Luhman K L, Busk H A, Naylor T, and King R R 2014 *Astrophys. J.* **787** 107
[13] Yu J L, De Grijs R, and Chen L 2011 *Astrophys. J.* **732** 16
[14] Arnold B, Goodwin S P, Griffiths D W, and Parker R J 2017 *Mon. Not. R. Astron. Soc.* **471** 2498
[15] Goodwin S P, and Whitworth A P 2004 *Astron. Astrophys.* **413** 929
[16] Priyatikanto R, Kouwenhoven M B N, Arifyanto M I, Wulandari H R T, and Siregar S 2016 *Mon. Not. R. Astron. Soc.* **457** 1339
[17] Kroupa P 2002 *Science* **295** 82
[18] Haghi H, Zonoozi A H, and Taghavi S 2015 *Mon. Not. R. Astron. Soc.* **315** 2153
[19] Portegies-Zwart S F, McMillan S L W, van Elteren A, Pelupessy F I, and de Vries N 2013 *Comput. Phys. Commun.* **183** 456
[20] Barnes J, and Hutt P 1986 *Nature* **324** 446
[21] Fujii M, Iwasawa M, Funato Y, and Makino J 2007 *Publ. Astron. Soc. Japan* **59** 1095
[22] Martinez-Barbosa C A, Jilkova L, Portegies-Zwart S, and Brown A G A 2016, *Mon. Not. R. Astron. Soc.* **457** 1062
[23] Schonrich R, Binney J, and Dehnen W 2010 *Mon. Not. R. Astron. Soc.* **403** 1829

# An Application of The Markov Chain Monte Carlo (MCMC) Method to Open Cluster Membership Determination

**Y A Hidayat, M I Arifyanto, Aprilia and M I Hakim**

[1]Department of Astronomy, Faculty of Mathematics and Natural Sciences, Institut Teknologi Bandung, Indonesia

Email: yudarhd@gmail.com

**Abstract.** The determination of membership in open clusters will be more difficult than determination membership of the globular clusters. This is due to the location of open clusters which located on the galactic disk. From that location, when observing open clusters, it will be observed foreground and background stars also. Determination of membership in cluster could be done by calculating the probability of a star becoming a cluster member. We used double elliptic bivariate Gaussian function to represent the population of cluster and field stars in the region. The function has eleven parameters and solved by using Markov Chain Monte Carlo (MCMC). The Metropolis-Hastings algorithm is used in this works to obtain the probability of open cluster membership. We used astrometric data from GAIA Data Release 2 for two open clusters which their membership are to be determined i.e. IC 1590 and NGC 2262. Our result is nearly in accordance with the MWSC II catalogue. After the cluster membership is obtained, the physical parameters of the two clusters using isochrones fitting from [1] are the distance(pc) 2200 and 2366, reddening(mag) 0.728 and 0.830, distance modulus(mag) 12.210 and 12.438 also log age 6.895 and 8.995, respectively for IC 1590 and NGC 2262.

***Keywords:*** *Markov Chain Monte Carlo (MCMC) and Open Cluster Membership Determination*

## 1. Introduction

The Study determining the membership of clusters is really important to learn about stellar evolution and Galaxy. A star cluster is any physically related group of stars that formed together, they remain together due to their mutual gravitational attraction. Star clusters are divided into two types which are open cluster and globular cluster. This paper will be focused on determining of open clusters because open clusters are located on the galactic disk. From that information, the field stars; foreground and background stars will be observed as well. So determining the membership of open clusters becomes important for identifying the actual members of the cluster. The determination of membership of open clusters can be done by calculating the probability of a star becoming a member of the cluster. To calculate the probability, we use double elliptic bivariate Gaussian function from [2] which contains have eleven parameters. In this work, Markov Chain Monte Carlo (MCMC) is applied to separate member stars from the field stars using Metropolis-Hastings algorithm. To ensure the parameters obtained by MCMC, quality of fit (D) parameter should be calculated. If the obtained value of D is close to 0, then the value from eleven parameters fit to the model. This probability is compared from MWSC II (Milky Way Stellar Cluster II) from [3].

## 2. Data and Methods

Proper motion data from GAIA Data Release 2 is used as the astrometric data to determine the membership of IC 1590 and NGC 2262. As for the photometry data, we use magnitude of the stars in G band from GAIA Data Release 2 and also J, H band from 2MASS. These photometry data is used to construct CMD. Data of probability from MWSC II catalogue is used as the comparation to the result. The data from this research has been selected with the criteria:

i) Parallax error < 0.15 (mas)
ii) $-15.0 < \mu_x < 10.0$ with error $\mu_x < 0.3$ (mas/yr)
iii) $-15.0 < \mu_y < 20.0$ with error $\mu_y < 0.3$ (mas/yr)

The stars of the cluster can be identified using double elliptic bivariate Gaussian equation

$$\Phi(\mu_x, \mu_y) = (1 - n_f)\Phi_c(\mu_x, \mu_y) + n_f\Phi_f(\mu_x, \mu_y)$$

$$= \frac{1 - n_f}{2\pi\sigma_{xc}\sigma_{yc}\sqrt{1 - \rho_c^2}}\exp\left\{-\frac{1}{2(1 - \rho_c^2)}\left[\begin{array}{c}\left(\frac{\mu_x - \mu_{x0c}}{\sigma_{xc}}\right)^2 + \left(\frac{\mu_y - \mu_{y0c}}{\sigma_{yc}}\right)^2 \\ -2\rho_c\left(\frac{\mu_x - \mu_{x0c}}{\sigma_{xc}}\right) + \left(\frac{\mu_y - \mu_{y0c}}{\sigma_{yc}}\right)\end{array}\right]\right\}$$

$$+ \frac{n_f}{2\pi\sigma_{xf}\sigma_{yf}\sqrt{1 - \rho_f^2}}\exp\left\{-\frac{1}{2(1 - \rho_f^2)}\left[\begin{array}{c}\left(\frac{\mu_x - \mu_{x0f}}{\sigma_{xf}}\right)^2 + \left(\frac{\mu_y - \mu_{y0f}}{\sigma_{yf}}\right)^2 \\ -2\rho_f\left(\frac{\mu_x - \mu_{x0f}}{\sigma_{xf}}\right)\left(\frac{\mu_y - \mu_{y0f}}{\sigma_{yf}}\right)\end{array}\right]\right\}. \quad (1)$$

This equation depends on eleven parameters namely $n_f$ is the proportion of field stars, $\mu_{x0c}, \mu_{y0c}, \sigma_{xc}$ and $\sigma_{yc}$are the proper motion centroids and the dispersions or standard deviations for the cluster, respectively; $\mu_{x0f}, \mu_{y0f}, \sigma_{xf}$ and $\sigma_{yf}$ are the proper motion centroids and the dispersions or standard deviations for the field, respectively; $\rho_c$ and $\rho_f$ are the cluster and the field proper motion correlation coefficients. To derive the eleven parameters, the MCMC method calculate a random probability of posterior that is

$$P(\theta|D, M) \propto P(D|\theta, M)P(\theta|M), \quad (2)$$

Where $P(D|\theta, M)$ is likelihood function and $P(\theta|M)$ is prior or range value of eleven parameters. Each of parameter, data and model is denoted as $\theta, D$ and $M$. Bayes rules is used to determine the probability of member stars,

$$P_i(C|\mu_{xi}, \mu_{yi}) = \frac{(1 - n_f)\Phi_c(\mu_{xi}, \mu_{yi})}{n_f\Phi_f(\mu_{xi}, \mu_{yi}) + (1 - n_f)\Phi_c(\mu_{xi}, \mu_{yi})}, \quad (3)$$

if the value of membership probability $> 50\%$ then it is identified as a member star and if the value of membership probability $< 50\%$ then it is identified as a field star. The MCMC method is very depended on the number of iterations, the greater number of iterations performed, the closer the value of eleven parameters to convergence of each parameters. Therefore, the value of each parameters obtained from mean and standard deviation. On the results obtained, then after the value obtained from eleven parameters, based on [4] to confirm from value of each parameter can be calculated from D parameter or quality of fit

$$D = \frac{(\mu_{x0cMC} - \mu_{x0cLIT})^2}{(\sigma_{xcMC})^2 + (\sigma_{xcLIT})^2} + \frac{(\mu_{y0cMC} - \mu_{y0cLIT})^2}{(\sigma_{ycMC})^2 + (\sigma_{ycLIT})^2}, \quad (4)$$

where MC is value from MCMC method and LIT is value from references.

If the value is very small or that means close to 0, the value obtained is considered good. To confirm that the probability value of the parameters is greater than 50% Vector Point Diagram (VPD) and Color-Magnitude Diagram (CMD) are done by performing isochrone fitting of [1] on physical parameters of MWSC II. The greater the probability value, the smaller the number of clusters members. After getting the probability value > 50% for each star (member stars), we choose the data near the peak of parallax value for further analysis.

## 3. Results and Analysis

The MCMC method used in this work is applied for the two open clusters. The initial and prior values are listed in the table 1. The best parameters values are obtained after iterating 1,000,000 times with step of 0.001 for IC 1590 and NGC 2262. The value obtained is the mean value with standard deviation for eleven parameters. The parameters values are listed in the table 2.

**Table 1.** Initial and prior for IC 1590[a] and NGC 2262[b]

| Parameter | Initial[a] | Prior[a] | Initial[b] | Prior[b] |
|---|---|---|---|---|
| $n_f$ | 0.5 | $0.0 < n_f < 1.0$ | 0.5 | $0.0 < n_f < 1.0$ |
| $\mu_{x0c}$ | $-1.736 \pm 0.081$ | $-4.0 < \mu_{x0c} < 0.5$ | $0.189 \pm 0.031$ | $-0.4 < \mu_{x0c} < 0.7$ |
| $\mu_{y0c}$ | $-0.954 \pm 0.048$ | $-2.5 < \mu_{y0c} < 0.5$ | $0.118 \pm 0.017$ | $-0.4 < \mu_{y0c} < 0.4$ |
| $\sigma_{xc}$ | $0.345 \pm 0.031$ | $0.0 < \sigma_{xc} < 9.0$ | $0.061 \pm 0.001$ | $0.0 < \sigma_{xc} < 2.0$ |
| $\sigma_{yc}$ | $0.279 \pm 0.012$ | $0.0 < \sigma_{yc} < 6.0$ | $0.053 \pm 0.006$ | $0.0 < \sigma_{yc} < 1.3$ |
| $\rho_c$ | 0.015 | $-1.0 < \rho_c < 1.0$ | 0.015 | $-1.0 < 1\rho_c < 1.0$ |
| $\mu_{x0f}$ | $-1.634 \pm 0.037$ | $-4.0 < \mu_{x0f} < 2.0$ | $0.149 \pm 0.013$ | $-0.6 < \mu_{x0f} < 0.8$ |
| $\mu_{y0f}$ | $-0.999 \pm 0.016$ | $-3.0 < \mu_{y0f} < 1.0$ | $0.124 \pm 0.006$ | $-0.5 < \mu_{y0f} < 0.8$ |
| $\sigma_{xf}$ | $3.552 \pm 0.075$ | $0.0 < \sigma_{xf} < 12.0$ | $0.824 \pm 0.03$ | $0.0 < \sigma_{xf} < 4.0$ |
| $\sigma_{yf}$ | $2.614 \pm 0.031$ | $0.0 < \sigma_{yf} < 8.0$ | $1.292 \pm 0.024$ | $0.0 < \sigma_{yf} < 3.0$ |
| $\rho_f$ | 0.015 | $-1.0 < \rho_f < 1.0$ | 0.015 | $-1.0 < \rho_f < 1.0$ |

[a] the values for IC 1590
[b] the values for NGC 2262

**Table 2.** Parameter value using MCMC method for IC 1590[a] and NGC 2262[b]

| Parameter | IC 1590 | NGC 2262 |
|---|---|---|
| N | 2874 | 908 |
| $n_f$ | $0.304 \pm 0.016$ | $0.429 \pm 0.022$ |
| $\mu_{x0c}$ | $-1.613 \pm 0.046$ | $0.064 \pm 0.026$ |
| $\mu_{y0c}$ | $-0.975 \pm 0.018$ | $0.101 \pm 0.029$ |
| $\sigma_{xc}$ | $1.692 \pm 0.048$ | $0.506 \pm 0.027$ |
| $\sigma_{yc}$ | $1.133 \pm 0.029$ | $0.487 \pm 0.036$ |
| $\rho_c$ | $0.015 \pm 0.029$ | $-0.114 \pm 0.061$ |
| $\mu_{x0f}$ | $-0.317 \pm 0.186$ | $-0.415 \pm 0.112$ |
| $\mu_{y0f}$ | $-0.972 \pm 0.022$ | $-0.902 \pm 0.082$ |
| $\sigma_{xf}$ | $4.667 \pm 0.089$ | $2.678 \pm 0.086$ |
| $\sigma_{yf}$ | $3.634 \pm 0.122$ | $3.913 \pm 0.132$ |
| $\rho_f$ | $0.030 \pm 0.034$ | $-0.107 \pm 0.047$ |

The results of fitting using MCMC after cross-match between MWSC II and 2MASS are 30 for IC 1590 and 49 for NGC 2262. The number is the star with probability value $> 50\%$. The comparation number of member stars of probability between MCMC method and MWSC II Catalogue are listed in the table 3. The kinematic and photometric membership probabilities for stars in a cluster region of MWSC II catalogue is obtained from data-processing pipeline on the basis of stellar data from PPMXL and 2MASS. Based on the table 3, we can see the comparation number of member stars of each probability between MCMC method and MWSC II Catalogue. They differ due to the difference in the centre of the clusters used and the radius they encompass.

**Table 3.** Comparation number of member stars of each probability between MCMC method and MWSC II Catalogue after cross-match

| | **MCMC Method** | | **MWSC II Catalogue** | |
|---|---|---|---|---|
| **Probability (%)** | **IC 1590** | **NGC 2262** | **IC 1590** | **NGC 2262** |
| 50 | 30 | 49 | 18 | 52 |
| 70 | 25 | 45 | 11 | 34 |
| 90 | 19 | 36 | 5 | 36 |

The obtained value of each parameter is calculated by using parameter D based on the equation (4). The value obtained in parameter D has the greatest value on NGC 2262 and the smallest value on IC 1590 written on the table 4. The value obtained has been in the order less than $10^{-1}$ that gives the meaning that the value obtained by the method is considered very good for the determination of open clusters membership. Then for the member stars, we obtained CMD by performing isochrone fitting of [4] on physical parameters are listed on table 5, the J and H band are used for the isochrone fitting. The distribution probability of membership for IC 1590 and NGC 2262 are described on the figure 1. Parallax value of each clusters are displayed on the figure 2 and we use the value near the peak.

**Table 4.** D value for IC 1590 and NGC 2262

| | **IC 1590** | **NGC 2262** |
|---|---|---|
| Parameter D | 0.0184 | 0.0489 |

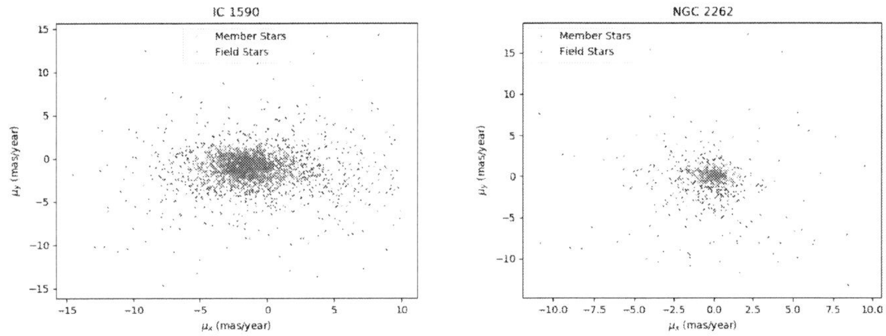


**Figure 1.** Vector Point Diagram (VPD) for IC 1590 (left) and NGC 2262 (right), the red color is representing member stars and the blue color is representing field stars

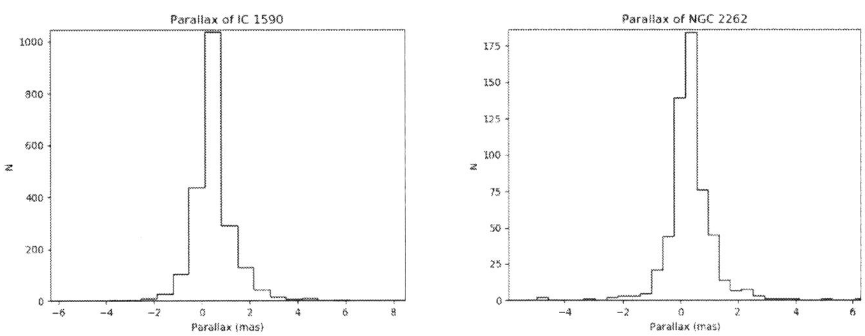


**Figure 2.** Histogram of parallax for IC 1590 (left) and NGC 2262 (right), the value near the peak is used for analysis

IOP Conf. Series: Journal of Physics: Conf. Series **1231** (2019) 012029 doi:10.1088/1742-6596/1231/1/012029

**Table 5.** The basic parameters for IC 1590 and NGC 2262 using isochrone fitting

| Name | Distance (pc) | Reddening (mag) | Distance Modulus (mag) | Log Age |
|---|---|---|---|---|
| IC 1590 | 2200 | 0.728 | 12.210 | 6.895 |
| NGC 2262 | 2366 | 0.830 | 12.438 | 8.995 |

The CMD of G, J and H band are very good and we can see the CMD from each open cluster is clearly on main sequences described by figure 3. Figure 4 describes the corner plot from each parameter. The parameter value obtained is the mean of the iteration after burn-in. The number of burn-in used in this work is 80% from the whole iteration. We check each value by evaluating the peak of the corner in every parameter and we get the value near the peak.

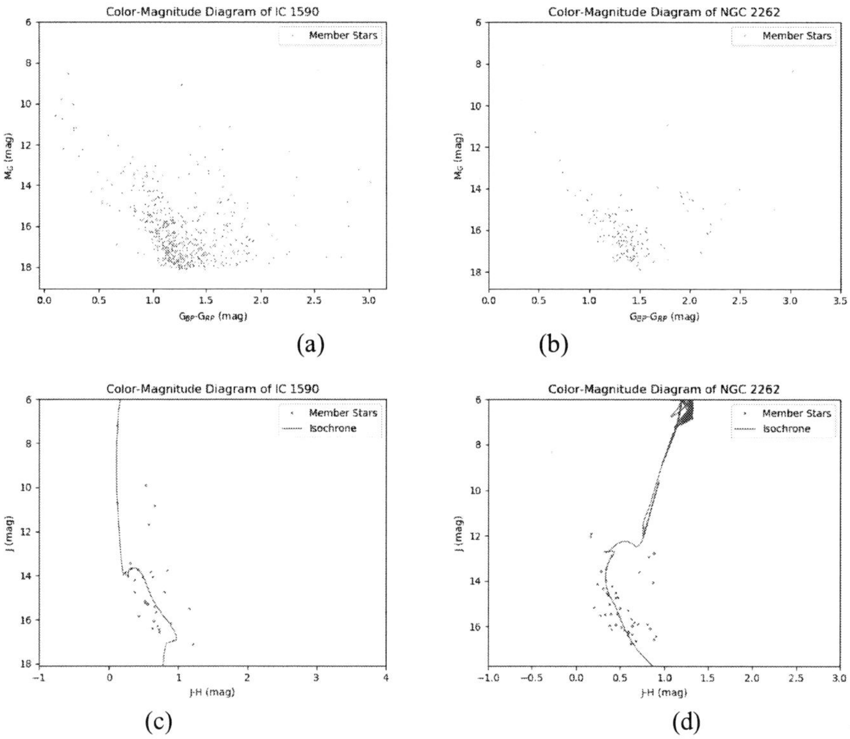


**Figure 3.** Color-Magnitude Diagram (CMD) using G band from GAIA Data Release 2 for member stars; probability value > 50% for (a) IC 1590 and (b) NGC 2262. Color-Magnitude Diagram (CMD) using J and H band from MWSC II after cross-match for member stars each open clusters. The red dots are member stars and the blue line is isochrone fitting from [4] for (c) IC 1590 and (d) NGC 2262

## 4. Conclusion and Discussion

The MCMC method can determine the open cluster membership by separating the stars around the star cluster into cluster and field stars. Using this method, the value of parameter D or quality of fit is obtained, with the largest value are 0.0489 for NGC 2262 and 0.0184 for IC 1595, that gives the meaning that the value obtained by the method is very good for the determination of open clusters membership. The amount of data and iteration is very influential on the time and convergence of the required value. The MCMC method is very good at determining the parameter value of double elliptic bivariate Gaussian function for a star cluster; it can be open clusters and globular clusters whose membership has not been determined. Furthermore, a precise of range value or prior is required to obtain a convergent parameter value. The advantages of MCMC method are usable for complicated distributions in high-dimensional spaces even when we don't know where the regions of high probability are, relatively easy to implement and fairly reliable to determine the parameter value from some functions.

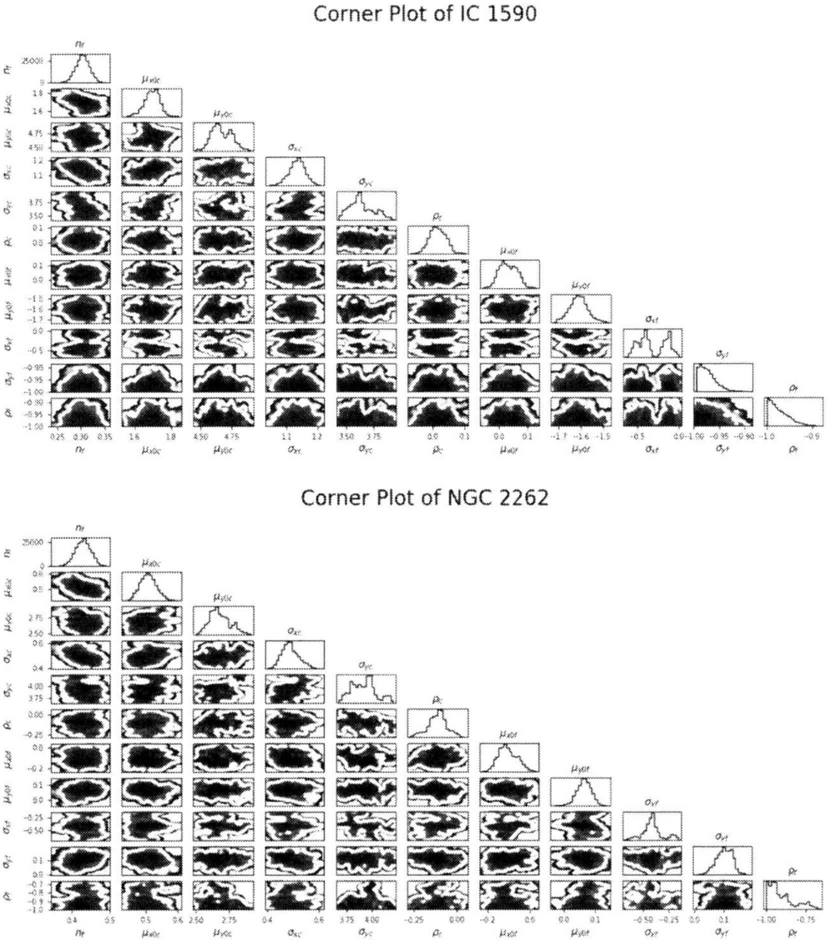


**Figure 4.** Corner plot from each parameters value for IC 1590 (upper) and NGC 2262 (lower). The parameter value obtained is the mean of steps after burn-in and the value is near the peak

**Acknowledgments**
We acknowledged Program Penelitian, Pengabdian Kepada Masyarakat, dan Inovasi (P3MI) for funding travel grant to SEAAN 2018

**References**

[1] Girardi L et al. 2000 *A&A* **141** 371-383
[2] Sabogal-Martinez B E *et al.* 2001 *Revista Mexicana de Astronomia y Astrofisica* **37** 105-114
[3] Kharchenko N V *et al.* 2013 *A&A* **53** 558
[4] Dias W S *et al.* 2014 *A&A* **79** 564
[5] Fitriana I K and M Ikbal Arifyanto 2016 *ISSEL. JoPCS* **771**
[6] Gaia Collaboration Brown A G A *et al.* 2018 *A&A* manuscript no. aa201833051
[7] Sharma S 2017 *Annu. Rev. Astron. Astrophys.* **55** 1-49

# The Magnetic Field Dynamics Equation of the Accreting and Rapidly Rotating Neutron Star in the ZAMO (Zero Angular Momentum Observers) Frame

**A Yasrina**[1*], **D Andra**[2]

[1]Department of Physics, Faculty of Mathematics and Natural Sciences, Universitas Negeri Malang
[2]Physics Education Study Program, Faculty of Teacher Training and Education, University of Lampung

Email: atsnaita.yasrina.fmipa@.um.ac.id

**Abstract.** The presence of accretion around neutron stars is assumed to be one of the causes of the decreasing of the magnetic field around neutron stars. Therefore, the equation that describes the relation between the decreasing of the magnetic fields and the accretion are required. The equation of the magnetic field decreasing describes the dynamics of the magnetic field around rapidly rotating neutron star. The relativistic magnetic field dynamics equation around a accreting and rapidly rotating neutron star in the ZAMO (Zero Angular Momentum Observers) frame has been formulated. The equation is derived from the rapidly rotating neutron star metric and the first and second of the relativistic Maxwell equations. The magnetic field dynamics equation is a differential equation in the radial, polar, and azimuthal parameter.

***Keyword**: Magnetic field, neutron stars, rapidly, ZAMO*

## 1. Introduction

Stars with mass $M_* \gtrsim 8M_\odot$, where $M_\odot$ denotes Solar mass, can undergo nucleosynthesis process or nuclear reactions. The nuclear reaction occurs in the form of burning silicon $^{28}Si$ to burning iron $^{56}Fe$. If the nucleosynthesis had finished, then the star collapsed and the supernova explosion occurred. After undergoing a supernova explosion, the state of the star depends on the thrown material. If a number of materials in the star's core after undergoing supernovae collapse, cool, reach an equilibrium and contain an abundance of degenerated neutrons, the star is called a neutron star [1]. In other words, neutron stars are produced by $M_* \gtrsim 8M_\odot$ massed massive stars that undergo supernova explosions [1]. The abundance of neutrons are obtained from the beta decay process, i.e., $+e^- \rightarrow n + v_e$ which occurs at high density.

A neutron star has a mass of $M_* \sim 1 - 2\, M_\odot$ and radius $R_* \approx 10 - 14$ km [2]. The maximum mass of neutron star is about $M_* \sim 1{,}5\, M_\odot$ with its radius is about $R_* \sim 3$ km [3]. The density of neutron star is about $\rho \simeq (2-3)\rho_o$, where $\rho_o = 2{,}8 \times 10^{14}$ g cm$^{-3}$ is the normal density of the nucleus [4]. The gravitational potential of a neutron star is about $E_{\text{grav}} \sim GM^2/R \sim 5 \times 10^{53}$erg $\sim 0{,}2\, Mc^2$, whereas the gravitational field at a neutron star's surface is about $g \sim GM/R^2 \sim 2 \times 10^{14}$cm s$^{-2}$ [4].

The very large density and gravitational fields of neutron stars indicate that neutron stars are relativistic objects. Therefore, to understand all of the various physical phenomena both inside and outside of neutron star and around it, it can be explained by the general relativity approach. General relativistic aspects are determined by the compactness parameters as shown in equation (1.1)

IOP Conf. Series: Journal of Physics: Conf. Series **1231** (2019) 012030 doi:10.1088/1742-6596/1231/1/012030

$$x_g = r_g/R, \text{ with } r_g = 2GM/c^2 \approx 2{,}95\ M_*/M_\odot\ , \tag{1.1}$$

where $r_g$, $G$, and $c$ denote the radius of Schwarzschild, gravitational constant, and the speed of light, respectively. In general, stars have a compactness parameter is about $x_g \ll 1$, for example, $x_g \sim 10^{-4}$ for white dwarfs, and $x_g \sim 10^{-6}$ for the main row star with $M_* \simeq M_\odot$ [5]. By substituting the mass, $M_* \sim 1{,}4\ M_\odot$, and radius, $R_* \approx 12$ km, of the neutron star into equation (1.1), then we obtain $x_g = 0{,}3$.

The fact that neutron stars are rapidly rotating astrophysical objects also requires us to do a general relativistic approach to consider various physical phenomena in neutron stars. The rapidly rotating neutron star is one of the astrophysical objects that is classified into a gravitational field class. The very rapidly rotating neutron star is called the pulsar [6]. Numerically, the explanation of rapidly rotating neutron star is the solution of Einstein's field equations. There are two main influences that differentiate the rotating relativistic star from its non-rotating pair, i.e., the form of a star is flattened by centrifugal force. Moreover, the local inertial frame is "pulled" by rotation from the source of the gravitational field. To study the "pull" of inertial frame in the spacetime of rotating neutron star can be assisted by ZAMO (Zero Angular-Momentum Observers).

The existence of pulsar is known from electromagnetic waves. There are two types of electromagnetic waves, i.e., radio waves and X-rays. The emission will show the magnetic field possessed by the neutron star [7]. If the pulses are radio pulsars, the magnitude of the neutron star magnetic field is $B \sim 10^8\text{G} - 10^{13}\text{G}$. Whereas the magnetic field of X-ray pulses is about $B > 10^{12}\text{G}$. The magnetic field is obtained from radio pulsars that emit infrared and accrete X-ray pulsars and LMXBs (Low Mass X-ray Binaries). The magnetic field inside the neutron star reaches of at least $\sim 10^{15}\text{G}$ [8] or perhaps up to $10^{18}$ G, while the pulsar magnetic field is about $B \sim 10^{15}\text{G}$ [9]. The neutron star magnetic field is the strongest magnetic field found in the universe [9].

Magnetic fields in neutron stars can decrease, for example, a neutron star that has a magnetic field of $\sim 10^{12}$ G becomes $\sim 10^8$ G in $5 \times 10^6$ years [10]. Neutron stars that experience a decrease in magnetic fields are found in multiple systems. A neutron star in a dual system shows a neutron star that performs the accretion process. Many astrophysicists therefore argue that the reduction of magnetic fields in neutron stars is due to the accreting neutron stars [11-17].

Cumming et al. have derived the equation that describes the decrease of the magnetic field in neutron stars due to accretion, for non-relativistic studies and non-rotating neutron star case. They obtained the equation of magnetic field dynamics related to the accretion velocity. Whereas the relativistic studies were introduced by Anderson et al. [18] that yielded the stationary electromagnetic fields in the Schwarzchild spacetime. The electromagnetic field is required to study the dynamics of magnetic field. The equation of the electromagnetic field can be obtained from Maxwell equations. Sengupta [19] stated that the obtained electric field of Schwarzschild metric in the neutron star was not the solution of Maxwell equations. Sengupta [20] has considered the velocity of Ohmic decay in the Schwarzschild spacetime. The Ohmic decay is assumed responsible for the cause of the magnetic field in the neutron stars. The studies related to the relativistic slowly rotating neutron stars were proposed by Muslinov and Tsygan [21]. While the study of the solution of Maxwell equations in spacetime for the slowly rotating neutron star with the absence of material around it has been conducted by Rezolla et al. [22]. In other words, they considered non-accretion neutron star. The considered case is slowly rotating neutron stars in the ZAMO (Zero Angular Momentum Observers) observers. These studies are required for describing the magnetic field dynamics for rotating neutron stars [22]. Previously, Atsnaita et al. [23] have also derived the equations of magnetic field dynamics for slowly rotating neutron star with the presence of material around it.

To describe the decrease of the magnetic field in neutron stars due to accretion, we need to derive the equations of magnetic field dynamics for rapidly rotating neutron stars. The equation is obtained from the formulation of Maxwell equations on neutron stars with the ZAMO frame.

## 2. Research methods

This research is an analytical theoretical-mathematical study. The research procedure to obtain the equation of magnetic field dynamics for the accreting rapidly rotating neutron stars in the ZAMO frame is shown in Fig. 2.1 below

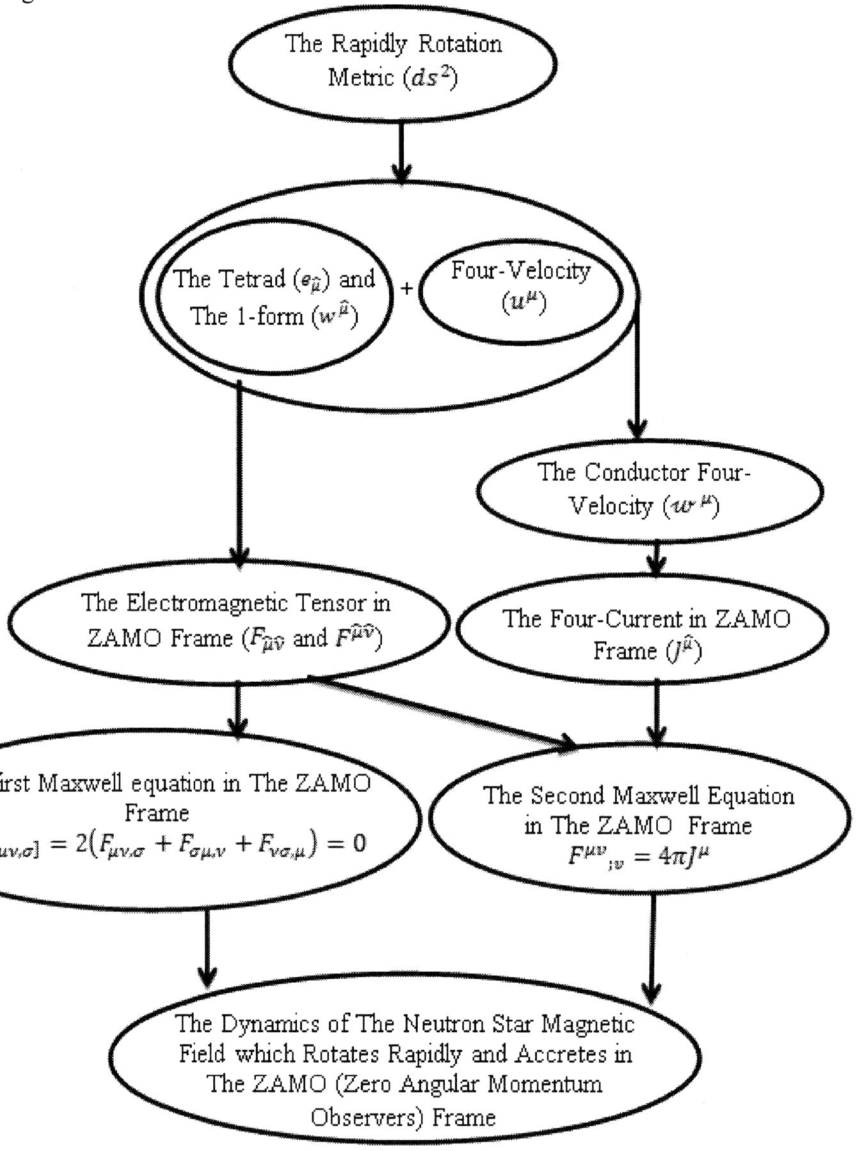


**Figure 2.1**. Research procedures to obtain the equation of magnetic field dynamics for the accreting rapidly rotating neutron stars in the ZAMO frame.

## 3. Results and discussion

### *3.1. The rapidly rotating metric, tetrad, 1-form, and 4-velocity component for relativistic neutron stars*

The metric for rapidly rotating neutron star in the coordinate system $\{ t, r, \theta, \varphi \}$ is given by

$$ds^2 = -e^{2\phi}dt^2 + e^{2\lambda}r^2 \sin^2\theta\,(d\varphi - \omega dt)^2 + e^{2\alpha}(dr^2 + r^2 d\theta^2), \tag{3.1}$$

where $\phi, \lambda, \omega, \alpha$ are the functions in the parameter of $r$ and $\theta$, respectively. The function of $\omega\ (r)$ is the angular velocity of the inertial reference frame [24]. The covariant metric component of rapidly rotating neutron stars is given by

$$g_{\mu\nu} = \begin{bmatrix} -(e^{2\phi} - e^{2\lambda}r^2\sin^2\theta\,\omega^2) & 0 & 0 & -e^{2\lambda}r^2\sin^2\theta\omega \\ 0 & e^{2\alpha} & 0 & 0 \\ 0 & 0 & e^{2\alpha}r^2 & 0 \\ -e^{2\lambda}r^2\sin^2\theta\,\omega & 0 & 0 & e^{2\lambda}r^2\sin^2\theta \end{bmatrix}. \tag{3.2}$$

The tetrad $\{e_{\hat{\mu}}\} = (e_{\hat{0}}, e_{\hat{r}}, e_{\hat{\theta}}, e_{\hat{\varphi}})$ carried by ZAMO observer are

$$e^{\nu}_{\hat{0}} = e^{-\phi}(1,0,0,\omega), \tag{3.3a}$$
$$e^{\nu}_{\hat{r}} = e^{-\alpha}(0,1,0,0), \tag{3.3b}$$
$$e^{\nu}_{\hat{\theta}} = e^{-\alpha}r^{-1}(0,0,1,0), \tag{3.3c}$$
$$e^{\nu}_{\hat{\varphi}} = \frac{e^{2\phi} - e^{2\lambda}r^2\sin^2\theta\omega^2}{e^{\phi}e^{\lambda}r\sin\theta}e^{-\phi}(0,0,0,1), \tag{3.3d}$$

The 1-form $\{w^{\hat{\mu}}\} = (w^{\hat{0}}, w^{\hat{r}}, w^{\hat{\theta}}, w^{\hat{\varphi}})$, corresponding to this tetrad are given by [25]

$$w^{\hat{0}}_{\nu} = \left(e^{2\phi} - e^{2\lambda}r^2\sin^2\theta\,\omega^2\right)^{\frac{1}{2}}(1,0,0,0), \tag{3.4a}$$
$$w^{\hat{r}}_{\nu} = e^{\alpha}(0,1,0,0), \tag{3.4b}$$
$$w^{\hat{\theta}}_{\nu} = e^{\alpha}r(0,0,1,0), \tag{3.4c}$$
$$w^{\hat{\varphi}}_{\nu} = e^{\lambda}r\sin\theta\,(-\omega,0,0,1). \tag{3.4d}$$

The 4-velocity vector for the rapidly rotating neutron star metrics are [25]

$$u^{\mu} = \frac{e^{-\phi}}{\sqrt{1-V^2}}(1,0,0,\omega), \tag{3.5a}$$
$$u_{\mu} = -\frac{e^{\phi}}{\sqrt{1-V^2}}(1,0,0,0), \tag{3.5b}$$

where

$$V^2 = e^{2\phi}\left[e^{2\lambda}r^2\sin^2\theta\ (\Omega - \omega)^2 + e^{2\alpha}(v^r)^2 + e^{2\alpha}r^2\left(v^{\theta}\right)^2\right]. \tag{3.6}$$

### *3.2. The 4-velocity vector of conductor*

The 4 -velocity vector of a conductor can be obtained from the equation below

$$\vec{w} \equiv \delta\vec{u}, \tag{3.7}$$

where $\delta$ denotes the existence of Eulerian perturbation [26]. The value of each component of velocity represents the perturbation is

$$\delta u^{\mu} = \Gamma\left(1, \delta v^{i}\right), \tag{3.8}$$

where

$$\delta v^{i} = e^{\nu}_{\hat{i}}\delta v^{\hat{i}}. \tag{3.9}$$

By substituting Eqs. (3.3a) - (3.3d) and Eq. (3.9) into Eq. (3.15), then we obtain the 4-velocity vector of the conductor, i.e,

$$\delta u^{\mu} = \Gamma\left(1, e^{-\alpha}\delta v^{\hat{r}}, e^{-\alpha}r^{-1}\delta v^{\hat{\theta}}, \frac{(e^{2\phi}-e^{2\lambda}r^2\omega^2\sin^2\theta)^{\frac{1}{2}}}{e^{\phi+\lambda}\, r\sin\theta}\delta v^{\hat{\varphi}}\right) = w^{\alpha}, \tag{3.10}$$

where $\Gamma$ is [26]

$$\Gamma = \left[-g_{00}\left(1 + \frac{g_{\alpha\beta}(\delta v^{\alpha}\delta v^{\beta})}{g_{00}}\right)\right]^{-\frac{1}{2}}. \tag{3.11}$$

If the considered perturbation is weak, then $\delta v^{\alpha}\delta v^{\beta}$ for $\alpha = \beta$ is very small. The metric components in equation (3.2) are not zero if $\alpha = \beta$, then equation (3.18) becomes

$$\Gamma \approx [-g_{00}]^{-\frac{1}{2}} \approx \left(e^{2\phi} - e^{2\lambda}r^2\omega^2\sin^2\theta\right)^{-\frac{1}{2}}. \tag{3.12}$$

If the neutron star is assumed as a conductor, then the relation $u^{\mu} = \Gamma(1,0,0,\Omega) = w^{\alpha}$ applies [26]. Therefore the 4-velocity vector of the conductor for the rapidly rotating neutron star is given by

$$\delta u^{\mu} = \Gamma\left(1,0,0, \frac{(e^{2\phi}-e^{2\lambda}r^2\omega^2\sin^2\theta)^{\frac{1}{2}}}{e^{\phi+\lambda}\, r\sin\theta}\delta v^{\hat{\varphi}}\right) = \Gamma(1,0,0,\Omega) = w^{\alpha}. \tag{3.13}$$

*3.3. The electromagnetic tensor in ZAMO frame*

The electromagnetic field tensor $F_{\alpha\beta}$ denotes the relation between the 4-vector of electric field $E^{\alpha}$ and magnetic field $B^{\alpha}$. The covariant component of the electromagnetic tensor component satisfies the equation

$$F_{\alpha\beta} \equiv 2u_{[\alpha}E_{\beta]} + \eta_{\alpha\beta\gamma\delta}u^{\gamma}B^{\delta}, \tag{3.14}$$

where $\eta_{\alpha\beta\gamma\delta}$ is pseudo-tensor. The pseudo-tensor expressed by the Levi-Civita symbol $\epsilon_{\alpha\beta\gamma\delta}$, are

$$\eta^{\alpha\beta\gamma\delta} = -\frac{1}{\sqrt{-g}}\epsilon_{\alpha\beta\gamma\delta}; \qquad \eta_{\alpha\beta\gamma\delta} = \sqrt{-g}\,\epsilon_{\alpha\beta\gamma\delta}, \tag{3.15}$$

Where

$$g \equiv \det|g_{\alpha\beta}| = -e^{2(\Phi+\Lambda)}r^4\sin^2\theta. \tag{3.16}$$

Therefore, the non-zero components of the covariant tensor of the electromagnetic field for the rapidly rotating neutron star are [25]

$$F_{00} = F_{11} = F_{22} = F_{33} = 0, \tag{3.17a}$$
$$F_{01} = -F_{10} = -YE_r - X\omega B^{\theta}, \tag{3.17b}$$
$$F_{02} = -F_{20} = -YE_{\theta} + X\omega B^{r}, \tag{3.17c}$$
$$F_{03} = -F_{30} = -YE_{\varphi}, \tag{3.17d}$$
$$F_{12} = -F_{21} = XB^{\varphi}, \tag{3.17e}$$
$$F_{13} = -F_{31} = -XB^{\theta}, \tag{3.17f}$$
$$F_{23} = -F_{32} = XB^{r}, \tag{3.17g}$$

where

$$X = \left(e^{2\alpha+\lambda}r^2\sin\theta\right)/\sqrt{1-V^2}, \tag{3.18}$$

and

$$Y = e^{\Phi}/\sqrt{1-V^2}. \tag{3.19}$$

Whereas the contravariant tensor components of the electromagnetic field are the inverse of the covariant tensor. The non-zero components of the contravariant tensor of the electromagnetic field for the rapidly rotating neutron star are

$$F^{00} = F^{11} = F^{22} = F^{33} = 0, \tag{3.20a}$$
$$F^{01} = -F^{10} = E^r/Y, \tag{3.20b}$$
$$F^{02} = -F^{20} = E^\theta/Y, \tag{3.20c}$$
$$F^{03} = -F^{30} = E^\phi/Y, \tag{3.20d}$$
$$F^{12} = -F^{21} = B_\varphi/X, \tag{3.20e}$$
$$F^{13} = -F^{31} = -(YB_\theta + X\omega E^r)/XY, \tag{3.20f}$$
$$F^{23} = -F^{32} = \left(YB_r - X\omega E^\theta\right)/XY. \tag{3.20g}$$

The covariant and contravariant tensor components of the electromagnetic field for the rapidly rotating neutron star are measured by ZAMO observer. They are obtained by changing the 4-vector of electric field $E^\alpha$ and magnetic field $B^\alpha$ in the ZAMO frame. The 4-vector of electric field $E^\alpha$ and magnetic field $B^\alpha$ in the ZAMO frame are given by

$$E^i = e^i_{\hat{\mu}} E^{\hat{\mu}}, \tag{3.21a}$$
$$E_i = g_{ii} e^i_{\hat{\mu}} E^{\hat{\mu}}, \tag{3.21b}$$
$$B^i = e^i_{\hat{\mu}} B^{\hat{\mu}}, \tag{3.21c}$$
$$B_i = g_{ii} e^i_{\hat{\mu}} B^{\hat{\mu}}. \tag{3.21d}$$

From the equation (3.22a) and (3.22b) then we get the 4-vector of electric field $E^\alpha$ for rapidly rotating neutron star in the ZAMO frame, i.e,

$$E_r = e^\alpha E^{\hat{r}}, \tag{3.22a}$$
$$E_\theta = e^\alpha r B^{\hat{\theta}}, \tag{3.22b}$$
$$E_\varphi = e^{\alpha-\phi} r \sin\theta \left(e^{2\phi} - e^{2\lambda} r^2 \omega^2 \sin^2\theta\right)^{\frac{1}{2}} E^{\hat{\varphi}}, \tag{3.22c}$$
$$E^r = e^{-\alpha} E^{\hat{r}}, \tag{3.22d}$$
$$E^\theta = e^{-\alpha} r^{-1} E^{\hat{\theta}}, \tag{3.22e}$$
$$E^\varphi = \frac{\left(e^{2\phi} - e^{2\lambda} r^2 \omega^2 \sin^2\theta\right)^{\frac{1}{2}}}{e^{\phi+\lambda} r \sin\theta} E^{\hat{\varphi}}. \tag{3.22f}$$

Whereas based on the equation (3.23c) and (3.23d) we can obtain the 4-vector of magnetic field $B^\alpha$ for rapidly rotating neutron star in the ZAMO frame as follow

$$B_r = e^\alpha B^{\hat{r}}, \tag{3.23a}$$
$$B_\theta = e^\alpha r B^{\hat{\theta}}, \tag{3.23b}$$
$$B_\varphi = e^{\alpha-\phi} r \sin\theta \left(e^{2\phi} - e^{2\lambda} r^2 \omega^2 \sin^2\theta\right)^{\frac{1}{2}} B^{\hat{\varphi}}. \tag{3.23c}$$
$$B^r = e^{-\alpha} B^{\hat{r}}, \tag{3.23d}$$
$$B^\theta = e^{-\alpha} r^{-1} B^{\hat{\theta}}, \tag{3.23e}$$
$$B^\varphi = \frac{\left(e^{2\phi} - e^{2\lambda} r^2 \omega^2 \sin^2\theta\right)^{\frac{1}{2}}}{e^{\phi+\lambda} r \sin\theta} B^{\hat{\varphi}}. \tag{3.23f}$$

By substituting equations (3.22a) - (3.22c) and (3.23d) - (3.23f) into equation (3.17a) - (3.17g), then we obtain the components of electromagnetic field covariant tensor for the rapidly rotating neutron stars measured in the ZAMO frame as follows

$$F_{\hat{0}\hat{0}} = F_{\hat{1}\hat{1}} = F_{\hat{2}\hat{2}} = F_{\hat{3}\hat{3}} = 0, \tag{3.24a}$$

$$F_{\hat{0}\hat{1}} = -F_{\hat{1}\hat{0}} = -\frac{e^{\phi+\alpha}}{\sqrt{1-V^2}}E^{\hat{r}} - \frac{\omega e^{\alpha+\lambda}}{\sqrt{1-V^2}} r\sin\theta\, B^{\hat{\theta}}, \quad (3.24b)$$

$$F_{\hat{0}\hat{2}} = -F_{\hat{2}\hat{0}} = -\frac{e^{\phi+\alpha}}{\sqrt{1-V^2}} rE^{\hat{\theta}} + \frac{\omega e^{\alpha+\lambda}}{\sqrt{1-V^2}} r^2\sin\theta\, B^{\hat{r}}, \quad (3.24c)$$

$$F_{\hat{0}\hat{3}} = -F_{\hat{3}\hat{0}} = -\frac{e^{\lambda}}{\sqrt{1-V^2}} r\sin\theta\left(e^{2\phi} - e^{2\lambda}r^2\omega^2\sin^2\theta\right)^{\frac{1}{2}}E^{\hat{\varphi}}, \quad (3.24d)$$

$$F_{\hat{1}\hat{2}} = -F_{\hat{2}\hat{1}} = \frac{e^{-\phi+2\alpha}}{\sqrt{1-V^2}} r\left(e^{2\phi} - e^{2\lambda}r^2\omega^2\sin^2\theta\right)^{\frac{1}{2}}B^{\hat{\varphi}}, \quad (3.24e)$$

$$F_{\hat{1}\hat{3}} = -F_{\hat{3}\hat{1}} = -\frac{e^{\alpha+\lambda}}{\sqrt{1-V^2}} r\sin\theta\, B^{\hat{\theta}}, \quad (3.24f)$$

$$F_{\hat{2}\hat{3}} = -F_{\hat{3}\hat{2}} = \frac{e^{\alpha+\lambda}}{\sqrt{1-V^2}} r^2\sin\theta\, B^{\hat{r}}. \quad (3.24g)$$

While the components of contravariant tensor of electromagnetic field for the rapidly rotating neutron stars measured in the ZAMO frame are obtained by substituting equations (3.22d) - (3.22f) and (3.23a) - (3.23c) into equation (3.20a) - (3.20f). The components of contravariant tensor of electromagnetic field for the rapidly rotating neutron stars measured in the ZAMO frame are

$$F^{\hat{0}\hat{0}} = F^{\hat{1}\hat{1}} = F^{\hat{2}\hat{2}} = F^{\hat{3}\hat{3}} = 0, \quad (3.25a)$$

$$F^{\hat{0}\hat{1}} = -F^{\hat{1}\hat{0}} = -e^{-\phi-\alpha}\sqrt{1-V^2}E^{\hat{r}}, \quad (3.25b)$$

$$F^{\hat{0}\hat{2}} = -F^{\hat{2}\hat{0}} = e^{-\phi-\alpha}r^{-1}\sqrt{1-V^2}E^{\hat{\theta}}, \quad (3.25c)$$

$$F^{\hat{0}\hat{3}} = -F^{\hat{3}\hat{0}} = \frac{\left(e^{2\phi}-e^{2\lambda}r^2\sin^2\theta\omega^2\right)^{1/2}}{e^{2\phi+\lambda}r\sin\theta}\sqrt{1-V^2}B^{\hat{\varphi}}, \quad (3.25d)$$

$$F^{\hat{1}\hat{2}} = -F^{\hat{2}\hat{1}} = -\frac{\left(e^{2\phi}-e^{2\lambda}r^2\sin^2\theta\omega^2\right)^{\frac{1}{2}}}{e^{\alpha+\phi+\lambda}r}\sqrt{1-V^2}B^{\hat{\varphi}}, \quad (3.25e)$$

$$F^{\hat{1}\hat{3}} = -F^{\hat{3}\hat{1}} = -\left[\frac{E^{\hat{\theta}}}{e^{\alpha+\lambda}r\, sin\theta} - \frac{\omega}{e^{\phi+\alpha}}B^{\hat{r}}\right]\sqrt{1-V^2}, \quad (3.25f)$$

$$F^{\hat{0}\hat{1}} = -F^{\hat{1}\hat{0}} = -\left[\frac{B^{\hat{\theta}}}{e^{\alpha+\lambda}r^2\sin\theta} + \frac{\omega}{e^{\phi+\alpha}}E^{\hat{r}}\right]. \quad (3.25g)$$

### 3.4. The 4 - current in ZAMO frame

The conduction current $j^{\mu}$ carried by electrons with electrical conductivity $\sigma$ satisfies Ohm's law, that is

$$j_{\mu} = \sigma F_{\mu\nu} u^{\nu}. \quad (3.26)$$

Therefore, using equations (3.17a) - (3.17g), (3.13), and (3.26) the covariant components of the conduction current are

$$j_0 = -\sigma\Gamma\, YE_{\varphi}, \quad (3.27a)$$

$$j_r = \sigma\Gamma\left(YE_r - XB^{\theta}(\Omega-\omega)\right), \quad (3.27b)$$

$$j_{\theta} = \sigma\Gamma\left(YE_{\theta} + XB^{r}(\Omega-\omega)\right), \quad (3.27c)$$

$$j_{\varphi} = \sigma\Gamma YE_{\varphi}, \quad (3.27d)$$

While the contravariant components of the conduction current satisfies the equation

$$j^{\mu} = g^{\mu\nu}j_{\nu}. \quad (3.28)$$

Therefore the contravariant components of the conduction current are

$$j^0 = e^{-2\phi}\sigma\Gamma\, YE_{\varphi}(\Omega+\omega), \quad (3.29a)$$

$$j^r = e^{-2\alpha}\sigma\Gamma\left[YE_r - XB^{\theta}(\Omega-\omega)\right], \quad (3.29b)$$

$$j^{\theta} = e^{-2\alpha}r^{-2}\sigma\Gamma\left[YE_{\theta} + XB^{r}(\Omega-\omega)\right], \quad (3.29c)$$

$$j^{\varphi} = \sigma\Gamma\, Y E_{\varphi}[\frac{(e^{2\phi}+e^{2\lambda}r^2\sin^2\theta\omega^2)}{e^{2\phi+2\lambda}r^2\sin^2\theta} + e^{-2\phi}\omega\Omega]. \quad (3.29d)$$

The 4-current density satisfies the equation

$$J^{\mu} = \rho_e \mathcal{w}^{\nu} + j^{\mu}, \quad (3.30)$$

where $\rho_e$ is charge density, so the components of 4-current density are

$$J^0 = \Gamma[\rho_e + e^{-2\phi}\sigma Y\, E_{\varphi}(\Omega-\omega)], \quad (3.31a)$$
$$J^r = e^{-2\alpha}\,\sigma\Gamma[Y\, E_r - XB^{\theta}(\Omega-\omega)], \quad (3.31b)$$
$$J^{\theta} = e^{-2\alpha}\, r^{-2}\sigma\Gamma[Y\, E_{\theta} + XB^{r}(\Omega-\omega)], \quad (3.31c)$$
$$J^{\varphi} = \rho_e\Gamma\Omega + \sigma\Gamma Y\, E_{\varphi}\left[\frac{(e^{2\phi}+e^{2\lambda}r^2\sin^2\theta\omega^2)}{e^{2\phi+2\lambda}r^2\sin^2\theta} + e^{-2\phi}\omega\Omega\right]. \quad (3.31d)$$

The relation between the 4-current density $J^{\mu}$ and the 4-current density in the ZAMO frame $J^{\hat{\iota}}$ is

$$J^{\mu} = e^{\mu}_{\hat{\iota}} J^{\hat{\iota}}. \quad (3.32)$$

The 4-current density in the ZAMO framework for rapidly rotating neutron stars are given by

$$J^{\hat{0}} = e^{\phi}\rho_e\,\Gamma + \sigma\, e^{-2\phi+\lambda} r\sin\theta\, Y\,(\Omega-\omega)E^{\hat{\varphi}}, \quad (3.33a)$$
$$J^{\hat{r}} = \sigma\Gamma\left[YE^{\hat{r}} - X\, e^{-2\alpha}r^{-1}B^{\hat{\theta}}(\Omega-\omega)\right], \quad (3.33b)$$
$$J^{\hat{\theta}} = \sigma\Gamma\left[YE^{\hat{\theta}} + X\, e^{-2\alpha}r^{-1}B^{\hat{r}}(\Omega-\omega)\right], \quad (3.33c)$$
$$J^{\hat{\varphi}} = \Gamma\left[\rho_e(\Omega-\omega)\frac{(e^{\phi+\lambda}r\sin\theta)}{(e^{2\phi}-e^{2\lambda}r^2\sin^2\theta\omega^2)^{\frac{1}{2}}} + \sigma\, Yr^2\sin^2\theta\, E^{\hat{\varphi}}\right], \quad (3.33d)$$

*3.5. Maxwell equation in a rapidly rotating spacetime*

The general form of the first pair of general relativistic Maxwell equation is given by

$$3!\,F_{[\alpha\beta,\gamma]} = 2(F_{\alpha\beta,\gamma} + F_{\gamma\alpha,\beta} + F_{\beta\gamma,\alpha}) = 0. \quad (3.34)$$

By substituting equations (3.24a) - (3.24g) into equation (3.34) then we get relativistic Maxwell first equation for rapidly rotating neutron stars in the ZAMO frame, namely as

$$-\frac{e^{\phi+\alpha}}{\sqrt{1-V^2}}r\left(E^{\hat{\theta}}\right)_{,\varphi} + \omega\frac{e^{\alpha+\lambda}}{\sqrt{1-V^2}}r^2\sin\theta\left(B^{\hat{r}}\right)_{,\varphi} + r\left(\frac{e^{\lambda}\sin\theta}{\sqrt{1-V^2}}(e^{2\phi}-e^{2\lambda}r^2\sin^2\theta\,\omega^2)^{\frac{1}{2}}E^{\hat{\varphi}}\right)_{,\theta}$$
$$+\frac{e^{\alpha+\lambda}}{\sqrt{1-V^2}}r^2\sin\theta\frac{\partial B^{\hat{r}}}{\partial t} = 0, \quad (3.35a)$$
$$-\frac{e^{\phi+\alpha}}{\sqrt{1-V^2}}\left(E^{\hat{r}}\right)_{,\varphi} - \omega\frac{e^{\alpha+\lambda}}{\sqrt{1-V^2}}r^2\sin\theta\left(B^{\hat{\theta}}\right)_{,\varphi} + \sin\theta\left(\frac{e^{\lambda}r}{\sqrt{1-V^2}}(e^{2\phi}-e^{2\lambda}r^2\sin^2\theta\,\omega^2)^{\frac{1}{2}}E^{\hat{\varphi}}\right)_{,r}$$
$$-\frac{e^{\alpha+\lambda}}{\sqrt{1-V^2}}r\sin\theta\frac{\partial B^{\hat{\theta}}}{\partial t} = 0, \quad (3.35b)$$
$$-e^{\phi+\alpha}\left(\frac{E^{\hat{r}}}{\sqrt{1-V^2}}\right)_{,\varphi} - \omega\, r\left(\frac{e^{\alpha+\lambda}}{\sqrt{1-V^2}}\sin\theta\, B^{\hat{\theta}}\right)_{,\theta} + e^{\phi+\alpha}\left(\frac{rE^{\hat{\theta}}}{\sqrt{1-V^2}}\right)_{,r} - \sin\theta\left(\omega\frac{e^{\alpha+\lambda}}{\sqrt{1-V^2}}r^2B^{\hat{r}}\right)_{,r}$$
$$+\frac{e^{2\alpha-\phi}}{\sqrt{1-V^2}}r\,(e^{2\phi}-e^{2\lambda}r^2\sin^2\theta\,\omega^2)^{\frac{1}{2}}\frac{\partial B^{\hat{\varphi}}}{\partial t} = 0. \quad (3.35c)$$

The relativistic Maxwell second equation satisfies the equation

$$F^{\mu\nu}{}_{;\nu} = 4\pi J^{\mu}. \tag{3.36}$$

Substituting equations (3.25a) - (3.25g) and (3.33a) - (3.33d) into equation (3.36), then we obtain the relativistic Maxwell second equation in the ZAMO frame that are

$$\left(e^{\alpha+\lambda} r^2 \sin\theta \sqrt{1-V^2} E^{\hat{r}}\right)_{,r} + \left(e^{\alpha+\lambda} r \sin\theta \sqrt{1-V^2} E^{\hat{\theta}}\right)_{,\theta} + \left(e^{-\phi+2\alpha} r \sqrt{1-V^2}\left(e^{2\phi} - e^{2\lambda} r^2 \sin^2\theta\, \omega^2\right)^{\frac{1}{2}} E^{\hat{\varphi}}\right)_{,\varphi} = 4\pi\Gamma\rho_e e^{2\phi+2\alpha+\lambda} r^2 \sin\theta + \frac{4\pi\Gamma e^{2\alpha+\lambda}}{\sqrt{1-V^2}} r^3 \sin^2\theta\, (\Omega-\omega) E^{\hat{\varphi}}, \tag{3.37a}$$

$$\left(-e^{\alpha+\lambda} r^2 \sin\theta \sqrt{1-V^2}\frac{\partial E^{\hat{r}}}{\partial t}\right)_{,r} + r\left(\sqrt{1-V^2} e^{\lambda} \sin\theta \left(e^{2\phi} - e^{2\lambda} r^2 \sin^2\theta\, \omega^2\right)^{1/2} B^{\hat{\varphi}}\right)_{,\theta} - e^{-\phi+\alpha} r \left(B^{\hat{\theta}}\right)_{,\varphi} - \omega e^{\alpha+\lambda} r^2 \sin\theta \left(E^{\hat{r}}\right)_{,\varphi} = \frac{4\pi\Gamma\rho_e e^{2\phi+2\alpha+\lambda}}{\sqrt{1-V^2}} r^2 \sin\theta\, E^{\hat{r}} - \frac{4\pi\Gamma e^{2\alpha+\lambda}}{\sqrt{1-V^2}} r^3 \sin^2\theta\, (\Omega-\omega) B^{\hat{\theta}}, \tag{3.37b}$$

$$\left(-e^{\alpha+\lambda} r \sin\theta \sqrt{1-V^2}\frac{\partial E^{\hat{\theta}}}{\partial t}\right)_{,r} - \sin\theta\left(\sqrt{1-V^2} e^{\alpha} r \left(e^{2\phi} - e^{2\lambda} r^2 \sin^2\theta\, \omega^2\right)^{\frac{1}{2}} B^{\hat{\varphi}}\right)_{,r} + e^{\phi+\alpha}\left(B^{\hat{r}}\right)_{,\varphi} - \omega e^{\alpha+\lambda} r \sin\theta \left(E^{\hat{\theta}}\right)_{,\varphi} = \frac{4\pi\Gamma\rho_e e^{2\phi+2\alpha+\lambda}}{\sqrt{1-V^2}} r^2 \sin\theta\, E^{\hat{\theta}} + \frac{4\pi\Gamma e^{\phi+2\alpha+\lambda}}{\sqrt{1-V^2}} r^3 \sin^2\theta\, (\Omega-\omega) B^{\hat{r}}, \tag{3.37c}$$

$$\left(-e^{-\phi+\alpha} r \left(e^{2\phi} - e^{2\lambda} r^2 \sin^2\theta\, \omega^2\right)^{\frac{1}{2}} \sqrt{1-V^2}\frac{\partial E^{\hat{\varphi}}}{\partial t}\right)_{,r} + \left(e^{\phi+\alpha} r B^{\hat{\theta}}\right)_{,r} - e^{\phi+\alpha}\left(B^{\hat{r}}\right)_{,\theta} - \left(\omega e^{\alpha+\lambda} r E^{\hat{r}}\right)_{,r} + r\,\omega\left(e^{\alpha+\lambda} E^{\hat{\theta}}\right)_{,\theta} = \frac{4\pi\Gamma\rho_e e^{2\phi+2\alpha+\lambda}}{\left(e^{2\phi}-e^{2\lambda} r^2 \sin^2\theta\omega^2\right)^{\frac{1}{2}}} r^3 \sin^2\theta\, (\Omega-\omega) + \frac{4\pi\sigma\Gamma e^{\phi+2\alpha+\lambda}}{\sqrt{1-V^2}} r^4 \sin^4\theta\, E^{\hat{\varphi}}, \tag{3.37d}$$

This equation corresponds to the slowly rotating neutron star case as conducted by Rezolla.

### *3.6. The dynamics equation of the magnetic field for the accreting and rapidly rotating neutron stars in the ZAMO frame*

As conducted by Rezolla [22], in this article we write the terms that contain $\omega$ and $\Omega$ as $\mathcal{O}(\Omega)$. Therefore the Maxwell second equation as shown by equations (3.37a) – (3.37d) can be written as follow

$$E^{\hat{r}} = \frac{\sqrt{1-V^2}}{4\pi\Gamma e^{2\phi+2\alpha+\lambda} r^2 \sin\theta}\left[e^{\alpha+\lambda} r^2 \sin\theta \sqrt{1-V^2}\frac{\partial E^{\hat{r}}}{\partial t} + r\left(\sqrt{1-V^2} e^{\lambda} \sin\theta \left(e^{2\phi} - e^{2\lambda} r^2 \sin^2\theta\, \omega^2\right)^{\frac{1}{2}} B^{\hat{\varphi}}\right)_{,\theta} - e^{-\phi+\alpha} r \left(B^{\hat{\theta}}\right)_{,\varphi}\right] + \mathcal{O}(\Omega). \tag{3.38a}$$

$$E^{\hat{\theta}} = \frac{\sqrt{1-V^2}}{4\pi\Gamma e^{2\phi+2\alpha+\lambda} r^2 \sin\theta}\left[\left(-e^{\alpha+\lambda} r \sin\theta \sqrt{1-V^2}\frac{\partial E^{\hat{\theta}}}{\partial t}\right)_{,r} - \sin\theta\left(\sqrt{1-V^2} e^{\alpha} r \left(e^{2\phi} - e^{2\lambda} r^2 \sin^2\theta\, \omega^2\right)^{\frac{1}{2}} B^{\hat{\varphi}}\right)_{,r} + e^{\phi+\alpha}\left(B^{\hat{r}}\right)_{,\varphi}\right] + \mathcal{O}(\Omega), \tag{3.38b}$$

$$E^{\hat{\varphi}} = \frac{\sqrt{1-V^2}}{4\pi\sigma\Gamma e^{\phi+2\alpha+\lambda} r^4 \sin^4\theta}\left[-e^{-\phi+\alpha} r\sqrt{1-V^2}\left(e^{2\phi}-e^{2\lambda}r^2\sin^2\theta\,\omega^2\right)^{\frac{1}{2}}\frac{\partial E^{\hat{\varphi}}}{\partial t} + \left(e^{\phi+\alpha} rB^{\hat{\theta}}\right)_{,r} - \left(e^{\phi+\alpha}B^{\hat{r}}\right)_{,\theta}\right] + \mathcal{O}(\Omega). \tag{3.38c}$$

Furthermore, by substituting equations (3.38a) - (3.38c) into Maxwell first equation, then we obtained the dynamic equation of the magnetic field for the accreting and rapidly rotating neutron star in the ZAMO frame, that are

$$\frac{\partial B^{\hat{r}}}{\partial t} = \frac{\sqrt{1-V^2}}{\left(e^{\alpha+\lambda}r^2\sin\theta\right)}\left\{\frac{e^{-\phi-\alpha-\lambda}}{4\pi\Gamma r^2\sin\theta}\left[-e^{\alpha+\lambda}r^2\sin\theta\sqrt{1-V^2}\left(\frac{\partial E^{\hat{r}}}{\partial t}\right)_{,\varphi} - \sin\theta\left(\left(\sqrt{1-V^2}e^{\alpha}r\left(e^{2\phi}-e^{2\lambda}r^2 sin^2\theta\,\omega^2\right)^{\frac{1}{2}}B^{\hat{\varphi}}\right)_{,r}\right)_{,\varphi} + e^{\phi+\alpha}\left(\left(B^{\hat{r}}\right)_{,\varphi}\right)_{,\varphi}\right] + \left(\mathcal{O}(\Omega)\right)_{,\varphi} - \omega\frac{e^{\alpha+\lambda}}{\sqrt{1-V^2}}r^2\sin\theta\left(B^{\hat{r}}\right)_{,\varphi} - \left(\frac{r\sin\theta}{4\pi\sigma\Gamma e^{2\phi+2\alpha}}\left(e^{2\phi}-e^{2\lambda}r^2\sin^2\theta\,\omega^2\right)^{\frac{1}{2}}\left[-e^{-\phi+\alpha}r\sqrt{1-V^2}\left(e^{2\phi}-e^{2\lambda}r^2\sin^2\theta\,\omega^2\right)^{\frac{1}{2}}\frac{\partial E^{\hat{\varphi}}}{\partial t} + \left(e^{\phi+\alpha}rB^{\hat{\theta}}\right)_{,r} - \left(e^{\phi+\alpha}B^{\hat{r}}\right)_{,\theta}\right] + \mathcal{O}(\Omega)\right)_{,\theta}\right\}, \tag{3.39a}$$

$$\frac{\partial B^{\hat{\theta}}}{\partial t} = \frac{\sqrt{1-V^2}}{\left(e^{\alpha+\lambda}r\sin\theta\right)}\left\{-\frac{e^{-\phi-\alpha-\lambda}}{4\pi\Gamma r^2\sin\theta}\left[-e^{\alpha+\lambda}r^2\sin\theta\sqrt{1-V^2}\left(\frac{\partial E^{\hat{r}}}{\partial t}\right)_{,\varphi} + e^{\alpha}r\left(\left(\sin\theta\sqrt{1-V^2}\left(e^{2\phi}-e^{2\lambda}r^2 sin^2\theta\,\omega^2\right)^{\frac{1}{2}}B^{\hat{\varphi}}\right)_{,\theta}\right)_{,\varphi} - e^{\phi+\alpha}\left(\left(B^{\hat{\varphi}}\right)_{,\theta}\right)_{,\varphi}\right] + \left(\mathcal{O}(\Omega)\right)_{,\varphi} - \omega\frac{e^{\alpha+\lambda}}{\sqrt{1-V^2}}r\sin\theta\left(B^{\hat{\theta}}\right)_{,\varphi} + \sin\theta\left(\frac{r}{4\pi\sigma\Gamma e^{2\phi+2\alpha}}\left(e^{2\phi}-e^{2\lambda}r^2\sin^2\theta\,\omega^2\right)^{\frac{1}{2}}\left[-e^{-\phi+\alpha}r\sqrt{1-V^2}\left(e^{2\phi}-e^{2\lambda}r^2\sin^2\theta\,\omega^2\right)^{\frac{1}{2}}\frac{\partial E^{\hat{\varphi}}}{\partial t} + \left(e^{\phi+\alpha}rB^{\hat{\theta}}\right)_{,r} - \left(e^{\phi+\alpha}B^{\hat{r}}\right)_{,\theta}\right] + \mathcal{O}(\Omega)\right)_{,\theta}\right\}. \tag{3.39b}$$

$$\frac{\partial B^{\hat{\varphi}}}{\partial t} = \frac{\sqrt{1-V^2}}{\left(e^{-\phi+2\alpha}r\left(e^{2\phi}-e^{2\lambda}r^2 sin^2\theta\,\omega^2\right)^{\frac{1}{2}}\right.}$$

$$\left\{e^{\phi+\alpha}\left(\frac{1}{4\pi\Gamma r^2\sin\theta}\left[-e^{\alpha+\lambda}r^2\sin\theta\sqrt{1-V^2}\frac{\partial E^{\hat{r}}}{\partial t} + e^{\alpha}r\left(\sin\theta\sqrt{1-V^2}\left(e^{2\phi}-e^{2\lambda}r^2 sin^2\theta\,\omega^2\right)^{\frac{1}{2}}B^{\hat{\varphi}}\right)_{,\theta} - e^{\phi+\alpha}r\left(B^{\hat{\theta}}\right)_{,\varphi}\right] + \mathcal{O}(\Omega)\right)_{,\theta} + \right.$$

$$\omega r\left(\frac{e^{\alpha+\lambda}}{\sqrt{1-V^2}}\sin\theta\, B^{\hat{\theta}}\right)_{,\theta} + e^{\phi+\alpha}\left(\frac{1}{4\pi\sigma\Gamma e^{2\phi+2\alpha} r\sin\theta}\left[-e^{\alpha+\lambda} r\sin\theta\sqrt{1-V^2}\,\frac{\partial E^{\hat{\theta}}}{\partial t} + e^{\phi+\alpha}\left(B^{\hat{r}}\right)_{,\varphi} - \sin\theta\left(e^{\alpha} r\sqrt{1-V^2}\left(e^{2\phi}-e^{2\lambda}r^2\sin^2\theta\,\omega^2\right)^{\frac{1}{2}} B^{\hat{\varphi}}\right)_{,r}\right] + \mathcal{O}(\Omega)\right)_{,r} + \sin\theta\left(\frac{\omega e^{\alpha+\lambda}}{\sqrt{1-V^2}} r^2\right)_{,r}\Bigg\}. \quad (3.39c)$$

## 4. References


[1] Lander S K 2010 *Thesis* (England: University Of Southamton)

[2] Potekhin A Y 2011 *astro-ph. SR*, 1235-1256.

[3] Shaphiro L S and Teukolsky S A 2004 *Black Hole, White Dwarfs, and Neutron stars* (Verlag: Wiley-VCH)

[4] Haensel P, Potekhin A Y, and Yakovlev D G 2007 *Neutron Stars 1 Equation of State and Structure* (New York: Springer)

[5] Istiqomah E L 2010 *Thesis* (Yogyakarta: Department of Physics, Faculty of Mathematics and Natural Sciences Universitas Gadjah Mada)

[6] Camenzind M 2007 Compact Objects in Astrophysics White Dwarfs, Neutron Stars, and Black Hole (Verlag Berlin Heidelberg: Springer)

[7] Reisenegger A 2003 Origin and Evolution of Neutron Stars Magnetic Fields *arXiv*:0307133v1

[8] Bhattacharya D 2002 *J Astrophys. Astr.* **23**, 67–72

[9] Reisenegger A 2007 *Atron. Nachr* **1** 5

[10] Zhang C M 1997 *Astron. Astropys* **330** 195

[11] Anzer U and Börner G 1979 *Astron. Astrophysics* 133-139

[12] Choundhuri A R and Konar S 2002 *arXiv*:0108229

[13] Cumming A, Zweibel E and Bildsten L 2001 *arXiv*:0102178

[14] Ho and Wynn C G 2011 *arXiv*:11024870v1

[15] Konar S and Bhattacharya D 1996 *R. Astron. Soc.* **284** 311

[16] Lovelace R V E and Romanova M M 2005 *The Astrophysical Journal* **625** 957

[17] Melatos A *et al.* 2001 *Astron. Soc. Aust.* 421

[18] Anderson J L and Cohen J M 1970 *Astropys. Space Science.* **9** 146.

[19] Sengupta S 1995 *ApJ.* **449** 224.

[20] Sengupta S 1997 *ApJ* **479** L133

[21] Muslinov A and Tsygan A I 1992 General Relativistic Electric Potential Drops Above Pulsar Polar Caps *MNRAS* **255** 61

[22] Rezzolla L, Ahmedov B J, and J. C. Miller 2004 General Relativitic Elektromagnetic Fields of a Slowly Rotating Magnetized Neutron Stars I. Formulation of the Equation MNRAS 1-19

[23] Yasrina A and Rosyid M F 2013 *Thesis* (Yogyakarta: Department of Physics, Faculty of Mathematics and Natural Sciences Universitas Gadjah Mada)

[24] Gregory B, Cook, Stuart L, Shapiro, Saul A and Teukolsky 1994 *The American Astronomical Society: Astrophysical Journal* **424** 828

[25] Yasrina A 2015 *Tensor Elektromagnetik Bintang Neutron yang Berotasi Cepat Diukur oleh Pengamat ZAMO* (Zero Angular Momentum Observers), ISBN 978-602-71273-1-9, A-1-A-5

[26] Rezzolla L, Ahmedov B J, and J. C. Miller 2004 Electromagnetic Fields in the Exterior of an Oscillating Relativistic Sta-I. General Expressions and Application to A Rotating Magnetic Dipole. MNAR. 1-21

# Photoionization modelling of the Wolf-Rayet planetary nebulae IC 4663 based on multiwavelength observations

**E Puspitaningrum[1] and H L Malasan[2,3]**

[1] Master Program in Astronomy, Faculty of Mathematics and Natural Sciences, Bandung Institute of Technology, Jalan Ganesha 10, Bandung 40132, Indonesia
[2] Department of Astronomy and Bosscha Observatory, FMIPA Institut Teknologi Bandung, Jalan Ganesha 10 Bandung 40132 Indonesia
[3]Lampung Astronomical Observatory, Institut Teknologi Sumatera, Way Hui, Lampung, Indonesia

Email: evaria.p@gmail.com

**Abstract.** Observations of the central star planetary nebulae (CSPN) IC 4663 revealed its spectral features similar to WN sequence of massive Wolf-Rayet (WR) stars. Because CSPN IC 4663 spectrum is dominated by broad He II and N V emission lines, it is classified as [WN3] spectral type. The similarity on the central stars composition of O(He) and IC 4663 provides evidence for evolutionary sequence [WN] → O(He). Evolution of CSPN IC 4663 followed evolution low-mass star, but experienced *born-again scenario* when it enters white dwarf phase. So it goes back to Asymptotic Giant Branch (AGB) phase. To investigate central star properties, the nebula and its evolutionary status, we constructed the spectral energy distribution (SED) model by using photoionization code CLOUDY based on data set from optic to far-IR wavelength (~3 – 100 $\mu m$) and include the *Postdam Wolf-Rayet* (PoWR) stellar model atmosphere. The comparison between the nebular abundances from photoionization model result and AGB model gives initial $3M_{\odot}$ star with Z = 0.02. The nebula ionized mass derived from the photoionization model also consistent with the prediction from AGB model.

***Keywords:*** *Photoionization modelling and Wolf-Rayet planetary nebulae IC 4663*

## 1. Introduction

Some of the central stars of planetary nebulae (CSPN) have H-deficient atmosphere and spectral features similar to massive Wolf-Rayet (WR) stars. This phenomenon can be explained by a scenario called "born again scenario" suggested by [4], when last helium thermal pulse occurs after the star left the thermally pulsing AGB (*Late Thermal Pulses*/LTP) or H-burning has almost come to the end on the early white-dwarf cooling branch (*Very Late Thermal Pulses*/VLTP). Observations of the CSPN IC 4663 revealed that it has [WN3] spectral features. The similarity on the composition of O(He) and IC 4663 central star suggested a secondary evolutionary sequence of H-deficient star [WN] → O(He) [6]. To gain insight about the progenitor star of this planetary nebulae (PN) and its evolutionary status, we constructed a photoionization model by using archival data from optic to far-IR ($\sim3 - 100\ \mu m$).

## 2. Data

For constructing a photoionization model, we looked for the observational data and its derivation that represents the ionizing sources and ionized gas from the archival data. We used spectra from [6] that were observed by using *Gemini Multi-Object Spectrograph* (GMOS) on Gemini South, with wavelength coverage of the spectra is from 3767 – 5228Å. We adopted the nebular abundances derived from [5] as input parameter into the model and used data from APASS, 2MASS, WISE, AKARI/IRC and IRAS to constraint the model. For dereddened $F(H\beta)$ of entire nebula, we used from [7] and get the value about -10.92. Distance (3.5 kpc) and diameter (14.4″) of the PN is adopted from [8].

## 3. Photoionization Modelling

Spectral energy distribution (SED) fitting between observational data and model is done by using CLOUDY C13.05 [3]. For the input parameters of ionizing source, we used Postdam Wolf-Rayet (PoWR) stellar model atmosphere [10] to derive the central star properties and the result is showed on Table 1.

We varied 14 parameters ion the model ($T_{eff}$, inner radius, hydrogen density, elemental abundances, dust mass fraction and filling factor) and showed these in Table 2. The spectra from [6] are used as constraints to optimize the model by using "optimize lines' method on CLOUDY. On the first model, we used $T_{eff} = 140000$ K and $\log g = 6.1$ adopted from [6]. Nevertheless, the higher excitation lines such as [Ne III] and [O III] are overestimated on this model. On the other hand, model with $T_{eff} = 125000$ K and $\log g = 6.0$ from PoWR result is consistent with the observation data. The best-fit model is the model with minimum reduced $\chi^2$ calculated from broadband fluxes, broadband flux densities and dereddened integrated flux of entire nebulae that is included on the input parameters. The $\chi^2$ value for the best-fit model is 3.75.

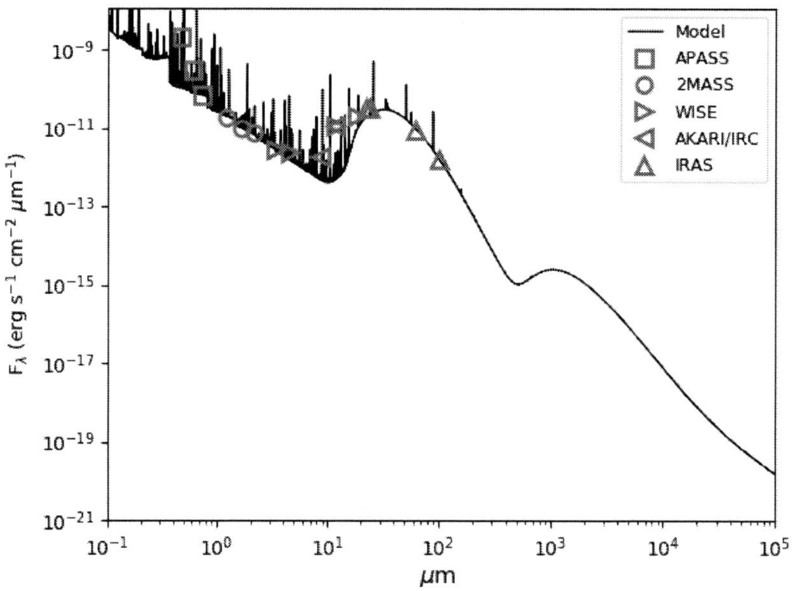


**Figure 1** Comparison of SED between the CLOUDY model and observational data of IC 4663.

**Table 1.** PoWR model atmosphere results for IC 4663

| Parameters | Value |
|---|---|
| $T_*$ | 125000 K |
| $v_\infty$ | 1600 km/s |
| $\log g$ | 6.0 cm/s$^2$ |
| $\log L/L_\odot$ | 3.5 |

**Table 2.** Parameters From CLOUDY Best-Fit Model

| Parameters | Value |
|---|---|
| $T_{\rm eff}$ | 125000 K |
| log g | 6.0 cm/s$^2$ |
| $\log L/L_\odot$ | 3.1 |
| $r_{\rm in}$ | 0.04$''$ |
| $n_{\rm H}$ | 1000 cm/s$^3$ |
| $m_{\rm d}/m_{\rm g}$ | 4.23(-4) $M_\odot$ |
| He/C/N/O/Ne/S/Cl/Ar | 11.08/8.55/8.4/8.7/8.4/7.2/5.38/7.13 |
| filling factor | 0.63 |

## 4. Results and Discussions

Figure 1 shows the plots between observed SED and the predicted SED from CLOUDY model. It is showed, however, that for the mid-IR part, the continuum does not show amorphous silicate bump although C/O < 1 indicates this PN is O-rich dust type. This should be confirmed by mid-IR spectroscopic observation. We do not find any mid-IR spectroscopic observation for this PN to constraint the model on the mid-IR emission line fluxes and dust continuum. Therefore, we cannot derived the dust-mass precisely and determine its dust type.

The nebular abundances of He/C/N/O/Ne/S from the model compared with the predictions from AGB model by [5] gives initially $3M_\odot$ star with Z = 0.02. From the model result, we also derived the gas-mass ($1.2M_\odot$). This result is also consistent with [5]. By using VLTP evolutionary tracks from [1], we get the core mass ($0.625M_\odot$), which also originated from a progenitor star with an initial mass $3M_\odot$. As a comparison, CSPN Abell 48 that had been classified as [WN5] by [9] is also suggested from $\sim 3M_\odot$ progenitor star [2]. But, Abell 48 is confirmed to be C-rich star which is different with IC 4663 that has C/O < 1. From the AGB nucleosynthesis prediction, the star with initial mass about $1.2 - 3M_\odot$ should be C-rich star. This can be confirmed by using mid-IR spectroscopic to see if IC 4663 is O-rich, C-rich or dual chemistry PN.

## 5. Conclusions

A new photoionization model for IC 4663 has been constructed by using CLOUDY and observation data from optical to far-IR. The input parameters for the central star properties are derived by using PoWR atmosphere model and give $\mathbf{T_{eff} = 125000\ K}$ and $\mathbf{\log g = 6.0}$ that are consistent with observation data. The comparison between nebular abundances from the model result with AGB model and born-again scenario evolutionary track suggested this PN is originated from $\mathbf{3M_\odot}$ progenitor star.

**References**

[1] Blöcker T 1995, A&A, **297**, 727
[2] Danehkar A, Todt,H, Ercolano B, and Kniazev A Y 2014, MNRAS, **439**, 3605
[3] Ferland G J, Porter R L, van Hoof, P A M, et al. 2013, Rev. Mexicana Astron. Astrofis., **49**, 137
[4] Iben I, Kaler J B, Truran, J W, and Renzini A 1983, ApJ, **264**, 605
[5] Karakas A, and Lattanzio J C 2007, PASA, **24**, 103
[6] Miszalski B, Crowther P A, De Marco O, Köppen J, Moffat A F J, Acker A, Hillwig T C 2012, MNRAS, **423**, 934
[7] Shaw R A and Kaler J B 1989, ApJs, **69**, 495
[8] Stanghellini L, Shaw R A and Villaver E 2008, ApJ, **689**, 194
[9] Todt H. et al 2013, MNRAS, **430**, 2302
[10] Todt H, Sander A, Hainich R, Hamann W-R, Quade M, and Shenar T 2015, A&A, **579**, A75

# Identification warp and flare milky way's disk in the third galactic quadrant

W Hidayat[1,*], M I Arifyanto[1], Aprilia[1], and M I Hakim[1]

[1]Department of Astronomy, Faculty Mathematics and Natural Sciences, Institut Teknologi Bandung

Email: wilhid@gmail.com

**Abstract.** Milky Way's disk has a feature called warp and flare. Warp is a bending phenomenon of the galactic disk. This feature is noticeable in the first and third galactic quadrant. Meanwhile, flare define as the increment of galactic disk scale high depends on galactocentric distance. Galactic disk tracers (dust, gaseous, and stars) can be used to explain the presence of both structures. The reddening of 10 young open clusters (OC) and 2985 early-type (OB) stars chosen in this work. Both components predicted to represent the galactic disk component where the warp and flare occurred. The Milky Way Star Cluster (MWSC) catalogue used to select the open clusters initial parameters. Then, we used proper motion data from GAIA DR2 combined with photometric data from Two Micron All Sky Survey (2MASS) catalogue to determine the open cluster membership and reddening. The open cluster memberships are selected using scatter algorithm. Then, the JHKs photometric data used to build the Two Colour Diagram (TCD) and Colour-Magnitude Diagram (CMD). We selected the reddening of OB stars data from Tycho 2 catalogue. The result is an increment of colour excess depends on galactocentric distance. This distribution could indicate the presence of warp and flare of the Milky Way's disk.

***Keywords:*** *flare milky way and third galactic quadrant*

## 1. Introduction

Warp define as bending phenomenon which is observed in the young population of galactic disk components such as stars, dust, and gaseous [1, 2]. The observation of Hydrogen Neutral (HI) emission from M33 revealed the structures of warp in another galaxy [3]. They found an anomaly of this data and make a hypothesis it came from the bending of the galactic plane. Further studies showed that the warp was detected in many spiral galaxies [4].

The Milky Way's galactic disks bend to the North Galactic Pole (NGP) and South Galactic Pole (SGP) in the first and third galactic quadrant, respectively [5]. Meanwhile, the star counts from Two Micron All Sky Survey (2MASS) catalogue constrain the warp started at 8.4 kpc from the centre of Milky Way. Therefore, the presence of the warp structure is one of an interesting problem in galactic structure study. The warp structure of Milky Way's disk accompanied by a structure named Flare. Flare define as the increment of galactic disk scale height depends on galactocentric distance [6]. In this work, we want to confirm the presence of both structures in Milky Way's disk. In section 2 we describe the data used in this work. Then we explain our method in section 3. At the end of this work, we describe our result concerning the presence of the warp and flare structure in Milky Way's disk.

## 2. Data

To support the goals of this work, the initial parameter of young OCs taken from the Milky Way Star Clusters (MWSC) catalogue [7]. 10 OCs located within longitudes $220^o < l < 240^o$ and $-15^o < l < 15^o$. Furthermore, we combined GAIA DR2 and 2MASS catalogue of stars that located in selected MWSC cluster's area. We describe the region as a circle with a radius (r_data) twice cluster's radius (r2). We picked the data which has relative error of parallax, proper motion RA (pmra) and proper motion declination (pmdec) less than 15%, and the photometric quality is "AAA". These criteria are used to minimize the uncertainty of kinematics and the photometric component of data.

**Table 1.** OC's initial parameters taken from MWSC 2013

| Name | RA (deg) | dec (deg) | l (deg) | b (deg) | r2 (deg) | r_ data (deg) | pmra (mas yr$^{-1}$) | pmdec (mas yr$^{-1}$) | pm error (mas yr$^{-1}$) | Distance Modulus | E(J-H) | Distance (pc) |
|---|---|---|---|---|---|---|---|---|---|---|---|---|
| NGC_2287 | 101.51 | -20.74 | 231.01 | -10.44 | 0.44 | 0.88 | -3.33 | -0.75 | 0.2 | 9.45 | 0.02 | 769 |
| NGC_2437 | 115.45 | -14.81 | 231.85 | 4.07 | 0.57 | 1.14 | -3.05 | 0.38 | 0.09 | 10.74 | 0.05 | 1375 |
| NGC_2383 | 111.17 | -20.94 | 235.26 | -2.46 | 0.12 | 0.25 | -6.84 | 2.83 | 0.63 | 11.16 | 0.07 | 1655 |
| NGC_2384 | 111.28 | -21.02 | 235.38 | -2.40 | 0.16 | 0.33 | -6.44 | 3.29 | 0.42 | 11.71 | 0.13 | 2074 |
| NGC_2421 | 114.06 | -20.61 | 236.27 | 0.07 | 0.16 | 0.32 | -2.92 | 4.81 | 0.3 | 11.84 | 0.15 | 2178 |
| DBSB_10 | 113.86 | -22.39 | 237.73 | -0.96 | 0.12 | 0.25 | -5.2 | 1.33 | 0.49 | 12 | 0.13 | 2362 |
| ASCC_32 | 105.52 | -26.56 | 237.94 | -9.61 | 0.32 | 0.65 | -2.33 | 4.25 | 0.35 | 9.45 | 0.02 | 767 |
| NGC_2362 | 109.68 | -24.96 | 238.18 | -5.55 | 0.26 | 0.52 | -2.6 | 2.34 | 0.32 | 10.74 | 0.05 | 1375 |
| Trumpler_7 | 111.82 | -23.96 | 238.21 | -3.35 | 0.18 | 0.36 | -4.1 | 4.19 | 0.28 | 11.2 | 0.1 | 1659 |
| NGC_2455 | 117.26 | -21.29 | 238.34 | 2.32 | 0.14 | 0.28 | -5.55 | 3.2 | 0.36 | 11.2 | 0.03 | 1711 |

In addition, we collected colour excess of 2985 early-type star (OB) from the Tycho 2 catalogue [8]. The OB stars position selected correspond to the galactic longitude of the OCs and another two regions as a complement.

## 3. Methodology

Tracing the warp signature from young open clusters (OC) is quite challenging due to OC's membership problems. The determination of clusters member in a specific area is necessary to find the intrinsic properties of OCs. As we know that the OC's member shares a similar age, metallicity, reddening, and proper motions. We used this description as the first assumption of our works.

### *3.1. Scatter Algorithm*

Scatter Algorithm describe as an algorithm which constructs density maps based on spatial distribution of the stars [9]. In this work, we would like to implement the density maps concept in open clusters proper motion area. First, we need to create a grid of the proper motion area. Then, we measure the density of each grid based on the presence of a star in the grid area and its surrounding. The main constant of this method is smoothing length (h), describes as the minimum distance of stars from the centre of the grid. If the stars located inside this range, it will contribute to the density of the designated grid.

$$h = \left(\frac{2}{N_{tot}}\right)^{\frac{1}{2}} s \tag{1}$$

$$s = \frac{2}{N_{tot}(N_{tot}-1)} \sum_{n=1}^{n=(N_{tot}-1)} \sum_{n'=(n+1)}^{n'=N_{tot}} \left[\left((x_n - x_{n'})^2 + (y_n - y_{n'})^2\right)^{\frac{1}{2}}\right] \tag{2}$$

The contribution of each star calculated by,

$$W(r) = \frac{3(h-r)}{\pi h^3} \quad \text{for } r < h \tag{3}$$

$$W(r) = 0 \quad \text{for } r > h \tag{4}$$

with $r^2 = (x - x_n)^2 + (y - y_n)^2$.

After we build the density map of the OC's area, we determine the OC's member as the stars located inside the circle with a radius of h. The centre of the circle is the location with the highest density in density maps.

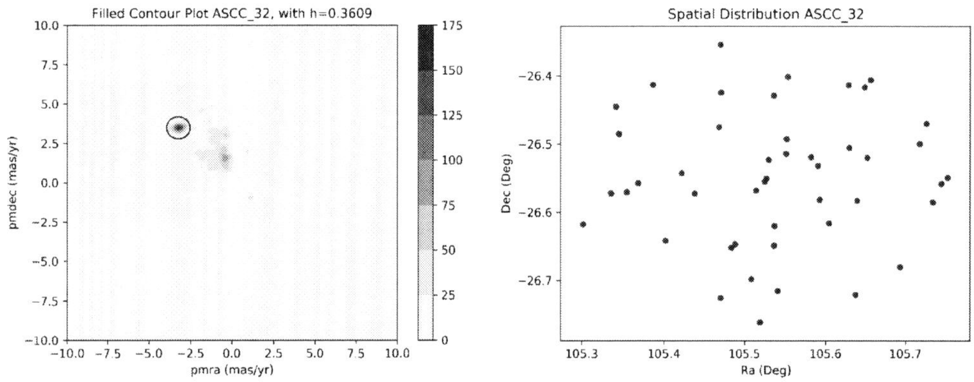


**Figure 1.** Left side showing the density maps of OCs in proper motion area. Right side showing the distribution of member stars in the equatorial plane

*3.2. Zero Age Main Sequences (ZAMs) Fitting*

The selected member of OCs which share similar proper motion properties plotted in CMD and TCD diagrams. We perform the fitting by eye method with ZAMs data to derive the colour excess and distance modulus of OCs. The ZAMs for 2MASS near-infrared photometry taken from the previous works [10]. Ratio of reddening $\frac{A_J}{A_V} = 0.276$; $\frac{A_H}{A_V} = 0.176$ [11], and $\frac{A_{K_S}}{A_V} = 0.118$ [12]. We calculated that $A_J = 0.669852\, E(J - H)$ and $E(J - H)/E(H - Ks) = 1.724$.

$$d = 10^{\left(\frac{\mu - A_J + 5}{5}\right)} \tag{5}$$

The distance of OCs calculated using equation (5) and we derived the error of each components. The data must fit as shown in figure 2.

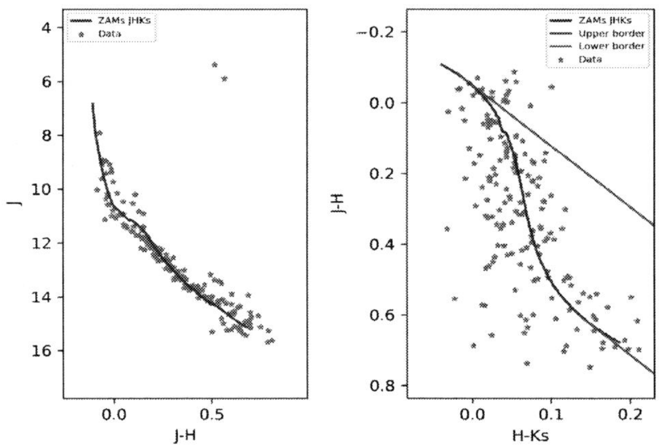


**Figure 2.** CMD and TCD fitting using near-infrared 2MASS ZAMs.

## 4. Result and Discussion

In this work, we derived the colour excess E(J-H) and distance modulus for constructing the graph of colour excess depend on galactocentric distance for 10 OCs. The determination of both parameters is challenging because as seen in figure 1, the spatial distribution of OC's member could not be determined as the clusters of stars directly. We need the information of its intrinsic parameter, such as proper motion, to decide the probable member of OCs. After selecting the member, we can see the OC's main sequences in figure 2. It makes the fitting by eye procedure are done more clearly for OC's members. The results of the OC's parameter determination procedure summarized in table 3. The centre of proper motion constant is inconsistent with the MWSC catalogue. Meanwhile, the distances modulus, distances, and colour excess E(J-H) meet a good agreement with the catalogue.

**Table 2.** The result of density based method and ZAMs fitting.

| Name | pmra (mas $yr^{-1}$) | pmdec (mas $yr^{-1}$) | h | Mod_fit | E(J-H)_fit | Distance_fit (pc) |
|---|---|---|---|---|---|---|
| NGC_2287 | -4.320 | -1.380 | 0.355 | $9.106 \pm 0.058$ | $0.012 \pm 0.001$ | $660.029 \pm 88.117$ |
| NGC_2437 | -3.840 | 0.420 | 0.249 | $10.756 \pm 0.074$ | $0.060 \pm 0.003$ | $1390.186 \pm 236.591$ |
| NGC_2383 | -1.620 | 1.920 | 0.222 | $11.378 \pm 0.054$ | $0.072 \pm 0.004$ | $1844.505 \pm 229.286$ |
| NGC_2384 | -1.620 | 1.920 | 0.225 | $11.989 \pm 0.037$ | $0.106 \pm 0.006$ | $2418.589 \pm 205.440$ |
| NGC_2421 | -3.120 | 3.120 | 0.230 | $11.667 \pm 0.047$ | $0.092 \pm 0.008$ | $2094.008 \pm 228.637$ |
| DBSB_10 | -1.620 | 1.560 | 0.280 | $11.933 \pm 0.038$ | $0.131 \pm 0.005$ | $2339.385 \pm 208.062$ |
| ASCC_32 | -3.240 | 3.480 | 0.361 | $9.411 \pm 0.081$ | $0.056 \pm 0.005$ | $749.513 \pm 139.598$ |
| NGC_2362 | -2.820 | 2.940 | 0.278 | $10.67 \pm 0.062$ | $0.076 \pm 0.003$ | $1334.853 \pm 191.954$ |
| Trumpler_7 | -2.520 | 3.360 | 0.246 | $11.311 \pm 0.043$ | $0.104 \pm 0.006$ | $1771.046 \pm 175.684$ |
| NGC_2455 | -2.760 | 2.700 | 0.261 | $11.222 \pm 0.044$ | $0.044 \pm 0.004$ | $1731.770 \pm 175.003$ |

If we construct the OCs and OB stars position in the cartesian galactic plane, we can see the plot in figure 3. We used the position of the Sun $x_0 = 8.3\ kpc, y_0 = 0\ kpc, and\ z_0 = 0.016\ kpc$ [13]. The number of OCs and OB stars in z negative relatively higher than z positive. It could be the first indication that the galactic disk components are following the warp in the third galactic quadrant that dominated in South Galactic Pole (SGP) directions.

IOP Conf. Series: Journal of Physics: Conf. Series **1231** (2019) 012032 doi:10.1088/1742-6596/1231/1/012032

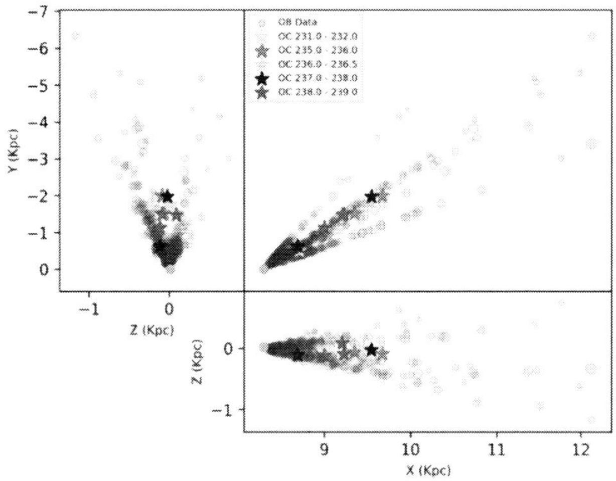


**Figure 3.** Position of OCs and OB in cartesian galactic plane. The size of blue circle represents the amount of colour excess J-H. Orange circle is the Sun position and the star represent the open clusters.

To clarify the existence of warp and flare structure, we construct the change of colour excess depend on galactocentric distance in figure 4. Generally, the colour excess of OCs and OB stars is higher below the galactic plane. The colour excess can represent the amount of dust in our galaxy that dominated below the galactic plane which indicate the presence of warp. Meanwhile, the value of E(J-H) in NGP directions that still increasing depend on the galactocentric distance which represent the structure of flare. The result of this work related to the previous study which found that the warp occurred in the third galactic quadrant [14]. For $238^o < l < 239^o$, this region has a higher amount of E(J-H) and could associate with the Canis Major over density in area of $l \approx 240^o$. Furthermore, the area of distance from 8.5 – 9 kpc show that the density of OB stars is higher in all longitude directions. This can be an indication of the existence of another structure called Gloud Belt.

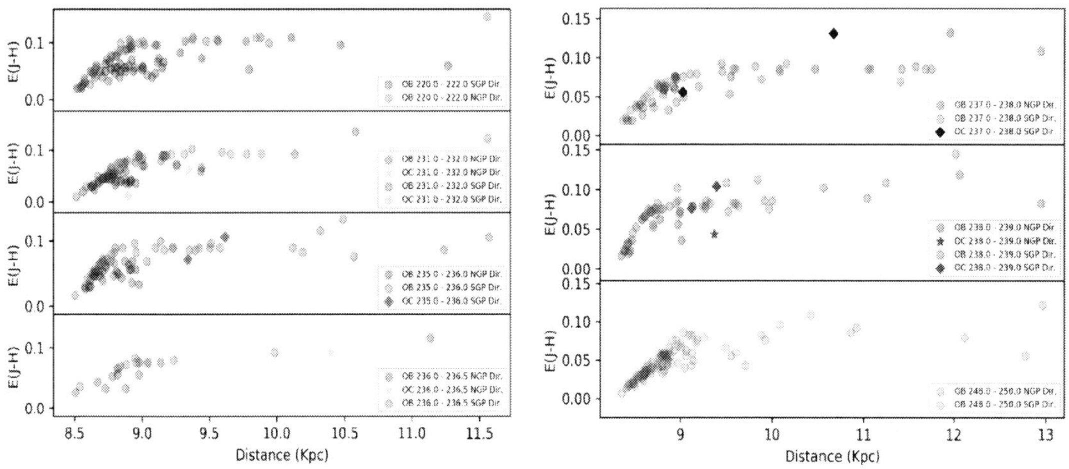


**Figure 4.** The change of colour excess depends on galactocentric distance of OCs and OB stars in 7 galactic longitude directions. The x axis is the distance of OCs and OB stars from our Sun. The y axis is the amount of colour excess E(J-H) for both components.

**Acknowledgement**
This work has made use of data from the European Space Agency (ESA) mission Gaia (https://www.cosmos.esa.int/gaia), processed by the Gaia Data Processing and Analysis Consortium (DPAC, https://www.cosmos.esa.int/web/gaia/dpac/consortium). Funding for the DPAC has been provided by national institutions, in particular, the institutions participating in the Gaia Multilateral Agreement. Besides that, this publication makes use of data products from the Two Micron All Sky Survey, which is a joint project of the University of Massachusetts and the Infrared Processing and Analysis Centre/California Institute of Technology, funded by the National Aeronautics and Space Administration and the National Science Foundation. Then, we acknowledged Program Penelitian, Pengabdian kepada Masyarakat, dan Inovasi (P3MI) for a funding travel grant to SEAAN 2018.

**References**
[1] Reyle C, D J Marshall, A C Robin and M Schultheis 2013 *Astron. Astrophys.* **495** 819-826
[2] Binney J and S Tremaine 1987 *Galactic Dynamics* (United State of America: Princeton University Press) chapter 6 pg 406-413
[3] Rogstad D H, M C H Wright and I A Lockhart 1976 *The Astrophis. Journal.* **204** 703-716
[4] Sánchez-Saavedra M L, E Battaner, A Guijarro, M Lopez-Corredoira and N Castro-Rodriguez 2003 *Astron. Astrophys.* **399** 457-467
[5] Kalberla P M W and J Kerp 2009 *Annu. Rev. Astro. Astrophys.* **47** 27-61
[6] Momany Y, S Zaggia, G Gilmore, G Piotto, G Carraro, L R Bedin and F de Angeli 2006 *Astron. Astrophys.* **451** 515-538
[7] Kharchenko N V, Piskunov A E, Schilbach E, Röser S and Scholz R D 2013 *Astron. Astrophys.* J., Lett. **558** A53
[8] Gontcharov G A 2016 *Astron. Letters.* **38** 694-706
[9] Gladwin P P, S Kitsionas, H M J Boffin and A P Whitworth 1999 *Mon. Not. R. Astron. Soc.* **302** 305-313
[10] Turner D G 2011 *RmxAA.* **47** 127-137
[11] Schlegel D J, D P Finkbener and M Davis 1998 *Astrophys. J.* **500** 525
[12] Dutra C M, Santiago B X and Bica E 2002 *Astron. Astrophys.* **381** 219-226
[13] Bobylev V V 2014 *Astrophysics* **57** 583-604
[14] Carraro G, R A Vázquez, E Costa, J A Ahumada and E E Giorgi 2014 *The Astron. J.* **149** 12

# Loop Quantum Gravity treatment of Charged Black Hole Thermodynamics

**Tirtho Daas**[1,*]

[1]Electronics and Communication Engineering, Institute of Engineering and Management, Kolkata, 700091, West Bengal, India

E-mail : tirthodaas@gmail.com

**Abstract.** This paper aims specifically at answering two contemporary requirements of a quantum theory of gravity, shown by using Loop Quantum Gravity (LQG) theory on the mathematics of black hole thermodynamics by (i) to quantise known classical General Relativistic formulations of black holes using LQG; and (ii) using LQG microstate spacelike sections (holonomies) on established quantum formulations of general relativity like Generalised Uncertainty Principle, to advocate for background independence in the mathematics of black hole horizon as the next step towards a viable quantum theory of gravity.

***Keywords:*** *Loop Quantum Gravity, Black hole thermodynamics, Reissner-Nordström, spin network*

## 1. Introduction

The Black hole is of fundamental importance to a study of the universe. It is believed by a consensus that the black hole mathematics is the test for any new theory of the universe: a proposed new theory of gravitation in cosmology is often tested against a black hole and the results compared. Literature have been found in the fields of forming a quantum approach to black hole thermodynamics pertaining to specific black holes, which however do not address the issue of micro scale quantum fluctuations which is intuitively expected from Quantum Field Theory which is amply discussed in Loop Quantum Gravity. Also, the theory of Loop Quantum Gravity does not consider higher dimensions, which have been strongly hinted to very likely be non-existent through contemporary independent research [7], because LQG theoretically is background independent, which is beautiful in that it is by far the most successful theoretical approach to quantising the background independent GR with the use of holonomies.

Considering a suitable candidate for the research demanded that a spherical, charged, non-rotating black hole be considered. The spherical shape of the black hole would ensure a simple formulation which can hence be expanded to suit non-spherical blackholes or systems of blackholes merging etc. A charge to the black hole would render the discussion generalised wherein the uncharged Schwarzschild blackhole would simply have in its formulation the charge nullified. A discussion on a rotation to the black hole has also been postponed to a separate subsequent project undertaking for the following reason: the spin knot geometry for the elementary quantum horizontal fluctuating area has to be revised for the gravitational shear in the vicinity of a rotating black hole due to its classical angular momentum. In such a case, the knot elements have to be reconstructed in view of the additional momentum.

This paper attempts and results in showing mathematically the first law of blackhole thermodynamics for a Reissner-Nordström charged spherical non-rotating black hole with quantum microstates embedded in it using Planck length spacelike holonomies along the horizon and their consequent effect on the minimum GUP 1 mass and Hawking Temperatures. The result is seen as a family of spin knot states which renders a way out from formalising along coordinates, i.e.: background independence. A further discussion is then added to argue for possible apparent dark matter candidature as a result of black hole remnants which exist due to the minimum mass requirement and are still active due to the horizon fluctuations as also advocacy of proof in light of possible theoretical understanding of very recent astronomical observational research data.

## 2. Charged Black Hole Thermodynamics: a non-quantum/averaged/course-grain view

### *2.1 The Reissner-Nordström metric*

This is a static solution to the Einstein-Maxwell field equations associated with the gravitational field of an uncharged, non-rotating, spherically symmetric massive body.

The metric (in spherical coordinates):

$$ds^2 = \left(1 - \frac{r_s}{r} + \frac{r^2{}_Q}{r^2}\right)c^2dt^2 - \left(1 - \frac{r_s}{r} + \frac{r^2{}_Q}{r^2}\right)^{-1} dr^2 - r^2 d\Omega^2 = g_{\mu\nu}dx^\mu dx^\nu \text{ (line element)}$$

$d\Omega^2$ being a 2-sphere defined by: $d\Omega^2 = d\theta^2 + \sin^2\theta d\phi^2$

The Schwarzschild radius: $r_s = \dfrac{2GM}{c^2}$

The characteristic length: $r^2{}_Q = \dfrac{Q^2G}{4\pi\varepsilon_0 c^4}$

The Reissner-Nordström black holes: The event horizon is located at where the component of the line element diverges: $0 = \dfrac{1}{g^{rr}} = 1 - \dfrac{r_s}{r} + \dfrac{r^2{}_Q}{r^2}$

solving into: $r_\pm = \dfrac{1}{2}\left(r_s \pm \left(r^2{}_s - 4r^2{}_Q\right)^{\frac{1}{2}}\right)$ [1]

This induces degeneracy when $2r_Q = r_s$ and when $2r_Q > r_s$ results in charge being more than the mass, which is physically not possible for the formation of a black hole. Conversely, such a possibility would result in a hypothetical naked singularity, refuted in supersymmetry [9] and which is also not feasible in Loop Quantum Gravity due to non-existence of spinfoam holonomies.

The ordinary first law of thermodynamics:

$dM = \dfrac{k}{8\pi}dA + \sum_{i=1}^{n} Y_i dy_i$ , where the last term on the right-hand side is related to the aggregate of work done externally on a black hole

Plugging in the substitution with the Hawking Temperature $T$ and charge $Q$, we have

$dM = TdS + \mu dQ$

Where, $S \equiv S_{BH} = \dfrac{k_B A}{4\hbar G}$ is the Bekenstein-Hawking entropy while $Q$ counts the charge pertaining to the chemical potential $\mu$ . Thus, following from above, we find

$$T = \left(\frac{\partial M}{\partial S}\right)_Q = \frac{dA}{dS}\left(\frac{\partial M}{\partial A}\right)_Q = \frac{dA}{dS} \times \frac{k}{8\pi}$$

which is a commonly known equation, derived by fixing Q as a constant.

Using Generalised Uncertainty Principle 1 [2] we note constraints on the position and momentum uncertainties

$$2r_+ \geq \Delta x_i \geq \frac{\hbar}{2}\left(\frac{1}{\Delta p_i} - \alpha l_{pl} + 2\alpha^2 l^2{}_{pl}\left(\Delta p_i\right)\right)$$

An approach stemming from the discussion in [2] gives the uncertainty in the energy $\Delta E \approx c\Delta p$. Considering the general notation $c$=1, we have

$T_H \approx \Delta p_i$ [2]

$$T_{H,\min}{}^{GUP1} = \frac{\left[\left(4r_+ + \alpha l_{pl}\right) - \left(\left(4r_+ + \alpha l_{pl}\right)^2 - 8\alpha^2 l_{pl}^2\right)^{\frac{1}{2}}\right]}{4\alpha^2 l_{pl}^2}$$

This gives the minimum value of the Hawking Temperature considering a section of the blackhole of Planck length.

*2.2 The Horizon Quantum fluctuations*

The above mathematics gives only a coarse grain description of the horizon. A closer look shows the presence of fluctuations on the horizon which are quantum mechanical. These fluctuations therefore are not given by the Reissner-Nordström metric but a time dependent nonsymmetric metric described in Quantum Loop Gravity. We shall now treat the surface of the black hole with a vigorous orthodox mathematics of Loop Quantum Gravity, namely the spin states which give rise to the loop and weave structures that are characteristic of this avenue of Quantum Gravity. The spacetime around the horizon is visualised as an ensemble of small (Planck scale) spacelike surfaces $\Sigma_t$, $t$ being time, and its geometry being a state $|s\rangle$ of spin knot s. Considering the horizon $h_t$ of the black hole to be comprised of a 2d surface, equivalent to the same spacelike surface $\Sigma_t$ .[3]

The required geometry $g_t$ is found by the intersections with the s-knot-s.

This is done by using the Area operator on the $|s\rangle$ (thus effectively an edge crossing the surface), the result is found calculated in established literature [5]

$$\hat{A}|s\rangle = \frac{G\hbar}{c^3}\sum_i \left[\frac{n_i\left(n_i+2\right)}{4}\right]^{\frac{1}{2}} |s\rangle$$

Putting the well-known relation $l_{pl} = \left(\frac{G\hbar}{c^3}\right)^{\frac{1}{2}}$ and the angular momentum variables $j = \frac{n}{2}$

$$A = l_{pl}{}^2 \sum_i \left[j_i\left(j_i+1\right)\right]^{\frac{1}{2}}$$

Considering the fundamental $j_i$ = 1, because these are boson [3]

$$A_1 = l_{pl}{}^2\left(2\right)^{\frac{1}{2}}$$

The dimension of a spin 1 network being 3, number of states of black hole,

$$N = 3^{\frac{A}{A_1}} = 3^{\frac{A}{l_{pl}^2 (2)^{\frac{1}{2}}}}$$

the entropy of the small spacelike segment under horizon fluctuations:

$$S_{hor.fluc} = k_B \ln N = k_B \frac{\ln 3}{l_{pl}^2 (2)^{\frac{1}{2}}} A$$

$$S_{hor.fluc} = k_B \ln N = k_B \frac{\ln 3}{l_{pl}^2 (2)^{\frac{1}{2}}} A$$

Being small, this can be written with respect to the whole black hole

$$\therefore dS_{hor.fluc} = k_B \ln N = k_B \frac{\ln 3}{l_{pl}^2 (2)^{\frac{1}{2}}} dA$$

Thus, the expression for the minimum mass of an element including the horizon fluctuations is a heavily revised first law of black hole thermodynamics

$$dM_{total,\min}{}^{GUP1} = T_{H,\min}{}^{GUP1} dS_{hor.fluc} + \mu dQ$$

Integrating this over the whole surface, we get

$$M_{total,\min}{}^{GUP1} = \frac{k_B \Omega \ln 3}{4(2)^{\frac{1}{2}} \left[\left(\frac{2(2)^{\frac{1}{2}}-1}{4}\right) \alpha l_{pl}{}^3\right]} \left( \left[\left(\frac{2(2)^{\frac{1}{2}}-1}{4}\right) \alpha l_{pl}\right]^2 + 8Q^2 \right)$$

with spin microstates of the black hole remnant being the full spectrum of all the interactions of the spin network $\Sigma_t$ with the surface $h_t$ including those which lie on the surface. This is because though the latter do not contribute to the mass/energy of the black hole, the spin states on the spacelike sections exist nonetheless.

i.e.: $\frac{1}{2}\sum_v \left[ 2 j_v^u \left( j_v^u + 1 \right) + 2 j_v^d \left( j_v^d + 1 \right) - j_v^t \left( j_v^t + 1 \right) \right]^{\frac{1}{2}}$ [5]

here, $j_v^u$ $\left( j_v^d \right)$ is the total spin with a positive(negative) sign and $j_v^t$ is the total spin of edges tangent to the surface at the vertex $v$.

## 3. Physical significance of the Result

We now compare the apparent physical significance of this result by comparing it to the supposed result in case these quantum fluctuations and their effects ignored. Thus, without considering horizon fluctuations $M_{\min}{}^{GUP1} = \frac{\Omega}{4\left[\left(\frac{2(2)^{\frac{1}{2}}-1}{4}\right) \alpha l_{pl}\right]} \left( \left[\left(\frac{2(2)^{\frac{1}{2}}-1}{4}\right) \alpha l_{pl}\right]^2 + 8Q^2 \right)$

Holding all other conditions and variables equal, the minimum mass, and therefore energy of the black hole remnant as seen in the result $M_{total,\min}{}^{GUP1}$ differs by a considerable magnitude from the latter $M_{\min}{}^{GUP1}$, in that the new result is a much larger minimum mass. This would result that a black hole remnant with its surface spin microstates considered thus should be more massive. This corroborates beautifully with two of the most recent and pertinent astronomical reports. Firstly, it could answer for the peculiar behaviour described as having a complex time-reversible wavelike structure. The effect is described in detail in the recent excellent paper by Hakkila et al. [6]. The reason was broadly discussed as a possible result of ejected material interacting with the surrounding medium which gives it an illusion of going back in time. We accept the reasoning and advocate that it is the added energy in $M_{total,\min}{}^{GUP1}$ of the black hole remnant that is interacting with the medium surrounding escaping radiation; thus, this can attempt to provide a theoretical answer. Secondly, it can account for the vast amounts of dark matter expected when black holes in themselves are a very unlikely candidate for dark matter as stated in Zumalacárregui M and Seljak U [8]. We reason that these intense fluctuations in gravity may often interfere constructively with fluctuations in adjacent spacelike segments, and thus give out pulses of strong gravitational waves which interact with the gravity of observable bodies, effectively acting as particles or dark matter.

## 4. Conclusion

We have hereby made a detailed discussion of the action of loop quantum gravity theory on a Reissner-Nordström black hole by first introducing the black hole class and making a coarse/averaged classical preliminary study. We then proceeded with the main motive by having considered small spacelike segments along the holonomies on the surface of the black hole and seen the microstate fluctuations, by considering the intersections of the horizon surface with the s-knot s state of the aforementioned spacelike segment. We then amalgamated this result with the result of the coarsely considered black hole and found the total mass of the blackhole remnant together with the mass due to the microstate fluctuations $M_{total,\min}{}^{GUP1}$ far exceeds (by orders of $\sim 10^{47}$) the mass without such considerations $M_{\min}{}^{GUP1}$. We concluded with arguing the significance of this result by trying to provide theoretical answers to two recent observations made but currently not theoretically accounted for.

**Acknowledgement**
I would begin by thanking the SEAAN 2018 Local Organising Committee for their support extended to me to present my research at the Conference and the academics and other participants present there for providing suggestions and comments. I would like to acknowledge Dr Miguel Zumalacárregui of the University of California, Berkeley, USA for kindly sharing with me the links to the data of his research and Mr Azrul Sulaiman Karim Pohan of the Institut Teknologi Sumatera, Lampung, Sumatra, Indonesia for giving his important insights and comments on this research.

**References**

[1] Xiang L and Wen X Q 2009 *JHEP* **910** 046

[2] Abbasvandi N. Soleimani M J, Radiman S, and Wan Abdullah W. A. T. 2016 *International Journal of Modern Physics A* **23** 1650129

[3] Rovelli C 2008 *Quantum Gravity* ed P V Landshoff *et al* (New York: Cambridge University Press) pp 299-317

[4] Rovelli C and Smolin L 1995 *Phys. Rev. D* **10** 5743-59

[5] Major S A 1999 A spin network primer *Am. J. Phys* **11** 972-80

[6] Hakkila J, Lesage S, McAfee S, Hofesmann E, Taylor C M and Cannon T *arXiv preprint arXiv:1804.10130*

[7] Pardo K, Fishbach M, Holz D E and Spergel D N 2018 *arXiv preprint arXiv:1801.08160*

[8] Zumalacárregui M and Seljak U *Phys. Rev. Lett.* **121** 141101

[9] Carter Brandon 1968 *Phys Rev* **174** 1559-71

# AUTHOR INDEX

**Institute of Physics**
Dirac House, Temple Back
Bristol BS1 6BE UK

ISSN: 1742-6588
ISBN 978-1-5108-9550-8

9 781510 895508

CURRAN ASSOCIATES INC.
proceedings.com